Computational Biology

The *Computational Biology* series publishes the very latest, high-quality research devoted to specific issues in computer-assisted analysis of biological data. The main emphasis is on current scientific developments and innovative techniques in computational biology (bioinformatics), bringing to light methods from mathematics, statistics and computer science that directly address biological problems currently under investigation.
The series offers publications that present the state-of-the-art regarding the problems in question; show computational biology/bioinformatics methods at work; and finally discuss anticipated demands regarding developments in future methodology. Titles can range from focused monographs, to undergraduate and graduate textbooks, and professional text/reference works.

Author guidelines: springer.com > Authors > Author Guidelines

For other titles published in this series, go to www.springer.com/series/5769

Ina Koch · Wolfgang Reisig · Falk Schreiber
Editors

Modeling in Systems Biology

The Petri Net Approach

Springer

Editors
Prof. Dr. Ina Koch
Institute for Computer Science
Johann Wolfgang Goethe University
Frankfurt
Robert-Mayer-Straße 11-15
60325 Frankfurt am Main
Germany
ina.koch@bioinformatik.uni-frankfurt.de

Prof. Dr. Falk Schreiber
Institute of Computer Science
Martin Luther University Halle-Wittenberg
Von Seckendorff Platz 1
06120 Halle
Germany
falk.schreiber@informatik.uni-halle.de

Prof. Dr. Wolfgang Reisig
Department of Computer Science
Humboldt-Universität zu Berlin
Unter den Linden 6
10099 Berlin
Germany
reisig@informatik.hu-berlin.de

ISSN 1568-2684
ISBN 978-1-4471-2598-3 ISBN 978-1-84996-474-6 (eBook)
DOI 10.1007/978-1-84996-474-6
Springer London Dordrecht Heidelberg New York

British Library Cataloguing in Publication Data
A catalogue record for this book is available from the British Library

Cover design: VTEX, Vilnius

Printed on acid-free paper

Springer is part of Springer Science+Business Media (www.springer.com)

Foreword

This is a book that looks at the universe of theoretical biology and biological systems analysis through an instrument that is a magnifying glass, and at the same time a telescope. The toolbox enables the chiseling of the tiniest molecular details as well as the embrace of the entireness of the world of living systems. And the elementary tools in this box are of surprising simplicity: places, transitions and directed arcs. Together with a concept of time and evolution this allows for concurrency, but not for continuity. Describing a system, be it chemical or biological, in the abstract automaton language of Petri nets can deliver a boring monotony or a fascinating holography.

It depends on the ingenuity of the designer. The Petri net approach offers not only the basic tools that Carl Adam Petri is said to have invented as a schoolboy, but in addition many extensions and expansions of those ideas, so that it becomes possible to combine quite a variety of modular elements (discrete, quasi-continuous, stochastic, spatially expanded, topologically involved, time-delayed, dimensionally separated, etc.) into one model without wreaking havoc in the abstract representation of the whole living system that is being simulated.

The book presented here aims to meet a challenge that previously has been largely avoided by the Petri net community. The community used to be content with the demonstration that Petri nets were able to simulate quite a large number of partial systems when their homogeneity permitted a rather restricted application of the Petri net methodology. The proof of principle used to say that Petri net models can simulate otherwise established theoretical descriptions. Here, the authors have undertaken the daunting task of explaining the whole realm of theoretical systems biology and of proving that it would all fit into the Petri net world. In this perspective, the Petri net world is extremely reductionist, condensed to a very limited abstract set of tools and interactions, but this very reductionism can help to reintegrate the perspective into a holistic picture.

This book will appeal to those who like to envisage life through a lens of mathematical tools, incorporated into the methodology of computer representation. The author of this foreword, for biographical reasons not having become a Petri net aficionado, can appeal only to the new generation of computational biologists: Do learn biological theory, do apply Petri net methods, and do show that this can yield rich harvests of interesting ideas. Use the Petri net tools as an incentive to generate exciting hypotheses about living systems!

Berlin
May 2010

Jens Georg Reich

Preface

Systems Biology is an emerging, multi-disciplinary field that has attracted increasing attention over the last few years. The rapid development of new experimental technologies in biology and medicine results in an enormous amount of biological data not only on sequences and structures, but also on their dependencies and interactions for many prokaryotes and eukaryotes. Thus, new data bases on gene expression, protein-protein interaction, and pathways have been developed. This quantity of data allows scientists to investigate molecular cell processes in a large scale manner. With the help of such newly available experimental data, many qualitative as well as quantitative in silico models have been constructed in order to obtain new insights into the behavior of biochemical systems, leading to a better understanding of molecular processes.

The lack of computational methods to explore such experimental data has led to an explosion of method development in this area. Methods in computational systems biology cover discrete, continuous, and stochastic techniques. Many of them are based on principles and algorithms which have been known for more than 20 years. The mathematical formalism of Petri net theory can encompass all of these techniques. For about 15 years, Petri net models of biochemical systems have been successfully developed, simulated, and analyzed. In the last five years, many papers have been published applying Petri net theory to different types of biochemical systems which model gene regulation, signal transduction and/or metabolism in biologically different application fields. This development provided the motivation to produce a book which explains Petri net foundations and reflects the main applications of Petri nets in molecular biology.

Who Should Read This Book?

This book intends to provide a comprehensive overview on recent applications of Petri nets in systems biology. The text has been designed to reach students, graduates, scientists with biological or medical background as well as with mathematical or computer science background, and also lecturers. The book aims to enable the interested reader to enter the field of modeling biochemical systems using Petri net concepts. The chapters have been divided into three parts, an introductory part, a methodological part, and an application-oriented part. The three introductory chapters comprise a general introduction on systems biology, biological foundations, and Petri net basics. As far as possible, we have unified the mathematical notations. We redraw the Petri nets in figures according to unifying drawing rules suggested by Wolfgang Reisig. Additionally, exercises are provided for each chapter, facilitating use of the book for lectures. Finally, cross references between chapters, a glossary and an index help in navigating the content and finding information quickly. We hope to spark interest not only in the application of Petri nets in modeling biochemical systems, but also in the fascinating and challenging fields of systems biology and of Petri nets.

How to Read This Book?

The book aims to introduce the latest research in Petri net applications in systems biology, but also to provide the necessary foundations of Petri net theory and of biochemical systems. The book can also be used as a textbook, as facilitated by the problems and their solutions provided for each chapter.

The book is organized into three parts each consisting of three to six chapters. The parts focus on theoretical foundations, basic modeling techniques, and special applications, respectively. Each chapter is self-contained and can be used as a unit of study. The first part introduces basic concepts. The second part discusses different modeling techniques, reflecting different levels of abstraction. The third part is dedicated to biochemical applications adopting a variety of different methods.

Part I introduces the field of systems biology, the foundations of biochemistry, and Petri net basics. This part should be read by newcomers to that field, but may be skipped, if the reader is already familiar with the foundations. Part I comprises three chapters. The first, Chap. 1, gives a general introduction into systems biology, also providing an overview of the main data resources, software tools, and visualization techniques. The second, Chap. 2, introduces the main biological principles of biochemistry comprising cell biology, metabolism, signal transduction, and gene expression. Chapter 3 explains basic principles of Petri nets, giving the necessary definitions and many examples.

Part II compiles basic Petri net modeling techniques for building and analyzing biological models. It starts with Chap. 4 on discrete modeling, comprising the classical Petri net modeling techniques as well as special new methods developed for application to biochemical systems. Discrete modeling techniques are particularly important for systems biology because often quantitative data is not available. Thus, a system's behavior can only be explored on the basis of its topology. One focus in the chapter concerns invariant analysis, which is in particular important for analyzing biochemical systems. Other special discrete modeling techniques have been developed to handle gene regulatory networks, which are described in Chap. 5, and with a special biological application in Chap. 12 in Part III.

In cases where some quantitative data is additionally available, hybrid modeling techniques, which combine discrete and continuous modeling, become suitable. These methods are introduced in Chap. 6. Many biochemical processes follow stochastic rules. Chapter 7 describes how stochastic systems can be modeled by Petri nets.

For continuous modeling, we have to know a critical amount of quantitative (kinetic) data. Methods applied in continuous modeling are mainly based on solving ordinary differential equation systems. These methods, which have been known for many years, differ in accordance with the underlying kinetics. Thus, for modeling enzymatic reactions, Michaelis–Menten kinetics is used, whereas for reflecting cooperativity, Hill kinetics is applied. Both concepts and the underlying mass action kinetics as well as their translation into Petri net formalism is explained in Chap. 8.

A new interesting application is fuzzy reasoning in Petri nets, which is introduced in Chap. 9.

Part III covers special applications to biochemical systems with regard to the type of the network according to its biological classification. Chap. 10 considers applications to metabolic networks, reflecting the relationship to stoichiometry-based methods, in particular to the concept of elementary modes. We continue in Chap. 11 with modeling of signal transduction pathways. The following chapter considers the modeling of gene regulation based on logical networks, and its conversion into Petri nets, using logical regulatory modules applied to development in segments of *Drosophila* embryos. Chapter 13 considers the modeling of feedback loops in the circadian clock of mammals using the hybrid Petri net approach. Additionally, Part III contains a special chapter on network prediction. The prediction of possible network structures from experimental data is a very challenging approach. Chapter 14 introduces such an approach using Petri net formalism.

Official FTP Site

http://www.springer.com/computer/bioinformatics/book/978-1-84996-473-9

FTP Site for Supplementary Material

http://pnbook.uni-frankfurt.de/

Berlin

Ina Koch
Wolfgang Reisig
Falk Schreiber

Acknowledgements

First of all, we would like to thank all the contributors for their excellent work and their suggestions on the structure and concept of this book. We are highly honored and grateful to Jens Georg Reich for writing the foreword. We are deeply grateful to Andrea Sackmann who compiled the LaTeX version, which involved checking consistency and completeness, transformation from Word into LaTeX for some chapters, and the drawing of the beautiful Petri net figures. We thank the Springer book series *Computational Biology* Editors-in-Chief, Andreas Dress and Martin Vingron, for giving us the opportunity to edit this book. We acknowledge the support, guidance, and patience of Simon Rees, Wayne Weeler, Catherine Brett, and the Springer Editorial Team, in particular Donatas Akmanavičius.

Contents

Part I Foundations

1 Introduction 3
Ina Koch and Falk Schreiber
1.1 Systems Biology 4
1.2 Models and Modeling 5
1.2.1 Analysis of Models 6
1.2.2 Model Development 8
1.2.3 Model Composition 9
1.2.4 Dynamic Behavior 9
1.3 Data Resources 10
1.3.1 Repositories of Gene Expression Data 11
1.3.2 Protein–Protein Interaction Databases 11
1.3.3 Pathway Databases 12
1.4 Visualization 13
1.4.1 Visualization Methods 13
1.4.2 Systems Biology Graphical Notation 14
1.4.3 Visualization Tools 15
1.5 Petri Nets in Biology 16
1.5.1 Motivation for Using Petri Nets 16
1.5.2 Petri Nets in Biology 16
1.6 Problems 17

2 Biochemical Fundamentals 19
Tiina Liiving, Syed M. Baker, and Björn H. Junker
2.1 Cell Biology 19
2.1.1 Cellular Organization 20
2.1.2 Mitochondria 21
2.1.3 Plastids 22
2.2 Metabolism 23
2.2.1 Metabolic Pathways and Networks 23
2.2.2 Metabolites 24

2.2.3 Enzymes . . . 24
2.2.4 Enzyme Inhibition . . . 25
2.2.5 Central Metabolic Pathways . . . 29
2.2.6 Metabolic Networks . . . 30
2.2.7 Regulation of Metabolism . . . 31
2.3 Gene Expression . . . 32
2.3.1 Transcription . . . 32
2.3.2 Translation . . . 32
2.3.3 Gene Regulation . . . 34
2.3.4 Gene Regulatory Networks . . . 34
2.4 Signal Transduction . . . 34
2.4.1 Hormones and Other Signaling Molecules . . . 35
2.4.2 Receptors . . . 36
2.5 Problems . . . 36

3 Petri Nets . . . 37
Wolfgang Reisig
3.1 Introduction . . . 37
3.2 An Introductory Example . . . 38
3.2.1 Models of Substances . . . 38
3.2.2 Models of Reactions . . . 38
3.2.3 Sequentially Composed Reactions . . . 39
3.2.4 Quantitative Aspects . . . 41
3.2.5 Alternative Composition . . . 41
3.2.6 Modeling the Interface . . . 42
3.2.7 Putting it All Together . . . 42
3.3 The Static Structure of Petri Nets . . . 43
3.3.1 Places . . . 43
3.3.2 Transitions . . . 43
3.3.3 Arcs . . . 44
3.3.4 Markings . . . 44
3.3.5 Static Net Structures . . . 44
3.4 Dynamic Behavior of Petri Nets . . . 45
3.4.1 Enabled Transitions . . . 45
3.4.2 Steps . . . 45
3.4.3 Step Sequences and Reachable Markings . . . 46
3.4.4 The Role of Infinity . . . 46
3.5 Analysis Techniques for Petri Nets . . . 47
3.5.1 Important Properties . . . 47
3.5.2 The Linear Algebra of Steps . . . 48
3.5.3 The Matrix of a Petri Net . . . 48
3.5.4 Place Invariants . . . 49
3.5.5 Transition Invariants . . . 51
3.5.6 Traps . . . 52
3.5.7 Syphons . . . 52

3.5.8 The Marking Graph 52
3.5.9 Concurrent Runs 53
3.6 Petri Nets as a Modeling Technique for Systems Biology 53
3.6.1 Discrete Steps 54
3.6.2 Local Cause and Effect 54
3.6.3 Invariance of Substance 55
3.6.4 Other Models of Dynamic Systems 55
3.7 References and Tools 55
3.8 Problems 56

Part II Modeling Techniques

4 Discrete Modeling 59
Andrea Sackmann
4.1 Modeling Concepts 59
4.2 Qualitative Analysis 62
4.3 P-invariants 63
4.4 T-invariants 64
4.4.1 Feasible T-invariants 65
4.4.2 MCT-sets 67
4.4.3 T-clusters 68
4.4.4 Mauritius Maps 69
4.5 Related Work 70
4.6 Software 71
4.7 Problems 71

5 Modeling Genetic Regulatory Networks 73
Richard Banks, Victor Khomenko, and L. Jason Steggles
5.1 Introduction 73
5.2 Synchronous Models of GRNs 75
5.2.1 Constructing a Qualitative Petri Net Model 76
5.2.2 Tool Support 80
5.2.3 Case Study: Nutritional Stress Response in *E. coli* 81
5.3 Asynchronous Models of GRNs 86
5.3.1 Signal Transition Graphs and Speed-Independent Circuits . 87
5.3.2 Refining Asynchronous Models of GRNs 91
5.3.3 Case Study: Lysis–Lysogeny Switch in Phage λ 94
5.4 Related Work 98
5.5 Summary 99
5.6 Problems 99

6 Hybrid Functional Petri Net with Extension for Dynamic Pathway Modeling 101
Ayumu Saito, Masao Nagasaki, Hiroshi Matsuno, and Satoru Miyano
6.1 Introduction 101
6.2 Pathway Modeling with Concept of Petri Net 103

6.2.1 Petri Net (PN) . . . 103
6.2.2 Representation of Biological Pathway . . . 103
6.2.3 Timed Petri Net (TPN) . . . 104
6.2.4 Continuous Timed Petri Net (CTPN) . . . 106
6.3 Hybrid Functional Petri Net with Extension . . . 106
6.4 Modeling MicroRNAs Double-Negative Feedback Loop in Gustatory Neurons of *C. elegans* . . . 114
6.5 Concluding Remarks . . . 119
6.6 Problems . . . 119

7 **Stochastic Modeling** . . . 121
Ivan Mura
7.1 Introduction . . . 121
7.2 Basic Concepts . . . 123
7.2.1 The Theoretical Basis of Stochastic Molecular Dynamics . 123
7.2.2 SPN Modeling of Stochastic Molecular Dynamics . . . 125
7.2.3 Beyond Pure Mass-Action . . . 130
7.3 Methods to Determine SPN Evolution . . . 132
7.3.1 Analytical and Numerical Approaches . . . 134
7.3.2 Simulation Approaches . . . 140
7.4 Examples of Modeling and Evaluation . . . 143
7.4.1 A Biochemical Oscillator . . . 143
7.4.2 A Protein Synthesis Network . . . 146
7.4.3 A Gene Regulatory Network . . . 147
7.5 Related Work . . . 148
7.6 Summary . . . 150
7.7 Problems . . . 150

8 **Quantitative Analysis** . . . 153
Jörg Ackermann and Ina Koch
8.1 Introduction . . . 154
8.2 Mass Action . . . 154
8.2.1 Zero-Order Reactions . . . 155
8.2.2 First-Order Reactions or Unimolecular Reactions . . . 155
8.2.3 Second-Order Reactions or Bimolecular Reactions . . . 157
8.2.4 Reversible Mass Action or Equilibrium Reactions . . . 158
8.2.5 Example: A Simple Predator Prey Model . . . 159
8.2.6 Steady States . . . 161
8.2.7 Stability Analysis . . . 161
8.2.8 Spatial Instability . . . 164
8.3 Michaelis–Menten Kinetics . . . 165
8.3.1 Enzyme Kinetics with Inhibitors . . . 167
8.4 Hill Kinetics . . . 169
8.4.1 Neutral Binding . . . 169
8.4.2 Cooperative Binding . . . 172

8.5 Tool Support 173
8.6 Problems 176
8.6.1 Continuous Versus Discrete Modeling 176
8.6.2 Stability Analysis of the Continuous Lotka–Volterra Model 177
8.6.3 Stability Analysis of the Discrete Lotka–Volterra Model . . 177
8.6.4 Michaelis–Menten Kinetics 178

9 Fuzzy Modeling 179
Lukas Windhager, Florian Erhard, and Ralf Zimmer
9.1 Introduction 179
9.2 Methods and Concepts 182
9.2.1 Ordinary Logic Reasoning 183
9.2.2 Fuzzy Sets 185
9.2.3 Fuzzy Logic Reasoning 187
9.2.4 Application to Petri Nets 192
9.3 Results 194
9.3.1 Design Principles 194
9.3.2 PNMA 197
9.3.3 Parameter Estimation 199
9.3.4 Application 200
9.4 Related Work 202
9.4.1 Common Network Motifs and Cell Cycle Model 202
9.4.2 Comparison of Fuzzy Logic to Multi-Class Discrete Logic . 203
9.4.3 Classification of Gene Expression Data 203
9.5 Summary 204
9.6 Problems 204

Part III Biochemical Applications

10 Topological Analysis of Metabolic and Regulatory Networks 209
Stefan Schuster and Björn H. Junker
10.1 Introduction 209
10.2 T-invariants 212
10.2.1 Modelling Steady States 212
10.2.2 Minimal T-invariants and Elementary Modes 213
10.2.3 Clustering of T-invariants 216
10.3 P-invariants 217
10.3.1 Definition and Biochemical Interpretation 217
10.3.2 Algorithms and Software Tools for Computing Minimal Invariants 219
10.4 Other Topological Properties 220
10.5 Related Work with Concrete Biological Examples 221
10.6 Conclusions 223
10.7 Problems 224

11 Analysis of Dynamical Models of Signaling Networks with Petri Nets and Dynamic Graphs . . . 225
Simon Hardy and Ravi Iyengar
11.1 Introduction . . . 225
11.2 Methods and Concepts . . . 227
11.2.1 From Ordinary Differential Equations to a Petri Net Model 228
11.2.2 From a Petri Net Model to an Interaction Graph . . . 230
11.2.3 From an Interaction Graph to an Influence Graph . . . 232
11.2.4 From an Influence Graph to Dynamic Graphs . . . 236
11.2.5 Tool Support . . . 238
11.3 Results . . . 238
11.3.1 Dynamic and Function of the Regulatory Motifs . . . 242
11.3.2 Lifetime of the Regulatory Motifs . . . 247
11.4 Related Work . . . 249
11.5 Summary . . . 249
11.6 Problems . . . 250

12 A Modular, Qualitative Modeling of Regulatory Networks Using Petri Nets . . . 253
Claudine Chaouiya, Hanna Klaudel, and Franck Pommereau
12.1 Introduction . . . 254
12.2 Logical Modeling of Regulatory Networks . . . 254
12.2.1 Logical Regulatory Graphs (LRGs) . . . 255
12.2.2 Logical Functions Representation Based on Decision Diagrams . . . 258
12.2.3 State Transition Graphs Associated to LRGs . . . 260
12.3 P/T Petri Net Representation . . . 261
12.3.1 Tools Support . . . 264
12.3.2 Related Works . . . 264
12.4 Modules, Their Composition and High-Level Petri Net Representation . . . 264
12.4.1 Interconnecting Logical Regulatory Modules . . . 265
12.4.2 High-Level Petri Net Representation . . . 267
12.4.3 Implementation . . . 269
12.5 Modeling Interconnected LRMs Using High-Level Petri Nets . . . 270
12.5.1 Interconnecting Occurrences of the Toy LRM . . . 270
12.5.2 The *Drosophila* Segment-Polarity Module . . . 273
12.6 Conclusions . . . 276
12.7 Problems . . . 278

13 A Case Study of HFPN Simulation: Finding Essential Roles of Ror Gene in the Interaction of Feedback Loops in Mammalian Circadian Clock . . . 281
Natsumi Mitou, Hiroshi Matsuno, Satoru Miyano, and Shin-Ichi T. Inouye
13.1 Introduction . . . 282

13.2 Modeling Molecular Circadian Oscillator in Mouse with Hybrid Functional Petri Net . . . 284
13.2.1 Circadian Clock Oscillator with Five Genes . . . 285
13.2.2 Construction of an HFPN Model . . . 285
13.3 Effect of the Two Interlocked Loops in Circadian Gene Regulations in Mouse . . . 288
13.3.1 A Problem of the Previous HFPN Model in an Oscillation Maintenance . . . 288
13.3.2 Modification in Parameters for Constituting Interlocked Loops . . . 288
13.4 Function of *Ror* in the Interlocked Model . . . 294
13.4.1 Introduction of *Ror* . . . 294
13.4.2 Inclusion of a Hypothetical Path: ROR Excites *Bmal* Only When PER/CRY Level Are Above the Threshold . . . 299
13.4.3 Simulation for *Rev-Erb* Knockout Mice . . . 302
13.5 Conclusions . . . 303
13.6 Problems . . . 304

14 Prediction of Network Structure . . . 307
Annegret Wagler
14.1 Introduction . . . 307
14.2 Methods to Predict Network Structures . . . 313
14.2.1 A General Combinatorial Reconstruction Approach . . . 313
14.2.2 An Approach for the Case of Monotone Data . . . 319
14.3 Results: A Reconstruction Algorithm . . . 326
14.4 Related Work . . . 332
14.5 Summary . . . 334
14.6 Exercises and Solutions . . . 335

Glossary . . . 337

References . . . 339

Index . . . 357

Contributors

Jörg Ackermann fluIT Biosystems GmbH, Mikroforum-Ring 1, 55234 Wendelsheim, Germany, joerg.ackermann@fluit-biosystems.com

Syed M. Baker Leibniz Institute of Plant Genetics and Crop Plant Research (IPK), Corrensstr. 3, 06466 Gatersleben, Germany, baker@ipk-gatersleben.de

Richard Banks School of Computing Science, Newcastle University, Newcastle upon Tyne, UK, Richard.Banks@ncl.ac.uk

Claudine Chaouiya Instituto Gulbenkian de Ciêencia, Rua da Quinta Grande, 6, 2780-156 Oeiras, Portugal, chaouiya@igc.gulbenkian.pt; TAGC Inserm U928, Campus de Luminy, case 928, 13288 Marseille, France

Florian Erhard Institut für Informatik, Ludwig-Maximilians-Universität München, Amalienstr. 17, 80333 Munich, Germany, florian.erhard@bio.ifi.lmu.de

Simon Hardy Department of Pharmacology and Systems Therapeutics, Mount Sinai School of Medicine, One Gustave L. Levy Place, New York, NY 10029-6574, USA, simon.hardy@mssm.edu

Shin Ichi T. Inouye Institute for Time Studies, Yamaguchi University, Yamaguchi 753-8511, Japan, inouye@c-able.ne.jp

Ravi Iyengar Department of Pharmacology and Systems Therapeutics, Mount Sinai School of Medicine, One Gustave L. Levy Place, New York, NY 10029-6574, USA, ravi.iyengar@mssm.edu

Björn H. Junker Leibniz Institute of Plant Genetics and Crop Plant Research (IPK), Corrensstr. 3, 06466 Gatersleben, Germany, junker@ipk-gatersleben.de

Victor Khomenko School of Computing Science, Newcastle University, Newcastle upon Tyne, UK, Victor.Khomenko@ncl.ac.uk

Hanna Klaudel IBISC, Université d'Evry, 523 place des terrasses, 91000 Evry, France, klaudel@ibisc.univ-evry.fr

Ina Koch Max Planck Institute for Molecular Genetics, Ihnestrasse 73, 14195 Berlin, Germany, ina.koch@molgen.mpg.de; Institute for Computer Science, Johann Wolfgang Goethe University Frankfurt am Main, Robert-Mayer-Strasse 11-15, 60325 Frankfurt am Main, Germany, ina.koch@bioinformatik.uni-frankfurt.de

Tiina Liiving Leibniz Institute of Plant Genetics and Crop Plant Research (IPK), Corrensstr. 3, 06466 Gatersleben, Germany, liiving@ipk-gatersleben.de

Hiroshi Matsuno Graduate School of Science and Engineering, Yamaguchi University, Yamaguchi 753-8512, Japan, matsuno@sci.yamaguchi-u.ac.jp

Natsumi Mitou Kyuden Business Solutions Co. Inc., Fukuoka 810-0004, Japan, natsumi.mitou@qdenbs.com

Satoru Miyano Human Genome Center, Institute of Medical Science, University of Tokyo, Tokyo 108-8639, Japan, miyano@ims.u-tokyo.ac.jp

Ivan Mura The Microsoft Research–University of Trento Centre for Computational and Systems Biology, Piazza Manci 17, 38123 Trento, Italy, mura@cosbi.eu

Masao Nagasaki Human Genome Center, Institute of Medical Science, University of Tokyo, Tokyo 108-8639, Japan, masao@ims.u-tokyo.ac.jp

Franck Pommereau IBISC, Université d'Evry, 523 place des terrasses, 91000 Evry, France, pommereau@ibisc.univ-evry.fr

Wolfgang Reisig Department of Computer Science, Humboldt-Universität zu Berlin, Unter den Linden 6, 10099 Berlin, Germany, wolfgang.reisig@informatik.hu-berlin.de

Andrea Sackmann Institute of Computing Science, Poznań University of Technology, Piotrowo 2, 60-965 Poznań, Poland, asackmann@cs.put.poznan.pl

Ayumu Saito Human Genome Center, Institute of Medical Science, University of Tokyo, Tokyo 108-8639, Japan, s-ayumu@ims.u-tokyo.ac.jp

Falk Schreiber Leibniz Institute of Plant Genetics and Crop Plant Research (IPK), Corrensstrasse 3, 06466 Gatersleben, Germany, schreibe@ipk-gatersleben.de; Institute of Computer Science, Martin Luther University Halle-Wittenberg, Von Seckendorff Platz 1, 06120 Halle, Germany, falk.schreiber@informatik.uni-halle.de

Stefan Schuster Biologisch-Pharmazeutische Fakultät, Lehrstuhl für Bioinformatik, Friedrich-Schiller-Universität Jena, Ernst-Abbe-Platz 2, 07743 Jena, Germany, stefan.schu@uni-jena.de

L. Jason Steggles School of Computing Science, Newcastle University, Newcastle upon Tyne, UK, L.J.Steggles@ncl.ac.uk

Annegret Wagler Magdeburg Center of Systems Biology (MaCS) and Institute for Mathematical Optimization, Otto-von-Guericke Universität Magdeburg, Universitätsplatz 2, 39106 Magdeburg, Germany, wagler@imo.math.uni-magdeburg.de

Lukas Windhager Institut für Informatik, Ludwig-Maximilians-Universität München, Amalienstr. 17, 80333 Munich, Germany, lukas.windhager@bio.ifi.lmu.de

Ralf Zimmer Institut für Informatik, Ludwig-Maximilians-Universität München, Amalienstr. 17, 80333 Munich, Germany, ralf.zimmer@bio.ifi.lmu.de

Part I
Foundations

Chapter 1
Introduction

Ina Koch and Falk Schreiber

Abstract This chapter gives a general introduction into the field of systems biology and the motivation for using Petri nets in this field. We consider modeling processes in the context of biological modeling approaches providing different examples. Starting from a general description of the purpose of a model and the modeling process, we cover the range from qualitative to quantitative modeling. We compile different modeling techniques at different abstraction levels, for example, at discrete, stochastic, and continuous levels. In this context, we introduce Petri nets and give the motivation for using Petri nets in particular for modeling biochemical systems. We describe the first applications of Petri nets in biology and give a brief overview of the progress made so far. Furthermore, we discuss the main public data resources for systems biology, giving an overview of microarray data repositories, protein–protein interaction databases, and pathway databases. Finally, we describe methods and tools for the visualization of biochemical systems and Petri net models.

I. Koch (✉)
Institute for Computer Science, Johann Wolfgang Goethe University Frankfurt,
Robert-Mayer-Strasse 11-15, 60325 Frankfurt am Main, Germany
e-mail: ina.koch@bioinformatik.uni-frankfurt.de

I. Koch
Max Planck Institute for Molecular Genetics, Ihnestrasse 73, 14195 Berlin, Germany
e-mail: ina.koch@molgen.mpg.de

F. Schreiber
Leibniz Institute of Plant Genetics and Crop Plant Research (IPK), Corrensstrasse 3,
06466 Gatersleben, Germany
e-mail: schreibe@ipk-gatersleben.de

F. Schreiber
Institute of Computer Science, Martin Luther University Halle-Wittenberg, Von Seckendorff
Platz 1, 06120 Halle, Germany
e-mail: falk.schreiber@informatik.uni-halle.de

I. Koch et al. (eds.), *Modeling in Systems Biology*, Computational Biology 16,

DOI 10.1007/978-1-84996-474-6_1,

1.1 Systems Biology

In biology, the science of living organisms, we want to understand life in its entirety. We examine all living things at different levels of abstraction and description. We are interested in distribution and classification, origin and evolution, structure and function of biological species. In order to really understand life we have to consider not only the single components, which are complicated enough, but also their interactions, again at different levels of abstraction. We have to move from the *reductionist paradigm*, where single components and single interactions have been identified and characterized, to the *holistic paradigm*. Following the holistic paradigm, components are not only pooled together, but rather integrated in such a way that all these components form a biological system, which reflects the dynamic behavior of life.

We can find different definitions of systems biology which express different aspects of the field. Instead of inventing a new one, representative for many others, we give four of them:

1. *A field that seeks to study the relationships and interactions between various parts of a biological system (metabolic pathways, organelles, cells, and organisms) and to integrate this information to understand how biological systems function* [134].
2. *A hypothesis-driven field of research that creates predictive mathematical models of complex biological processes or organ systems* [133].
3. *A discipline that aims at deciphering relationships between different parts of a biological system with the goal of understanding and predicting the behavior of the system as a whole* [121].
4. *Systems biology is a relatively new biological study field that focuses on the systematic study of complex interactions in biological systems thus using a new perspective (integration instead of reduction) to study them* [382].

The change to a holistic paradigm is only possible if data about system components and their interactions is available. The development of high-throughput data collecting techniques, for example, microarrays, protein chips, yeast two-hybrid screens etc., enables us to simultaneously interrogate cell components at any given time. Based on this data and depending on the abstraction level, we can consider various types of interaction networks, for example, protein–protein interaction networks, metabolic networks, signaling networks, transcriptional and gene regulatory networks, etc. For an introduction into biology including the description of these networks, see Chap. 2.

Kitano, one of the pioneers in systems biology, summarized in [193]:

> To understand biology at the system level, we must examine the structure and dynamics of cellular and organismal function, rather than the characteristics of isolated parts of a cell or organism. Properties of systems, such as robustness, emerge as central issues, and understanding these properties may have an impact on the future of medicine. However, many breakthroughs in experimental devices, advanced software, and analytical methods are required before the achievements of systems biology can live up to their much-touted potential.

For further reading on systems biology, see [11, 114, 186, 203, 290, 383].

Methodologically, we can divide the field of systems biology into experimental (wet lab) and theoretical and computational (dry lab) systems biology. Our book focuses on the theoretical and computational aspects in systems biology, and within that field on the application of Petri net theory to solve biological questions in biochemical networks. Before starting with the different aspects of Petri net modeling, let us formulate some general thoughts on models and modeling.

1.2 Models and Modeling

First of all, what is a model? Which purposes should be served by a model? Mathematical logic models are based on sets of axioms, which are always valid in the considered models. These models do not represent any reality outside of the axiom system. Other models reflect the actuality, often in an abstract and simplified way allowing for understanding, because actuality is too complex to be represented in its entirety. In biology, we are also accustomed to working with model organisms. For example, we use yeast and mouse as model organisms for humans in the wet lab because data can be obtained more easily, but also to avoid ethical questions.

Thus, models should reflect known behavior, but can contain also hypotheses in order to verify them. Models can be based on different levels of abstraction. For example, to represent protein, DNA or RNA sequences we use sequences of letters over a well defined alphabet. We describe chemical structures, e.g. metabolites, as chemical structure graphs. Additionally, we depict the complex protein structures by topological diagrams that are based on graph-theoretical descriptions [209, 252]. For network visualization, we apply wireframe schemes which also represent graphs. The underlying graphs of these descriptions can differ in the definitions of vertices and edges and their labels depending on of the biochemical network type. To represent dynamic properties, we apply different mathematical concepts at different levels of abstraction, so logical formalisms for Boolean modeling, discrete formalisms for Petri net modeling, differential equations for modeling continuous properties, and stochastic equations to express the stochastic properties of chemical reaction networks.

Secondly, why is modeling useful? Experimental observations cover many simple and complex processes, ranging from isolated enzymatic reactions through temporal processes in metabolic networks to patterns of gene expression and regulation. In many cases, we can not intuitively predict the behavior of even simple systems from experience. Thus, we also can not predict the behavior of complex dynamic processes with sufficient precision solely from experience. Here, we should develop a computational model, which can be easily modified. For example, we can easily extend a model or knockout components and then simulate the network behavior. Also, we can stretch or compress time scales. We can model quantities that are experimentally hidden.

Nevertheless, we have to keep in mind that models are not *right* or *wrong* at all, only with respect to specific aspects of the reality. We are developing models for different purposes addressing different aspects of a complex system at different levels of abstraction and, thus, resulting in different accuracies. To summarize these thoughts, we cite Box [45]:

> Essentially, all models are wrong, but some are useful.

Another advantage is that computer programs and special algorithms can be used for several systems independently from the actually modeled system. Finally, it is obvious that costs of modeling are much lower than for experiments. We can reduce animal experiments, and there is no risk for real living systems.

The purpose of a model is to give answers to specific questions. Some crucial questions regarding the model and the modeling process are the following:

- Is our knowledge about a network/pathway complete?
- How to decide whether the model is complete for the questions we want to address?
- Is our model consistent?
- Which dynamic properties can we infer only from network topology?
- How is the scalability of our model?
- How can we combine different modeling methods into one unique model?

Regarding biology following questions can arise:

- What happens in the cell at molecular level?
- How are the cross-links between different pathways?
- How is cellular response to environmental changes and stress regulated?
- How do gene regulation and signal transduction influence the metabolism and vice versa?
- How should a cell be treated to yield a high output of a desired product (Biotechnology)?
- Where should a drug operate to cure a disease?

The model has to predict the system's behavior in such a way that we yield precise results or output according to a special input. In many cases, the system itself might be treated as a black box.

To obtain this goal parts, structure, and relations of the object should be clear. The function of an object is to make the model nearly realistic. The model should exhibit a generality, that is, it should be applicable to many different objects and processes. Finally, the model should also be as simple as possible, for example, for the mathematical treatment.

1.2.1 Analysis of Models

The analysis of models generates a language or formalism that is able to describe complex processes. After collecting data, usually investigations start with the ordering and classification of data, for example, by the creation of databases. So, we select prototypes for processes that are common in some organisms or for a special cell type. We try to invent a systematics which covers the obtained knowledge. To be unique in terminology and to get a sound notion for computer representations, ontologies have been developed, for example, gene and gene product ontologies in GO [16]. For systems biology, the Systems Biology Ontology (SBO) [339] is under development.

For example, let us consider enzyme catalyzed reactions, for which a systematics E.C. (Enzyme Commission)-number can be assigned [108]. Enzymes may have one or two or more substrates and can catalyze different groups of reactions. Thus, we name enzymes according to their specific function. For example, the enzyme pyruvate kinase is a transferase (E.C. 2.) that transfers phosphorus-containing groups (E.C. 2.7.) with an alcohol group as acceptor (E.C. 2.7.1.). The full E.C. number is 2.7.1.40 and catalyzes the following reaction

$$ATP + Pyruvate \rightarrow ADP + Phosphoenolpyruvate \tag{1.1}$$

We can classify complex processes according to their direction, dynamic behavior, and abstraction level into reversible and irreversible, periodic and nonperiodic, deterministic and stochastic processes.

Reversibility: Every single chemical reaction is, in theory, reversible. It depends on the environmental conditions which direction is preferred, for example, on thermodynamic parameters such as temperature and pressure or on the availability of an enzyme. A reaction can be classified as irreversible if nearly all of its reactants are used to form products. Then, it is very difficult, even under extreme conditions, to reverse the reaction.

Periodicity: Periodic chemical reactions generate macroscopic patterns both in sequence and time. In nature, we know calcium oscillation as an important periodic process that controls a wide variety of cellular mechanisms, and is often organized into intracellular and intercellular calcium waves [353].

Determinism: Deterministic models produce the same output for the same starting condition. If a deterministic system is known at one time then it is known forever, that is, the system involves no randomness in the development of subsequent states. According to the level of abstraction, we distinguish between discrete and continuous deterministic models.

Stochasticity: Stochastic models are nondeterministic models. In stochastic systems, the development of subsequent processes is additionally determined by a random element, which is caused by the fact that collisions in a system of molecules take place in a random manner, leading to a probability distribution of system states. The random element acknowledges that resulting subsequent system states come from both, known and unknown causes.

Discreteness: In biology, discrete models consider the objects, for example, proteins or genes, as countable amounts. Processes take place according to discrete rules, that is, discrete objects react or will be translated as whole objects. In contrast, using continuous models we do not consider countable entities; we work with concentration of objects and reaction rates that depend on substance concentrations.

Other system-relevant categories include stability (stable or unstable), robustness (robust or fragile), and activity (active or inactive).

Stability: Biological systems have to be stable, performing a certain function in a specific changeable environment.

Robustness: Robustness is a crucial property of biochemical systems that enables the organism to flexibly respond to environmental perturbations or other changes. It is implemented in biological systems by redundancy, feedback loops, and structural stability [193].

Activity: It is characteristic of biochemical systems that parts of the system can be active or inactive depending on other activated or nonactivated components of the system. There are many possible mechanisms of activation and deactivation, for example, proteins can be activated or deactivated by phosphorylation or dephosphorylation, respectively.

Adequateness: Modeling the (biological) reality, the adequateness of models has to be considered. We have to translate the biological problem into a mostly mathematical model using a special mathematical formalism, which reflects those aspects of system's behavior we are interested in. So, we have always to find a compromise between adequateness (reflection of reality) and handling (simplicity).

1.2.2 Model Development

We consider model development as an iterative process of model extension and model verification with the aim of further improvement of the model. In particular, during data collection, model development is subjective and selective. As modeling people, we project our view and understanding of the biological problem onto the model. Thus, we have to define the scope and model boundaries.

We have to formulate the biological problem in such a way that it fits into a mathematical formalism. This is not easy, in particular for biology, because languages and thinking of biologists and mathematicians strongly differ. Moreover, biological knowledge is often uncertain and can change rapidly, even during the modeling process. Furthermore, the terminology is not unique. Thus, an object, for example, gene or metabolite, can have different names. To structure and control the vocabulary, various ontologies have been developed. For example, Gene Ontology (GO) [16] defines genes and gene product attributes. For other ontologies, see, for example, [26, 339, 364].

Usually, we start with a simple model, a sketch or a small set of chemical stoichiometric reaction equations or Ordinary Differential Equations (ODEs). After validation we extend the model, validate it again and so on, creating an iterative process of model refinement and improvement and model validation. Model verification and validation are crucial points in model development. We have to check whether the model reflects known system properties and behavior. Can the model explain the experimental observations? We have to carefully compare model results with experimental results. How we can achieve that? We solve the related (mathematical) problems. We try to define or estimate the parameters from experimental data, which is a complicated process, where many methods have been developed [17]. Further, we can check the model for consistency, applying different techniques such as T-invariant analysis or elementary mode analysis, see Chaps. 4 and 10. Once we have a verified and consistent model available, we are able to make predictions on the system's behavior, for example, by knockout analysis. But, we have to be aware that in the end the experiment will determine the correctness of the model.

1.2.3 Model Composition

The structure of a system is defined by its variables, parameters, constants, and boundaries. The boundaries define the interface to the surroundings of the system. Variables, parameters, and constants define quantities of the system. *Variables* are modifiable quantities, for which the model establishes a relation. A quantity with fixed value is called a *constant*, such as the Avogadro's number $N_A = 6.02214179(30) \times 10^{23}$ mole^{-1} that defines the number of molecules per mole. A *parameter* is a fixed value for a certain considered state, but it can change when the system switches to another state. The parameter value is changeable and depends on measurements.

For example, in the formula of the Michaelis–Menten equation

$$v = \frac{V_{max}[S]}{K_M + [S]} \tag{1.2}$$

v (reaction velocity) and $[S]$ (substrate concentration) are variables, and V_{max} (maximal velocity) and K_M (Michaelis–Menten constant) are the parameters. For details of Michaelis–Menten kinetics in this book, see Chap. 8.

It depends on the model, whether a quantity is a variable or parameter. For example, enzyme concentration is mostly considered as a parameter, but it can become a variable in a refined model that includes the dependence of enzyme concentration from gene expression or protein degradation.

The set of variables, describing the system completely, form the *state variables*. The *dimension* of a system is defined by the number of independent state variables. Depending on the number of variables, we call a system *underdetermined*, if we have too few variables, or if we have too many variables, then the system is *overdetermined* and probably contradictory.

1.2.4 Dynamic Behavior

The dynamic behavior (*kinetics*) describes the changes of system states over time. A *system state* is described by a set of variables at a given time point. One state should contain enough information to predict the behavior in all future times. Each model defines what it means by the state of the system. A state can be represented in different ways. Let us consider a simple monomolecular reaction:

$$A \xrightarrow{k} B \tag{1.3}$$

Deterministic processes with discrete change of system states in time can be represented by Boolean models. They consider the presence or activity of B at time $t+1$ in dependence on the presence or activity of A at time t:

$$B(t+1) = f\big(A(t)\big) \tag{1.4}$$

For example, gene regulation can be expressed as a Boolean model, distinguishing between the states where the gene is expressed (1) or not expressed (0).

Deterministic models with continuous change in time can be expressed by ODEs. ODE-models consider concentrations, for example, of mRNA or of metabolites. Then, in (1.3) the concentration of A decreases and of B increases in the time interval dt

$$\frac{dB}{dt} = k \cdot A \tag{1.5}$$

Nondeterministic models exhibit a random component and consider discrete changes of system states in time. They can be represented by stochastic models using a current probability distribution of numbers of molecules. The probability of transformation of molecule A into molecule B in a time interval dt is

$$P(a-1, t+dt|a, t) = k \cdot A \tag{1.6}$$

with a as number of molecules of type A.

When the values of all state variables remain constant in time the system resides in a *stationary state* (or *steady state* or *fixed point*). Metabolic systems reach after a sufficiently long time such a steady state, which can be regarded as an asymptotic behavior. Oscillatory or chaotic regimes are other types of asymptotic behavior.

Petri nets represent a formalism that can work at all these levels of description. Moreover, the combination of discrete and continuous modeling in one model is possible using hybrid Petri nets.

In this book, Chaps. 4, 10, 11 consider discrete Petri nets. Chapters 5 and 12 describe applications using Boolean and Petri nets. Hybrid modeling is explained in Chaps. 6 and 13. Continuous and stochastic approaches are discussed in Chaps. 7 and 8 .

Before considering Petri net applications to systems biology, let us say some words on data resources that are necessary for the initial modeling step.

1.3 Data Resources

Similarly to the development of databases in bioinformatics, we could follow a rapid increase of new databases for systems biology driven by the exhaustive effort for exploring biochemical systems using high-throughput technologies all over the world.

Table 1.1 Repositories of microarray gene expression data

Name	URL
ArrayExpress [292]	http://www.ebi.ac.uk/microarray-as/ae/
GEO [46]	http://www.ncbi.nlm.nih.gov/geo/
CIBEX [165]	http://cibex.nig.ac.jp/index.jsp

For developing biochemical systems, data on chemical reactions, enzymes, gene expression, proteomics, protein–protein interactions (PPIs) and on pathways is necessary. In the following, we want to give a very brief overview of the main existing data resources. We restrict ourself to five databases giving references for further reading; for a review, see [287].

1.3.1 Repositories of Gene Expression Data

To store gene expression data and to provide a free distribution, the Microarray Gene Expression Data (MGED) society [262] recommends the repositories given in Table 1.1. These data repositories use standards developed by MGED, for example, the Minimum Information About a Microarray Experiment (MIAME) [47] and the Microarray Gene Expression Markup Language (MAGE-ML) [367].

ArrayExpress (version 9.5) contains a curated subset of more than 1,000 experiments, 27,586 essays and 5,067 conditions. It answers queries for condition-specific gene expression patterns and searches for biologically interesting genes/samples.

Gene Expression Omnibus (GEO) serves as a public repository for a wide range of high-throughput experimental data including microarray-based experiments measuring mRNA, miRNA, genomic DNA (arrayCGH, ChIP-chip, and SNP), and protein abundance, as well as non-array techniques such as serial analysis of gene expression (SAGE) and mass spectrometry (MS) peptide profiling. It also provides online facilities for data browsing, querying, and retrieval.

CIBEX is another public database for microarray data using the MIAME standard. It covers more than 51 experiments, 93 arrays, and 1,771 hybridizations. For further information, see [27, 44].

1.3.2 Protein–Protein Interaction Databases

Protein–protein interactions (PPI) are important for the systems-level understanding of biological processes. Experimental methods have been developed to determine and explore PPI. Thus, in the last years many databases of PPI have been created, see Table 1.2.

The Protein Protein Interaction Database (BIND) is part of the BOND (Biomolecular Object Network Database) database. BINDplus contains more than 200,000

Table 1.2 Protein–protein interaction databases

Name	URL
BINDplus [9]	http://www.thomsonreuters.com/products_services/scientific/BINDplus
DIP [332]	http://dip.doe-mbi.ucla.edu/dip/Main.cgi
MINT [425]	http://mint.bio.uniroma2.it/mint/Welcome.do
STRING [366]	http://string.embl.de/
pSTIING [286]	http://pstiing.licr.org/

curated biomolecular interactions and complexes including more than 60,000 unique gene identifiers, 1,500 organisms, and 7,555 gene ontology terms.

The Database of Interacting Proteins (DIP) is another curated database of experimentally determined interactions between proteins.

The Molecular INTeraction database (MINT) stores experimentally verified protein–protein interactions. MINT contains more than 110,944 interactions of 29,213 proteins.

Search Tool for the Retrieval of Interacting Genes/Proteins (STRING) is a database of known and predicted protein interactions. The database currently covers more than 2,483,276 proteins from 630 organisms.

pSTIING (Protein, Signalling, Transcriptional Interactions & Inflammation Networks Gateway) is a database of protein–protein, protein–lipid, protein–small molecules, and ligand–receptor interactions, receptor-cell type information, transcriptional regulatory and signal transduction modules relevant to inflammation, cell migration, and tumor genesis. It is based on curated information from the literature, biochemical experiments, functional essays and in vivo studies across human, rat, mouse, fly, worm, and yeast.

1.3.3 Pathway Databases

Pathway databases are essential in systems biology. Over the last few years, many pathway databases have been developed. Table 1.3 gives some important pathway databases. One of the first pathway databases is KEGG (*Kyoto Encyclopedia* of *Genes and Genomes*), the development of which started in 1995.

KEGG consists of several subdatabases of genes and proteins (KEGG GENES), chemical substances (KEGG LIGAND), interaction and reaction networks (KEGG PATHWAY), and hierarchies and relationships of various biological objects (KEGG BRITE). It is still the largest pathway database, covering more than 94,339 pathways in KEGG PATHWAY, 4,490,246 genes in 99 eukaryotes + 826 bacteria + 61 archaea in KEGG GENES, 15,430 compounds, 8,903 drugs, 10,969 glycans, 7,907 reactions, 11,339 reactant pairs in KEGG LIGAND, and 21,479 hierarchies in KEGG BRITE.

Table 1.3 Important pathway and pathway-containing databases

Name	URL
KEGG [180]	http://www.genome.jp/kegg/
REACTOME [176]	http://www.reactome.org/
MetaCyc [57]	http://metacyc.org/
EcoCyc [183, 184]	http://ecocyc.org/

Other databases are specific for certain organisms. REACTOME is a curated database of core pathways and reactions in human biology. The curated human data are used to infer orthologous events in 22 nonhuman species including mouse, rat, chicken, puffer fish, worm, fly, yeast, two plants, and *E. coli*.

EcoCyc and MetaCyc are databases of nonredundant, experimentally elucidated metabolic pathways. EcoCyc contains data of metabolic pathways of the bacterium *Escherichia coli* K-12 MG1655. MetaCyc covers data from different organisms. It contains more than 1,200 pathways from at least 1,600 different organisms.

1.4 Visualization

The graphical representation of biological processes in general and of Petri nets representing these processes in particular helps in understanding them and is essential to make biological sense of much of the complex data. Such pictures of networks are called network diagrams. A network diagram representing biological processes or Petri nets consists of a set of elements (called nodes or vertices) and their connections (called arcs or edges) which usually have a defined appearance and are placed in a specific layout.

Drawings of Petri nets as well as of biological networks have been done manually for a long time. Typical examples can be found on posters, in textbooks on biochemistry (e.g., [224]) and in information systems such as KEGG [180]. Such drawings are often created manually long before their use by the end-user and provide a view of the data defined by the creator. Some navigation or even animation of the dynamics of Petri nets may be supported in electronic systems, but the layout of the network is fixed and hence this type of visualization is called static visualization. Nowadays, automatic visualization and interactive exploration methods are desired and dynamic visualization, that is the generation of a Petri net image or other network diagrams on demand at the time it is needed, is that state-of-the-art.

1.4.1 Visualization Methods

There exist three main classes of algorithm for the automatic layout of networks: force-directed methods [119], layered methods [380], and orthogonal methods [37].

The general idea of force-directed methods is to simulate a system of physical forces such as spring or magnetic forces on the network and compute a distribution of the objects with low forces/energy. Layered (also called hierarchical) algorithms, which work for Petri nets and directed networks, consist of the following four phases: 1. removal of cycles in the network by changing the direction of some edges temporarily, 2. assignment of vertices to layers such that all edges have the same direction, 3. permutation of vertices within layers to reduce edge crossings, and 4. assignment of coordinates to vertices and bend points of edges. Orthogonal methods place the vertices of the graph on a grid and use horizontal and vertical segments to represent the edges.

In addition to these main methods, a number of other layout techniques exist, which are either infrequently used or relatively straight forward to implement, but often unsatisfactory in their result. Examples are grid layout, which places all vertices on a grid, circular layout which puts all vertices on a circle, visibility representations which uses horizontal line segments for vertices and vertical ones for edges, and matrix representations which shows the adjacency matrix of a graph. It should be also noted that most layout algorithms consider vertices as uniformly sized. To obtain good drawings of networks or Petri nets containing vertices of different size or shape, extended methods are necessary. Furthermore, most layout algorithms have been developed for drawings in two dimensional space. However, some methods can be easily extended to three dimensions such as force-directed methods.

The question whether a layout of a network is a good one or not is an aesthetic question and hard to answer. However, there are several commonly agreed aesthetic criteria and quality measures which a good layout should fulfill, for example, a low number of edge crossings, no vertex–vertex or vertex–edge overlaps, and emphasizing symmetries. Furthermore, there are several application specific requirements to fulfill, for example, emphasizing the main direction or clustering of parts of a Petri net if they belong together. There is always a trade-off between different criteria, that is, fulfilling one criteria often prevents the layout method from attaining another one.

Although some solutions for visualizing the dynamics or properties of Petri nets have been proposed (e.g., [58, 201]), the layout of Petri nets is usually done with standard algorithms as described previously. However, these automatic layouts have drawbacks, such as that specific placement constraints are not taken into account and that interactive network layouts with similar subsequent drawings are not sufficiently supported. A recently proposed method [107, 344] could be adapted to Petri nets to overcome these problems. For Petri net visualization tools, see Chaps. 6 and 8.

1.4.2 Systems Biology Graphical Notation

Networks in biology are often visualized in many different ways. However, uniform network nomenclatures with a limited number of easily recognizable symbols are

Table 1.4 Visualization tools for biochemical networks

Name	URL
GenMAPP [86]	http://www.genmapp.org/
VANTED [177]	http://vanted.ipk-gatersleben.de/
BioLayout [110]	http://www.biolayout.org/
Cytoscape [358]	http://www.cytoscape.org/
CellDesigner [120]	http://www.celldesigner.org/

common in many other areas such as engineering sciences and physics. Such strict specification builds the basis for unlimited and unambiguous scientific communication. In biology, several attempts have been made to introduce such nomenclatures, but none of them have been able to become a standard. Recently, all previous efforts were combined within the framework of a consortium developing a uniform notation for the representation of biological systems: the Systems Biology Graphical Notation (SBGN) [229]. SBGN comprises a small number of easily recognizable elements (glyphs) and can be applied for the representation of various kinds of biological networks. It combines three languages, (1) Process Description, (2) Entity Relationships, and (3) Activity Flow. They allow the representation of different levels of granularity.

Process Description is the most detailed presentation of biochemical systems which focus on the mechanistics of the underlying processes. It consists of entity pool nodes, process nodes, connecting arcs, and some other elements. There is a similarity between entity pool and process nodes of SBGN and places and transitions of Petri nets, respectively, thus SBGN maps and Petri nets can be translated into each other.

1.4.3 Visualization Tools

Several tools have been developed for editing and visualizing data of biochemical systems (for comparisons, see, [204, 379]). Some databases, such as MINT, pSTI-ING and STRING use their own visualization tools. Table 1.4 gives some examples for visualization programs of biological network data.

GenMAPP visualizes maps of biological pathways and groups of genes. It also provides programs to globally analyze gene expression or genomic data in the context of pathway maps and gene ontology terms.

VANTED facilitates editing graphs, which may represent biological pathways or functional hierarchies. Experimental datasets can be mapped onto the graph elements, networks can be automatically, layouted and statistic functions for evaluation of the data can be applied. The VANTED extension SBGN-ED also supports all three types of SBGN maps.

Cytoscape visualizes molecular interaction networks and biological pathways, and integrates these networks with annotations, gene expression profiles, and other data.

BioLayout Express3D is another tool for the visualization and analysis of networks derived from biological systems. BioLayout has been designed for the visualization of large network graphs in two- and three-dimensional space.

CellDesigner is a structured diagram editor for drawing gene-regulatory and biochemical networks. CellDesigner uses standards for representing models of biochemical and gene-regulatory networks. Networks are able to link with simulation and other analysis packages through the Systems Biology Workbench.

1.5 Petri Nets in Biology

Petri nets were defined by Carl Adam Petri in his dissertation thesis in 1962 [299]. The graphical representation, together with the rules for coarsening and refinement, and the application to chemical processes were already proposed in August 1939 by Petri [300]. The aim was to define a mathematical formalism for representing and analyzing causal systems with concurrent processes. To represent concurrent processes, two types of vertices were introduced, separating the active ones (the transitions) from the passive ones (the places). Tokens as discrete objects implement the dynamics in the system. The Petri net foundations are described in Chap. 3.

1.5.1 Motivation for Using Petri Nets

The motivation for using Petri nets to model biochemical systems comes mainly from the fact that biochemical systems exhibit many concurrent reactions, similar to concurrent processes in technical systems. The intuitive description of chemical processes coupled with the possibility to simulate and analyze token movement, representing substance or information flow in biochemical systems, facilitates the idea to use Petri net formalism for systems biology.

Using a discrete representation allows us to explore the main aspects of the possible system's behavior without any knowledge of kinetic data, on the basis of network topology only. Moreover, possible system states can be analyzed, finding deadlocks, traps, and siphons, which can be interpreted in a biological sense. The animation of Petri net models provides additional insights into a system's behavior. The possibility to involve time, while still working at a discrete level, offers interesting facilities for estimating systems dynamics without kinetic data [308]. Finally, the well-known stochastic and kinetic approaches can be easily involved into Petri net formalism [208].

1.5.2 Petri Nets in Biology

The application to biochemical systems started with the papers in 1993 and 1994 by Reddy et al. [314] and by Hofestädt [162], respectively. Reddy et al. introduce a method of representation of metabolic pathways as Petri nets, and illustrate some useful properties, for example, liveness, reachability, reversibility, fairness, and invariant properties. As an example, they model fructose metabolism as a Petri net. Hofestädt describes the metabolic process depending on expressed genes as a Petri net. He gives examples for modeling biosynthesis, protein biosynthesis, and cell communication processes, considering the isoleucine biosynthesis in *E. coli*.

In the meantime, many very different applications have been published, ranging from various types of Petri nets to many diverse biological applications. Thus, colored Petri nets [124], hybrid Petri nets [253–255], continuous Petri nets [163], stochastic Petri nets [141], and fuzzy Petri nets [410] have been applied to model biochemical systems.

These techniques have been used to analyze metabolic systems, for example, for modeling the glycolysis and pentose phosphate pathway (PPP) [162, 249, 314, 407], the main carbon metabolism in *Tuberosum solanum* (potato) tubers [211], and others. Signal transduction pathways, such as in apoptosis [153, 254], the pheromone response [327], the filamentous and the HOG-pathways in *Saccharomyces cerevisiae* (baker's yeast), and the EGFR pathway [288] have been modeled. Also assembly processes of complexes, for example, of the human spliceosomal subunit U1 [190] and of the whole spliceosome have been modeled [41] as Petri net.

Gene regulatory networks have been considered in various papers [96, 253–255], for example, the modeling processes in Duchenne Muscular Dystrophy [147]. The use of Petri nets in medical applications plays a major role because, in particular, for medicine there is a lack of in vivo and in vitro kinetic data due to experimental difficulties and ethical reasons. Moreover, in contrast to few kinetic data we have lots of qualitative data which arise with the high-throughput technologies, for example, huge amount of gene expression data. Thus, discrete modeling is particularly of advantage to obtain at least some insights into possible system behavior. Examples for modeling systems of medical interest are the Petri net models of Duchenne Muscular Dystrophy [147] and of iron homeostasis in humans [328].

Our book compiles many of these approaches providing an overview of recent research activities in that field. We try not only to give an impression of the wide range of applications, but also to explain them including their extensions of Petri net techniques.

1.6 Problems

1.1 What is new about systems biology compared to classical biology?

1.2 What is the purpose of a model?

1.3 How can models be classified?

1.4 What are the differences between discrete, continuous, and stochastic modeling?

1.5 Give examples for main data resources in modeling biochemical systems.

1.6 Why are Petri nets suitable to model biochemical systems and why, in particular, for modeling medical systems?

Chapter 2
Biochemical Fundamentals

Tiina Liiving, Syed M. Baker,
and Björn H. Junker

Abstract Living organisms are among the most complex phenomena in our world. To describe, model, and simulate living organisms or at least parts thereof, formal descriptions such as Petri nets are needed. As the focus of this book is the use of Petri net theory in biology, the readership will be very diverse. Thus, this chapter is meant to provide a general introduction to biology, especially those areas that will be modeled with the use of Petri net approaches throughout this book. The experienced biochemist might want to skip this chapter, but for computer scientists and readers from similar fields this chapter contains important fundamentals.

2.1 Cell Biology

A cell is the smallest highly organized basic life form from which all living organisms are built. Thus, a cell might be called the "building block of life". The word *cell* comes from the Latin word *cellula* which means *small room*. Each cell is self-contained and self-maintaining and has its own set of instructions for carrying out all of its activities.

Cells can be categorized in two main types, eukaryotic and prokaryotic, which principally differ from each other by having or not having a membrane enclosed nucleus. The names of the cell types indicate this property as in Greek *karyose* means *kernel*, *pro* means *before*, and *eu* means *true* or *good*. Eukaryotic cells have

T. Liiving · S.M. Baker · B.H. Junker (✉)
Leibniz Institute of Plant Genetics and Crop Plant Research (IPK), Corrensstr. 3,
06466 Gatersleben, Germany
e-mail: junker@ipk-gatersleben.de

T. Liiving
e-mail: liiving@ipk-gatersleben.de

S.M. Baker
e-mail: baker@ipk-gatersleben.de

I. Koch et al. (eds.), *Modeling in Systems Biology*, Computational Biology 16,
DOI 10.1007/978-1-84996-474-6_2,

membrane enveloped compartments called organelles. Only Bacteria and Archaea have prokaryotic cells and all other organisms have eukaryotic cells.

2.1.1 Cellular Organization

All cells consist of a cytoplasm surrounded by a plasma membrane, also called plasmalemma, as a border. The plasma membrane can be surrounded by a cell wall. Cells of plants, algae, and fungi have cell walls, whose main contents are cellulose or chitin. Prokaryotes have different cell wall composition than eukaryotes; a bacterial cell wall is built up mainly of peptidoglycan and an Archae cell wall of protein. Animal cells and protozoans lack a cell wall, instead they usually have some other type of covering. In most cases, animal and protozoan cells are surrounded by an extracellular matrix, which has the same basic function as the cell wall, but is more flexible. The extracellular matrix, which consists of the space between cells, is filled with polysaccharides and proteins [221]. The cytoplasm contains membrane-surrounded organelles, and the cytosol, which is the space between the organelles. Membranes in prokaryotic and eukaryotic cells are very similar and are composed of two layers of lipids into which the proteins are incorporated.

There is great variation in the size of a cell. It is organism and tissue specific and also depends on the cell's developmental stage. While most prokaryotic cells are usually up to some micrometer in diameter, eukaryotic cells can reach up to 100 μm in diameter. A typical prokaryotic cell has a simple internal structure and no membrane-surrounded organelles (see Fig. 2.1). It is thought that prokaryote cells have to remain small in order to keep the metabolism and substance diffusion in high rate. Metabolism of prokaryotes takes place largely in the cytosol and the substrates can diffuse very quickly over the cell. Organelles in a typical eukaryotic cell are the nucleus, mitochondria, the endoplasmatic reticulum, the Golgi apparatus, lysosomes and peroxisomes. Usually cell organelles have flexible shape and size (about 1 to 5 μm in diameter). Some special organelles can be only found in plants, such as a

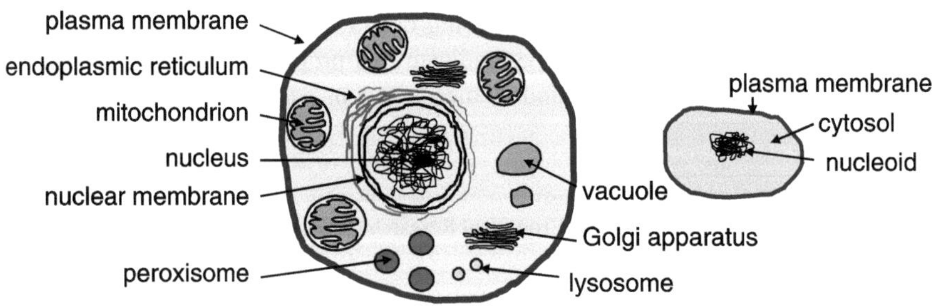

Fig. 2.1 Schematic structure of a typical eukaryotic (*left*) and prokaryotic cell (*right*). The eukaryotic cell contains the cytosol, which is surrounded by a plasma membrane and in turn contains membrane-surrounded organelles. The typical prokaryotic cell does not contain differentiated organelles

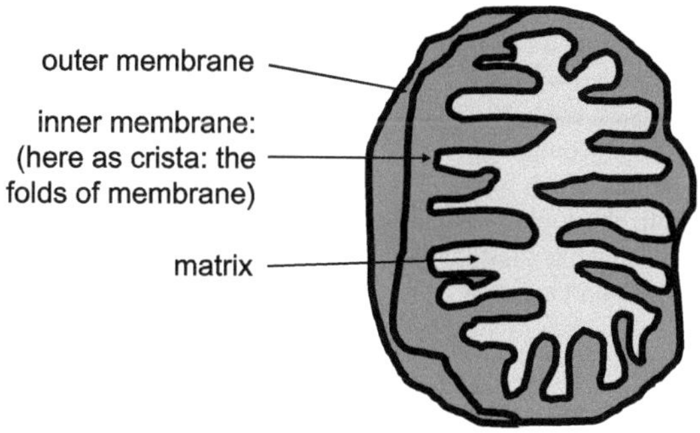

Fig. 2.2 Simplified structure of a mitochondrion: two membranes surround the matrix, in which the main metabolic steps occur. The inner membrane forms folds called cristae, which plays an important role in ATP synthesis

big central vacuole and plastids. In the next sections, the central role of mitochondria and plastids will be discussed.

For a deeper understanding of cell biology, the reader might refer to detailed textbooks [6].

2.1.2 *Mitochondria*

In any organism, energy is needed for most vital functions. Mitochondria are the main place for energy production in all eukaryotic cells and there are usually several hundred mitochondria in one cell. A mitochondrion is an approximately 0.5 to 1 μm long organelle, surrounded by two layers of membranes. The inner membrane is usually folded, these folds are called *cristae*. Inside of mitochondria there is a cytosol like compartment called the *matrix*.

The chemical reactions in an organism usually take place in the presence of adenine triphosphate (ATP), which contains chemical energy in the form of covalent bonds between phosphates. ATP consists of one adenine, one ribose (sugar) molecule, and three phosphate groups, which can be detached to provide energy for other chemical reactions. Most of the ATP is generated in mitochondria; plant cells are an exception, in which the ATP can also be produced in chloroplasts when exposed to light. Nicotine adenine dinucleotide phosphate (NADP), which is a reducing agent important for many metabolic reactions, is also produced in the mitochondrion.

Beside the production of ATP the mitochondria has many other functions in cellular metabolism. Important metabolic pathways located in mitochondria are the citric acid cycle and oxidative phosphorylation. Mitochondria participate in the metabolism of several essential substances like amino acids, steroids, heme groups, and iron-sulphur (Fe-S) clusters. Furthermore, heat production and storage of calcium ions are regulated by mitochondria.

Figure 2.2 describes a very simplified picture of mitochondria.

For better understanding of the mitochondria and its functions, the reader is referred to recent review articles of [235].

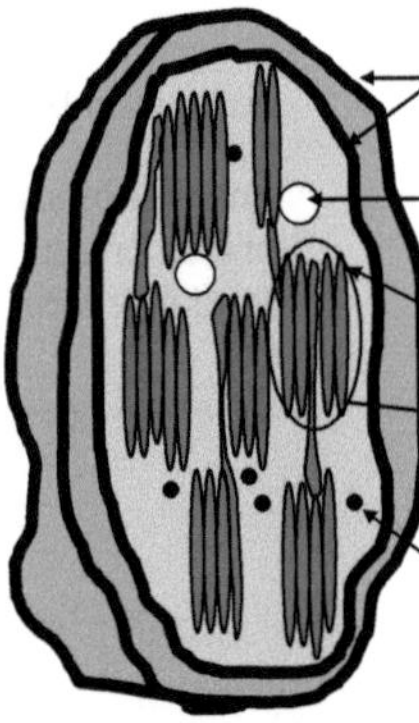

Fig. 2.3 Simplified structure of chloroplast: a chloroplast is surrounded with two membranes; contain chlorophyll containing thylakoids, which form granum; chloroplast stroma usually contains plastoglobuli and some starch grains

2.1.3 Plastids

Plastids are organelles exclusively found in the plant kingdom. Plastids are crucial to plant functionality and develop from proplastids to generate different plastid forms. Proplastids are generally much smaller than derived plastids and do not contain pigments (e.g., chlorophyll). Several metabolic pathways take place only in plastids, such as synthesis and storage of starch, and synthesis of some pigments (e.g., carotenoids). Undoubtedly, the most important function of plastids is the process of photosynthesis, which takes place in special plastids, called chloroplasts. By photosynthesis, light energy is converted to chemical energy. Therefore in addition to mitochondria, plastids are an important place for energy production in plants.

Depending on their morphology and function, plastids have the ability to differentiate, or redifferentiate. Differentiation therefore depends on the developmental stage of the organism, the specific tissue and on environmental impulses. Plastids can be grouped to chloroplasts, chromoplasts and leucoplasts according to their main functions and the accumulating of substances. Leucoplasts contain no pigments and to this group belong amyloplasts (starch accumulation), elaioplasts (oil accumulation), proteinoplasts (protein accumulation) and combinations of these. Number and size of plastids in a plant cell are similar to mitochondria, that is, up to several hundred plastids each with a size of some micrometers. Plastids are surrounded by two membranes called the *envelope*.

In chloroplasts (see Fig. 2.3), photosynthesis takes place and therefore they are present in all photosynthetic tissues and organs such as leaves, green stems, cotyledons and hypocotyls, unripe fruits as well as seed coats and embryos. In one photosynthetic cell, there can be a few to hundreds of chloroplasts. Similar to mitochondria, in chloroplasts the inner membrane is folded forming structures called *thylakoids*, to which chlorophyll is bound. Here, the process of harvesting light energy takes place. Inside of the plastids there is a cytosol like compartment called *stroma*, in which the actual carbon fixation from carbon dioxide and synthesis of the basic units for carbohydrate takes place.

Chromoplasts are red-, orange- and yellow-colored plastids containing relatively high levels of carotenoid pigments. Carotenoids are located also in chloroplasts, where they play an essential role by stabilizing chlorophyll. Chromoplasts often develop from chloroplasts, but may also be formed from proplastids and amyloplasts.

Amyloplasts are plastids specialized for storage starch accumulation and are found in roots, tubers, seeds, and other storage tissues.

For further reading about plastids, we refer the reader to detailed textbooks [40, 51].

2.2 Metabolism

Metabolism comprises the set of reactions that occur in living organisms for the production and degradation of organic compounds needed for an organism's vital functions. Metabolism is essential for life and transforms the input substrates into required products.

Metabolism can be divided into two categories: catabolism and anabolism. Catabolism breaks down organic matter to harvest energy in cellular respiration and also thereby can produce substrates for anabolic reactions. The chemical energy produced is in the form of ATP. Anabolism is the process of building new compounds from the basic units, which requires energy input from the outside or from catabolic reactions.

Cell metabolism is usually divided into primary and secondary metabolism. Primary metabolism can be defined as all of the processes essential for growth and development of an organism. Primary metabolism is considered a complex network of carbohydrate, fatty acid, protein and nucleic acid metabolic pathways and is largely similar in all organisms. Carbohydrate metabolism and energy metabolism may be called as a central metabolism, because they involve the production of basic structures for the other metabolic pathways. Primary metabolism also supplies the substrates for secondary metabolism. Secondary metabolism in contrast is a term for pathways that produce metabolites not absolutely essential for survival of the organism (see also Sect. 2.2.2).

2.2.1 *Metabolic Pathways and Networks*

The chemical reactions of metabolism are organized into metabolic pathways, in which one chemical is transformed into another by the help of enzymes. Enzymes are crucial to metabolism because they allow organisms to drive desirable but thermodynamically unfavorable reactions by coupling them to favorable ones and by lowering the activation energy. Pathways may differ between organisms, but some basic metabolic pathways are conserved between different organisms.

Many metabolic pathways are linear, that is, they begin with a specific substrate and end with a specific product. Some pathways, such as the citric acid cycle, are cyclic, that is, the end product can be again used for starting this specific metabolic pathway. Metabolic pathways usually have several chemical reactions, which means they generate several intermediates (metabolites which are not end product of the pathway) and the metabolic pathway can be connected to other pathways through these intermediates. Furthermore, pathways are connected to each other as a metabolite from one pathway can be a substrate for the next pathway. A collection of pathways is called the metabolic network, which is relatively dense due to many connections between different pathways.

2.2.2 Metabolites

A metabolite is a small molecule produced within the cell and participating in metabolic reactions. A primary metabolite is essential for growth, development, and/or reproduction of an organism. Amino acids are primary metabolites being the basis for proteins and relevant for many biochemical processes. Many important primary metabolites belong to the class of sugars like glucose and fructose, or contain sugar chains. A secondary metabolite has some other, less vital function. Examples of secondary metabolites are antibiotics, pigments, and hormones.

2.2.3 Enzymes

As most chemical reactions are relatively slow, there is a need for catalysts. Enzymes are proteins that catalyze (i.e., increase the rates of) chemical reactions without being consumed, and are therefore essential for metabolism. Almost all processes in a biological cell need enzymes to occur at significant rates. Like all catalysts, enzymes work by lowering the activation energy (the energy that is required to activate a process) for a reaction, thus dramatically increasing the rate of the reaction. Most enzyme reaction rates are orders of magnitude faster than those of comparable spontaneous reactions.

Enzymes are selective for their substrates and speed up only up to a few reactions. The set of enzymes produced in a cell determines which metabolic pathways occur in that cell. Without enzymes, metabolism would neither progress through the same steps, nor be fast enough to serve the needs of the cell. An enzyme-catalyzed reaction starts by binding the substrate at a special place of the enzyme, called the active site. The active site usually is shaped in a particular way to allow interactions with the substrate, which results in binding of the substrate [212]. In many cases, the substrate also changes shape slightly as it enters the active site. After the enzyme has catalyzed the reaction, the new product is released. Figure 2.4 describes this process.

In a cell, chemical reactions can be regulated by several enzymes. Enzymes that catalyze the same chemical reaction but differ in their amino acid sequence are called *isozymes* (also *isoenzymes*). In some cases, slightly different enzymes are formed from one gene, in this case the two proteins are called *isoforms*. Both

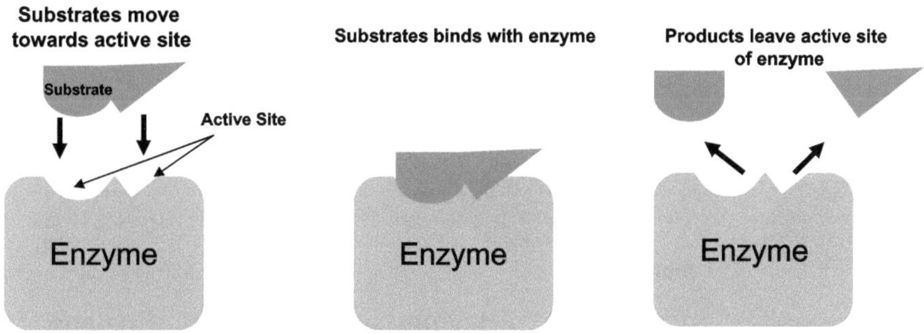

Fig. 2.4 Basic working process of an enzyme. The substrate binds to the enzyme in a site called the active site. After binding of the substrate, the enzyme catalyzes the reaction to create the product. The product is then released from the active site

isozymes and isoforms may display different kinetic parameters and/or regulatory properties. The parallel presence of isozymes/isoforms is needed to regulate the cell metabolism according to the needs of a given tissue or organ and/or to meet the needs of a developmental stage. Furthermore, different isozymes might be targeted to different cell compartments (cell organelles or cell parts) and thus allow compartment-specific regulation of the respective enzyme activity.

Enzymes can act alone or with the help of several factors, such as metal ions or organic molecules (called cofactors). The velocity of enzymatic reactions is affected by other molecules, environmental conditions, as well as the substrate and product concentrations. These regulatory factors are called inhibitors or activators, depending on the direction of the regulation (up or down).

2.2.4 Enzyme Inhibition

Molecules which bind to the enzyme and thereby decrease its activity are called inhibitors. Inhibition can be reversible or irreversible, depending on the inhibitor. The major role in regulation of synthesis may lay in reversible inhibition, in which enzyme activity is the same after removal of the inhibitor. There are basically three types of inhibition: competitive, noncompetitive and uncompetitive inhibition.

In competitive inhibition, the substrate has to compete with the inhibitor for binding to the active site. Commonly, the chemical structure of the competitive inhibitor resembles the chemical structure of the substrate. The degree of inhibition of the specific reaction in the cell depends on substrate and inhibitor concentrations. The principle of competitive inhibition is outlined in Fig. 2.5.

In noncompetitive inhibition, the inhibitor binds to some place other than the active site of the substrate in the enzyme and reshapes the enzyme in a way that the active site for a substrate is changed, and thus the substrate can not bind to the enzyme (see Fig. 2.6). In other words, noncompetitive inhibition can be described as

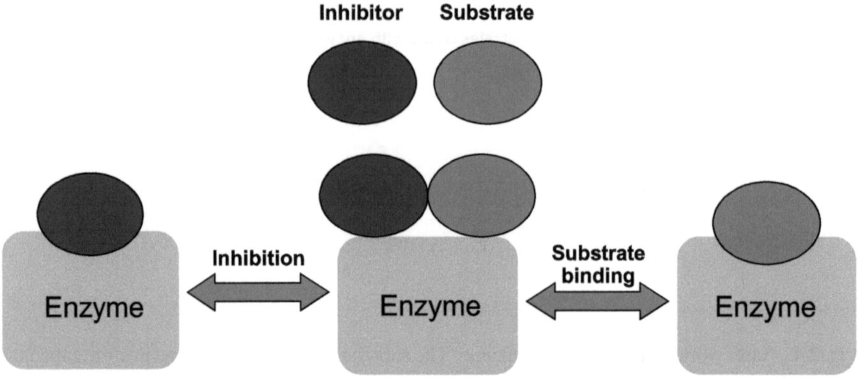

Fig. 2.5 Working procedure of competitive inhibition. In this type of inhibition, both the inhibitor and the substrate compete for the same active site of the enzyme. Both of them have the capability to bind with the enzyme in the active site. This creates a competitive environment between the substrate and the inhibitor to bind with the enzyme

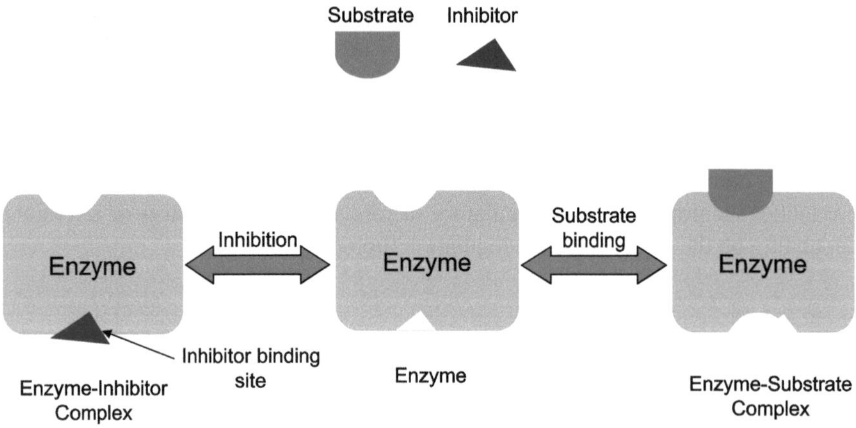

Fig. 2.6 Working process of noncompetitive inhibition. In this type of inhibition, the inhibitor binds to some place other than the active site. This place is called the allosteric site. By binding to the allosteric site, the inhibitor changes the structure and shape of the enzyme, so that the substrate can no longer bind properly with the enzyme and thus reduces the maximum rate of the chemical reaction

allosteric inhibition. Allostery (from the Greek "other site") means that the regulatory binding site of the inhibitor and the substrate binding site (the active site) are physically separate.

An uncompetitive inhibition means that the inhibitor binds to its regulatory site in the enzyme after the substrate has bound to the active site forming an inactive substrate-inhibitor-enzyme complex. For a deeper understanding of enzyme inhibition, the article [324] is suggested.

2.2.4.1 Enzyme Kinetics and Activity

Enzyme activity describes the use of the substrate or the amount of formed products. The SI unit for enzyme activity is katal (kat), which is the conversion of one mole substrate into a new product in one second (mol s^{-1}). The activity can also be expressed in enzyme units (U), which describe the amount of the enzyme that catalyzes the conversion of one micro mole of substrate per minute (1 U = 1 μmol min^{-1}).

$$1\ \mathrm{U} = \frac{1}{60}\ \text{micro katal} = 16.67\ \text{nano katal}$$

The term enzyme kinetics refers to the study of the speed, also called rate or velocity, of an enzyme-catalyzed reaction. Reaction velocity depends on the substrate concentration as well as the environment, especially on temperature, pressure and pH value of the surrounding medium. The velocity of enzymatic reactions describes the speed of the reaction, which is composed of 3 steps: substrate binding to the enzyme, the process of production of the new product, and release of the product. The following equation describes this process in a simplified way:

$$E + S \rightarrow ES \rightarrow E + P$$

The equation describes an irreversible reaction in which E is the enzyme, S is the substrate, ES is the enzyme-substrate complex, and P is the product.

The reaction velocity v is equal to the rate of formation of P or the rate of reduction of S. The following equation expresses the relationship between the reaction velocity and the change of concentration of substrate and product with time.

$$v = -\frac{d[S]}{dt} = \frac{d[P]}{dt}$$

A graph of product concentration vs. time is given in Fig. 2.7. The time-dependent behavior of the product concentration can be divided into three stages.

It is difficult to fit a curve to a graph of product as a function of time, as it ignores the transient phase and assumes that the reaction is irreversible. Therefore, the enzyme velocity is typically described as a function of substrate concentration as depicted in Fig. 2.8.

The simple velocity function shown above can be described by the Michaelis–Menten equation

$$v = \frac{V_{\max}[S]}{K_m + [S]}$$

K_m (Michaelis-constant) is (roughly) an inverse measure of the affinity, i.e. the strength of binding between the enzyme and substrate. The lower the K_m, the greater is the affinity, which means that lower concentrations of substrate are needed to achieve a certain velocity of the turnover of substrate (the amount of product produced per unit time). K_m is measured as the substrate concentration at half of maximum velocity ($v_{\max}/2$). Maximal velocity is the maximum rate of the reaction, which occurs when the enzyme is completely saturated with substrate.

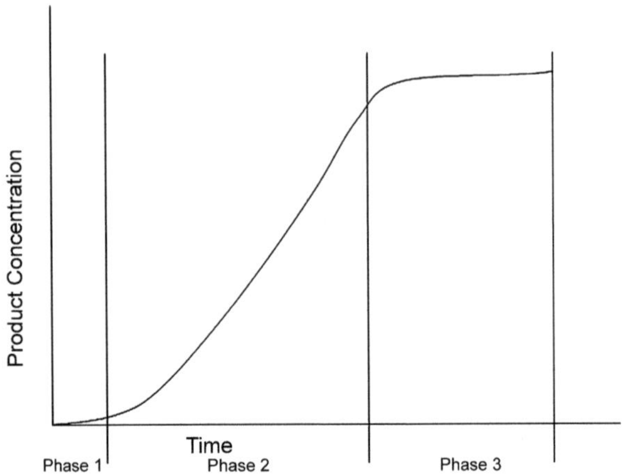

Fig. 2.7 Time-course of product formation in a typical enzymatic reaction. The graph shows the change of product concentration over time. With the passing of time, the rate of product accumulation also increases. This product accumulation can be divided into three phases. At the start of the reaction, the product concentration is relatively low (phase 1). Then for an extended period of time, the product concentration increases in a nearly linear manner with time (phase 2). During the last periods the enzyme is saturated, so the curve starts to level off (phase 3)

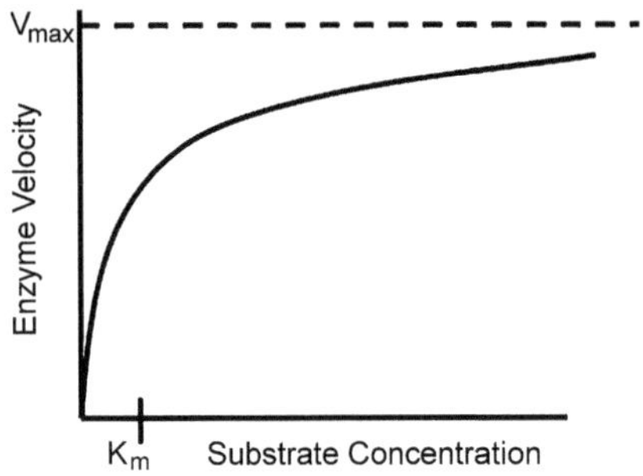

Fig. 2.8 Substrate concentration versus reaction rate of a typical enzymatic reaction. V_{max} is the maximal possible reaction rate. K_m is the substrate concentration at which the reaction reaches $1/2V_{max}$. This graph can be described with the Michaelis–Menten equation, see text

Inhibition has an effect on the v_{max} and/or K_m values. For *competitive inhibition*, in the presence of an inhibitor, a higher substrate concentration is required to achieve the same velocities that were reached in its absence. So while v_{max} can still be reached if sufficient substrate is available, one-half v_{max} requires a higher $[S]$ than before and thus the apparent value of K_m is larger than without inhibition . For *noncompetitive inhibition*, enzyme molecules that have been bound by the inhibitor are taken out of the reaction so that the reaction rate is reduced for all values of $[S]$, including v_{max} and one-half v_{max}, but K_m remains unchanged because the active site of those enzyme molecules that have not been inhibited is unchanged. For further reading on enzyme kinetics, the reader is referred to a textbook [38].

2.2.5 Central Metabolic Pathways

Central metabolic pathways are usually considered as pathways for the synthesis of carbohydrates, proteins and fatty acids. These are essential pathways as they produce energy and metabolites which are the starting point for many further products. Common central metabolic pathways are glycolysis, the citric acid cycle, the pentose phosphate pathway, and fatty acid synthesis, each of which will be described in detail below. Central metabolic pathways can be differently defined for plants and animal cells, or for heterotrophic and autotrophic cells. Autotrophic cells can produce organic substances from nonorganic elements (e.g., photosynthesis in plants). Heterotrophic organisms or cells can not produce organic substances from nonorganic elements and are therefore dependent on autotrophs. Plants are *mixotroph organisms*; they contain both, autotrophic and heterotrophic cells.

2.2.5.1 Glycolysis

Glycolysis is an essential part of energy generation in a cell and takes place in the cytosol. Glycolysis is the metabolic pathway that converts glucose, which is the main energy source of most of the organisms, into energy. The energy released in this process is used to form the high energy compound ATP and the reductant NADH (reduced nicotinamide adenine dinucleotide). In addition, the end product pyruvate is a relevant starting point for many other metabolic pathways.

The overall process of all steps of glycolysis is:

$$\begin{aligned} &\text{Glucose} + 2\ \text{NAD}^+ + 2\ \text{P}_i + 2\ \text{ADP} \\ &\quad\rightarrow 2\ \text{pyruvate} + 2\ \text{NADH} + 2\ \text{ATP} + 2\ \text{H}^+ + 2\ \text{H}_2\text{O} \end{aligned}$$

For a more detailed introduction into glycolysis, the reader might refer to the review article of [305].

2.2.5.2 The Citric Acid Cycle

The citric acid cycle, also named tricarboxylic acid (TCA) cycle or Krebs cycle, is important as an energy generator, and it is involved in the chemical conversion of carbohydrates, fats and proteins into carbon dioxide and water. In eukaryotic cells, the citric acid cycle takes place in the mitochondrial matrix. Pyruvate, which is the end product of glycolysis, is transported from the cytosol into the mitochondrial matrix. There it is transformed by the enzyme pyruvate dehydrogenase under addition of Coenzyme A into acetyl-CoA and CO_2. Coenzyme A carries a thiol group and reacts with carboxylic acids to form thioesters, thus functioning as an acyl group carrier. Acetyl-CoA is the primary substrate entering the citric acid cycle.

The citric acid cycle can be called the second step in respiration. The citric acid cycle oxidizes acetyl-CoA to carbon dioxide, and produces energy carriers like ATP and GTP (guanosin triphosphate, generated in animal cells) and molecules relevant to redox reactions, such as NADH. Beside that the intermediates of the citric acid cycle have a big relevance as they are substrates for many other biosynthetic pathways, for example, the synthesis of several amino acids.

In a process called oxidative phosphorylation, the reducing equivalents that are generated from TCA cycle activity are used by the electron transport chain to drive the synthesis of ATP. For this, the protons and electrons from NADH are transported at the mitochondrial inner membrane. The resultant potential across the membrane is used to drive ATP synthesis.

2.2.5.3 The Oxidative Pentose Phosphate Pathway

The oxidative pentose phosphate pathway (OPPP) has the role to produce the reductant NADPH and metabolic intermediates for several biosynthetic pathways including synthesis of nucleotides, aromatic amino acids, and phenylpropanoids. In most cells, the OPPP takes place in the cytosol, but in plants it is mainly localized in plastids. In plants, NADPH is also synthesized by photosynthesis, the OPPP is only essential for nonphotosynthetic cells.

The overall sum reaction of oxidative pentose phosphate pathway is:

$$\begin{aligned} &3\ \text{glucose-6-phosphate} + 6\ \text{NADP}^+ \rightarrow 6\ \text{NADPH} + 6\ \text{H}^+ \\ &\quad + 2\ \text{fructose-6-phosphate} + \text{glycerinaldehyde-3-phosphate} + 3\ \text{CO}_2 \end{aligned}$$

For a better understanding of the role of OPPP, the review of [214] is suggested.

2.2.5.4 Fatty Acid Synthesis

Fatty acids are usually synthesized in the cytosol. However, in plants their synthesis also takes place in the chloroplast stroma in photosynthetically active cells. Fatty acids constitute an energy storage form, but also play an important role in cell structure as a membrane compound. Storage lipids are formed by esterification of glycerol with up to three fatty acids.

Fatty acids are synthesized from acetyl-CoA, which in eukaryotic cells is synthesized in the mitochondrion, and thus has to be transported to the cytosol in order to be available for fatty acid synthesis. This is achieved by exporting citrate to the cytosol and cleaving it into oxalacetate and acetyl-CoA.

2.2.6 Metabolic Networks

The functions of a living cell are regulated by different networks of interacting biochemical components. How these molecules are connected to each other and what are their influences on the activity of each of the reactions under diverse physiological conditions is a central issue in understanding cellular organization. As mentioned

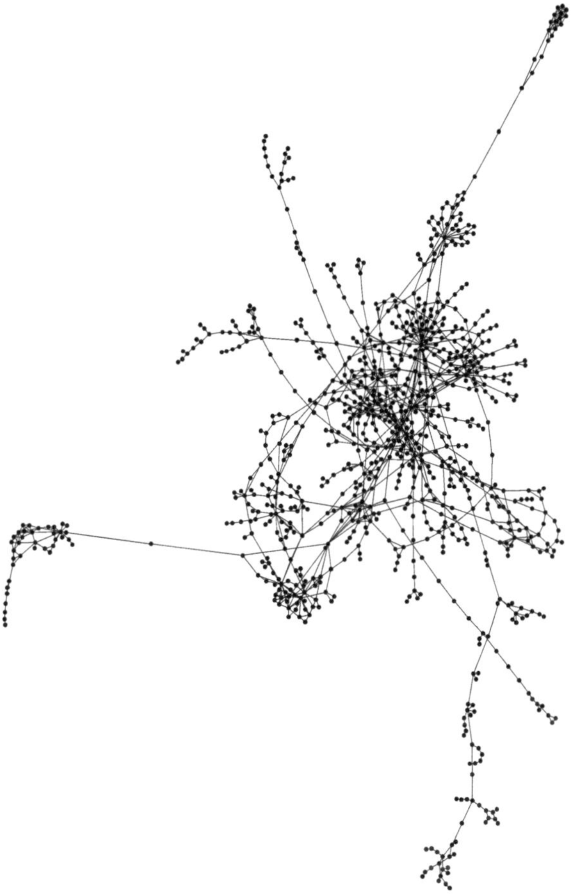

Fig. 2.9 Schematic view of a metabolic network. The nodes in the network are the metabolites and the edges between them are the enzymes. The figure was created using the software VANTED (Visualization and Analysis of Networks containing Experimental Data) [177]

before, the metabolic pathways can be connected to each other and produce starting substrates or intermediates for other metabolic pathways. All the pathways together form a large metabolic network. This metabolic network comprises the chemical reactions of metabolism as well as the regulatory interactions that guide these reactions. Figure 2.9 shows a major component of the metabolic network in the model plant *Arabidopsis thaliana*.

2.2.7 Regulation of Metabolism

A cell contains a large number of molecules. To ensure that every molecule is produced in the correct amounts at the time required, the cell must have a control device for their production and consumption. Metabolism is regulated through several factors, which can be both source-based and environment-based (temperature, light). In addition metabolism also depends on the developmental stage of the organism as well as the tissue function. This means that the current needs of the cell influences the metabolism. In the simplest case, the regulation of a metabolic reaction is achieved by controlling two issues: (1) the availability of enzymes or (2) the activity of the enzyme. More complicated mechanisms of metabolic regulation comprise, but are not limited to, redox regulation, allosteric regulation, and feedback mechanisms.

2.3 Gene Expression

Gene expression is the mechanism by which proteins are produced from DNA (deoxyribonucleic acid). The main role of DNA in the cell is the storage of long-term information. DNA contains literally the information needed to construct and maintain all components of cells, as well as to mediate all regulatory processes. DNA is a molecule composed of a chain of four different types of nucleotides (also called bases) named adenine (A), thymine (T), guanine (G), and cytosine (C). DNA has a double helix structure and each strand runs opposite to the other which makes them anti-parallel. The backbone of the DNA consists of the sugar ribose and phosphate groups.

Different parts of DNA chain have various roles. The DNA segments that carry the genetic information are called genes. Other segments of DNA sequences have structural purposes, or are involved in regulating the use of this genetic information.

The whole process of gene expression can be simply divided into two major stages: transcription and translation (see Fig. 2.10).

2.3.1 Transcription

By transcription, RNA is generated from DNA. RNA is the abbreviation of ribonucleic acid, which is a single stranded long chain of nucleotides. Instead of the base thymine it contains the base uracil (U).

Transcription is the process of the transfer of genetic information from the archival copy of DNA to RNA. The gene sequence is copied to produce a complementary nucleotide RNA strand called messenger RNA (mRNA), as it carries a genetic message from the DNA to the protein-synthesizing machinery (ribosomes) of the cell.

Transcription is a highly regulated process which is guided by the enzyme RNA polymerase. Transcription is regulated by transcription factors which suppress or promote binding of the RNA polymerase onto a specific DNA sequence. Transcription factors are short protein sequences and affect the transcription by binding to DNA.

2.3.2 Translation

Translation is the second large stage in gene expression. In this stage peptides, that is, amino acid chains, are created by decoding the information contained in the

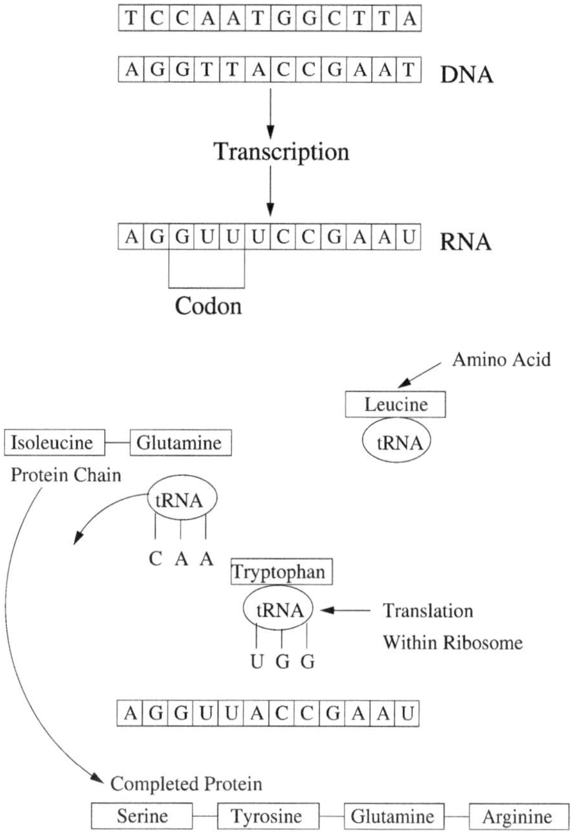

Fig. 2.10 Basic steps of gene expression. This is divided into two phases, transcription and translation. In transcription, double stranded DNA is converted in to a single stranded RNA. In translation, the RNA enters the ribosome, where it encounters the complementary codon of tRNA (transfer RNA) that carries the amino acid. Amino acids then connect with each other to create a peptide chain which is called protein

mRNA. It is mediated by the ribosomes which are located in the cytosol. This process uses an mRNA sequence as a template to guide the synthesis of a chain of amino acids that form a protein. In this process, the tRNA (transport RNA) carrying one amino acid per molecule is transported to a protein complex called the ribosome. In the ribosome, the tRNA binds specifically to a triplet of three mRNA nucleotides, which is called the codon. This is mediated by the fact that each tRNA contains an anticodon that matches a specific codon. The amino acids carried by the tRNA are joined together as a chain according to the codon chain. The polypeptide chain is converted into a three-dimensional, functional protein by specific folding procedures aided by chaperones. Furthermore, different or identical proteins can bind together to form protein complexes.

There are $4^3 = 64$ different codon combinations possible with a triplet codon of three nucleotides, but there are only 20 amino acids to code for. Thus, multiple codons might be used to encode the same amino acid. Furthermore, in the mRNA there is one start codon (AUG), which specifies the position where the peptide synthesis starts, and three stop codons (UAA, UAG, or UGA), which specify the end of peptide synthesis.

2.3.3 Gene Regulation

While some housekeeping genes are expressed all of the time and in all cells, most genes are turned on or off in a specific manner. These genes are specific to different tasks and are only required at a specific time and in specific cell types. Furthermore, some genes are expressed in certain tissues only at a certain developmental stage of the organism.

For example, the arrival of a hormone may turn on (or off) certain genes in that cell. So producing a protein at a wrong place or at a wrong time will disrupt the whole mechanism. Genes are turned off by transcription factors if there is no need for the protein that they encode or turned on when the environment changes and the proteins are once again needed. These mechanisms prevent a waste of energy.

2.3.4 Gene Regulatory Networks

Gene (or genetic) regulatory networks (GRNs) regulate the interactions between genes and more precisely the expression of genes in specific amounts and at a specific time and place. Very simplified, a GRN consists basically of a signaling pathway, a target gene and gene products. The studies of GRN therefore are focused on gene transcription to RNA, translation and protein formation as well as the interactions of these processes and products.

GRNs can be divided into many types, but two of them are basic: transcription factor network and gene expression network. The transcription factor network involves the regulatory mechanism which affects the first step of gene expression: the transcription of DNA into RNA. This network consist of transcription factor genes, transcription factors, and of regulatory molecules which regulate the transcription factor binding to a special DNA sequence called the promoter. The transcription factor network is regulated by external and cellular signals. Under gene expression, the network is meant to produce the functional gene product. The mRNAs and proteins are products of gene expression. These products can interact with each other as well as with transcription factors and therefore belong to a gene regulatory network.

For further reading and understanding of GRNs and the challenges of investigating GRNs, there are several reviews like [13, 181] and [247].

2.4 Signal Transduction

A cell has to react to the changes taking place inside itself, in surrounding cells and outside the organism. One of the most important functions of cell signaling is to control and maintain a physiological balance (called homeostasis) within the body. The reaction is controlled by handing over the signal to a receptor which leads to reactions called signal transduction. Activation of different signaling pathways leads to diverse physiological responses, such as cell proliferation, cell death, cell differentiation, and changes in metabolism.

Signals can be molecular as well as sensory in response to environmental changes such as light, pressure and temperature. Molecular signals can be simple elements, usually in the form of ions, complex inorganic as well as organic molecules. The start signal is called the stimulus, effector or elicitor. The signals are received by receptors specific to the signal and these receptors pass this message to a messenger.

There are many different signal transduction pathways which involve enzymatic reactions to activate the correct response. Signaling pathways in cells interact with each other, as there are many signals received at the same time. The signaling proteins and secondary messengers inhibit or increase the signal transduction or gene expression.

Signal transduction pathways are different depending on the signal and the receptor and usually consist of several steps of transferring the information. Usually, one signal can mediate many reactions; this is called a signal-transduction cascade, through which the response reactions can be regulated more precisely. In this process, the signal is passed over from one signaling protein to another. Signal transduction ends with a molecule which can activate the gene expression. This could be a transcription factor, by which many genes can be activated at the same time.

The term signal transduction network refers to a complex of all reactions (including interactions) in signaling from receptors to final targets that mediate the specific gene expression.

One of the best known signaling transduction systems is the mitogen activated protein kinase (MAPK) signaling pathway [85], which regulates many genes and therefore is important for several cellular processes and development. The MAPK functions as a cascade of kinases: one kinase phosphorylates (adds phosphate) the next kinase in order to propagate the signal and the last kinase phosphorylates the target protein.

In a cell there are four main groups of proteins involved in signal transduction: protein kinases, protein phosphatases, guanosine triphosphatases, and adapter proteins. Kinases and phosphatases regulate giving over the signal by adding or removing phosphate, respectively. The adapter proteins act as linkers or binders.

A clear understanding of the signal transduction pathways and the signal transduction networks is hard to achieve, because there are many participants and cross talk between the transduction pathways as well as between gene expression levels. The problem of understanding signaling pathways is well reviewed by [102].

2.4.1 Hormones and Other Signaling Molecules

There are many internal and external signaling molecules, which belong to different chemical groups. Hormones belong to signaling molecules, which are transported from one cell to another. The main role of the hormones is to regulate the growth and development of an organism. Many hormones belong to the chemical group of steroids, which is a very common group of signaling substances in mammals. In plants, there are hormone like substances called phytohormones. Furthermore, single elements like calcium or potassium play an important part of the cell signaling system. Signaling with ions is usually based on changes in the cell's electrical potential and concentration of ions.

Most external signals can not penetrate the cell, these are typically large molecules such as proteins, peptides and amines. Some signals which can penetrate the cell are light, steroids, gases, some hydrophilic molecules, and ions.

2.4.2 Receptors

Receptors are proteins specialized to detect signals and are usually membrane bound. Signals can enter the cell through the plasma membrane bound receptor proteins, ion channels, diffusion, or by active transport through the plasma membrane with the help of transport proteins.

The receptor usually gets modified by the signal and this conformation initiates signal transduction. Receptors can generate a chemical signal inside the cell by interacting with one or more proteins.

2.5 Problems

2.1 What is the role of mitochondria and plastids in a cell?

2.2 What is metabolism? Describe different categories of metabolism.

2.3 What is the role of metabolites and enzymes in a metabolic network?

2.4 How do enzymatic reactions take place?

2.5 What is described by the Michaelis-Menten equation?

2.6 What are common central metabolic pathways? Describe the role of central metabolic pathways.

2.7 What are the processes that take place in a gene expression?

2.8 Describe in your own words the role that a receptor could play for metabolism.

Acknowledgements We greatfully acknowledge Dr. Hart Poskar for his valuable comments and the German Federal Ministry of Education and Research (BMBF) for funding under grant 0315295.

Chapter 3
Petri Nets

Wolfgang Reisig

Abstract This chapter introduces the basic notions and notations, as employed in the rest of this book. This includes the distinct modeling of substances and reactions, together with their logical connection. We explain the graphical representation of Petri nets, as well as the "token flow mechanics", representing dynamic behavior. We show how sequential, alternative and concurrent occurrences of reactions are modeled, including quantitative aspects of stoichiometric reactions. We present a number of specific analysis techniques for Petri nets, including place invariants and transition invariants, based on systems of linear equations. Further analysis techniques such as traps and siphons exploit the graph-based structure of nets. The use of Petri nets as a modeling- and analysis technique for biological process is illustrated by a fraction of the combined glucose and pentose phosphate pathway.

3.1 Introduction

In Sect. 3.2, we present the concepts of Petri nets, one by one, by means of an evolving example. We finally arrive at a model of a nontrivial part of the glycolysis- and the pentose phosphate pathway in erythrocytes. This example has been chosen for readers not familiar with Petri nets at all, but familiar, to some extent, with biochemical processes.

As Petri nets are an utterly intuitive formalism, engaging only a fairly small set of concepts, this introductory example suffices for a good grasp of the principles of Petri nets.

Section 3.3 explains the static structure of Petri nets and the general principles of using distinguished components in adequate Petri net models of real systems.

W. Reisig (✉)
Department of Computer Science, Humboldt-Universität zu Berlin, Unter den Linden 6,
10099 Berlin, Germany
e-mail: wolfgang.reisig@informatik.hu-berlin.de

I. Koch et al. (eds.), *Modeling in Systems Biology*, Computational Biology 16,
DOI 10.1007/978-1-84996-474-6_3, © Springer-Verlag London Limited 2011

Based on Sect. 3.3, Sect. 3.4 describes how Petri nets are used to model dynamic behavior.

A faithful model of a system can be used to predict properties of the system from consequences drawn out of the model. Section 3.5 describes those techniques, mainly based on linear-algebraic arguments.

Finally, Sect. 3.6 puts Petri nets into perspective of other modeling techniques for systems biology.

3.2 An Introductory Example

We present the basic concepts of Petri nets, one by one, by means of a biologically inspired example.

3.2.1 Models of Substances

Biological systems process chemical substances. In a Petri net model of such systems, a substance is depicted as a circle (called a *place*). For example, *glucose* is a substance occurring in red blood cells; thus, a Petri net model of red blood cells includes

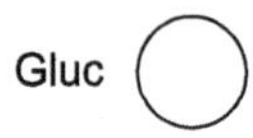

where Gluc is just a shorthand for glucose. Likewise, ATP (adenosine triphosphate), ADP (adenosine diphosphate) and G6P (glucose-6-phosphate) occur in red blood cells, and are consequently depicted as

respectively.

3.2.2 Models of Reactions

Substances are processed in stoichiometric reactions. For example, Glucose and ATP are turned into G6P and ADP. Stoichiometric reactions are usually written as "equations", for example,

$$\mathsf{Gluc} + \mathsf{ATP} \rightarrow \mathsf{G6P} + \mathsf{ADP} \tag{3.1}$$

for the above example. Here, all stoichiometric coefficients are equal to one, that is, one mole of Gluc and one mole of ATP react to one mole of G6P and one mole of ADP. Such a reaction is in a Petri net depicted as a box (called a *transition*), for example,

Hexokinase is the name of the enzyme that catalyzes this reaction. Here, we assume that the enzyme is available in sufficient amount in the cell. In addition, a substance required for a reaction is linked to the box by an *ingoing* arrow, such as

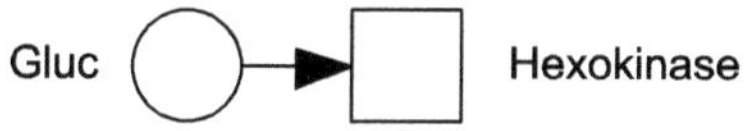

(3.2)

A substance occurring as a result of a reaction is likewise depicted, for example,

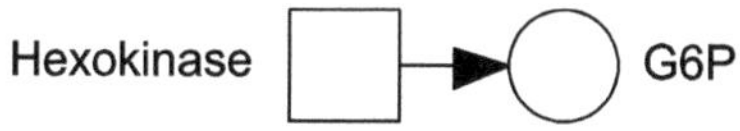

(3.3)

with an arrow going out of the box. The direction of arrows reflects the intuition of substance flow, just as the direction of the arrow in (3.1). With (3.2) and (3.3), we have seen the principles to represent equations such as (3.1): Just compose *all* substances affected by the reaction. Thus, the Petri net representation of (3.1) is

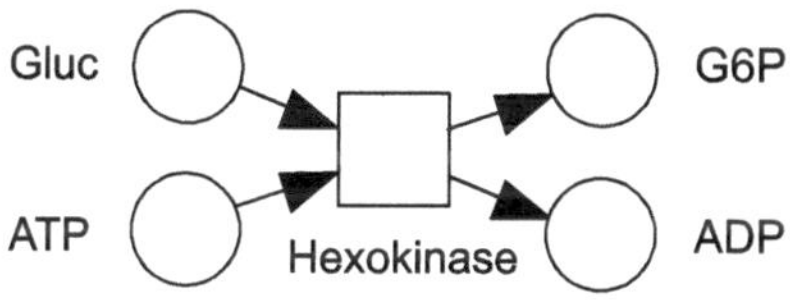

(3.4)

3.2.3 Sequentially Composed Reactions

As a further, fairly trivial example, the reaction of phosphoglucose isomerase, that is, the stoichiometric equation

$$\text{GGP} \rightarrow \text{F6P}$$

is depicted in Petri nets as

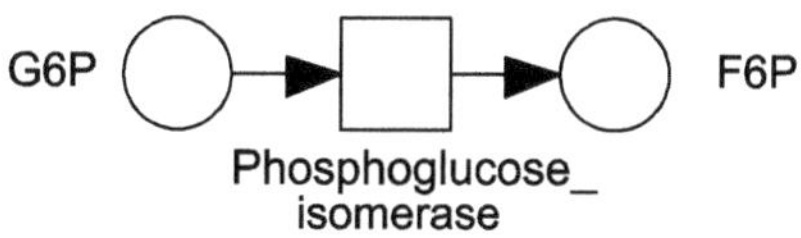

(3.5)

We observe that G6P, as produced in (3.4), is consumed in (3.5). Consequently, (3.4) and (3.5) can be composed, resulting in

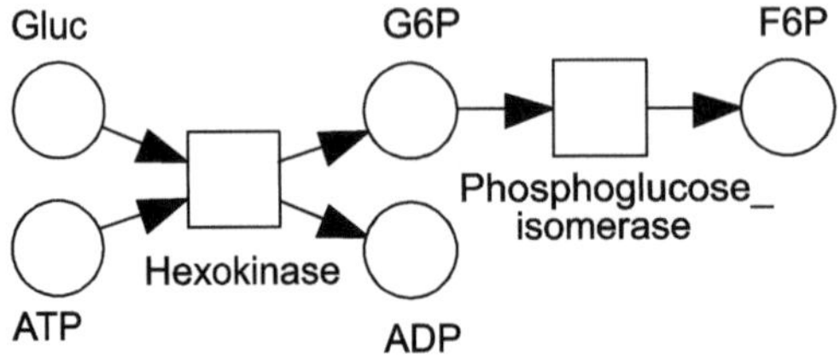

(3.6)

A graphical representation such as (3.6) not only gives *structural* information on substances and reactions, but also provides *quantitive* aspects, as well as information on *states*: In a *state* of the blood cell where one mole of Gluc and one mole of ATP are available, (3.6) shows that eventually one mole of ADP and one mole of F6P are produced. Information on states are represented as dots (called *tokens*) in places. The state described above is in (3.6) depicted as

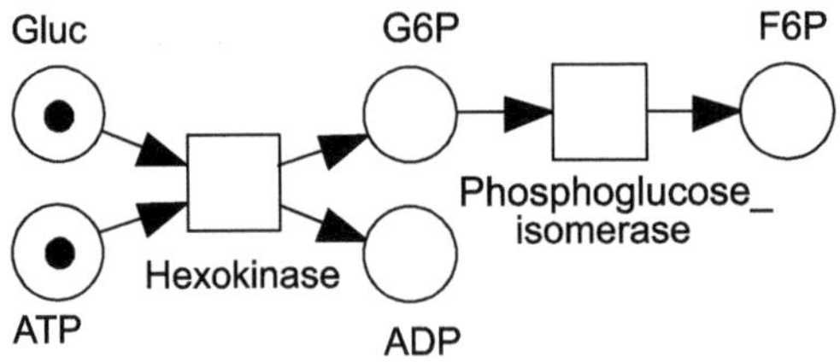

Hexokinase will occur in this state, yielding the intermediate state

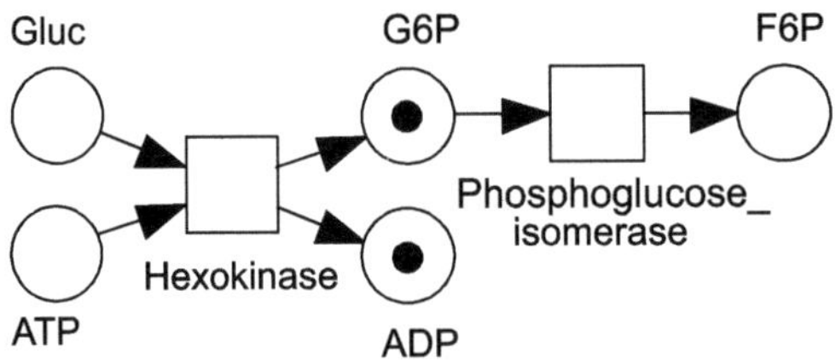

and eventually the state

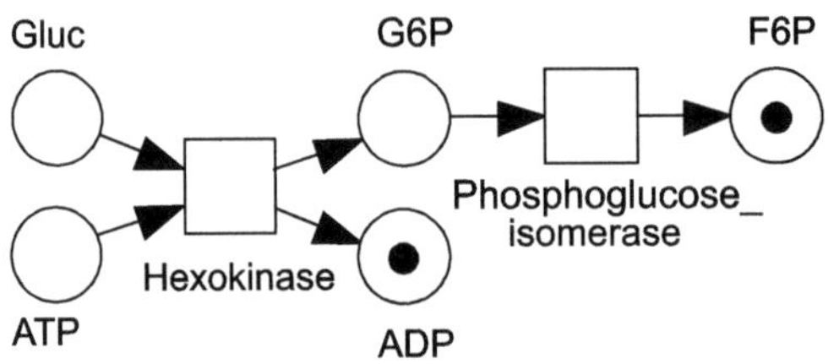

Intuitively formulated, a reaction can occur if each arrow ending at the corresponding box (transition) starts at a place that holds a token. Those tokens are removed upon the reaction's occurrence, and a new token is produced at each place linked to the transition by an outgoing arrow.

3.2.4 *Quantitative Aspects*

Quantitative aspects are even more crucial in the example of the G6P_dehydrogenase reaction. Its stoichiometric equation reads

$$\mathsf{G6P} + 2\,\mathsf{NADP}^+ \;\rightarrow\; \mathsf{Ru5P} + 2\mathsf{NADPH}.$$

This reaction requires *two* moles of NADP^+ (nicotinamide adenine dinucleotide, oxidized form) and produces *two* moles of NADPH (nicotinamide adenine dinucleotide, reduced form). The corresponding Petri net representation is

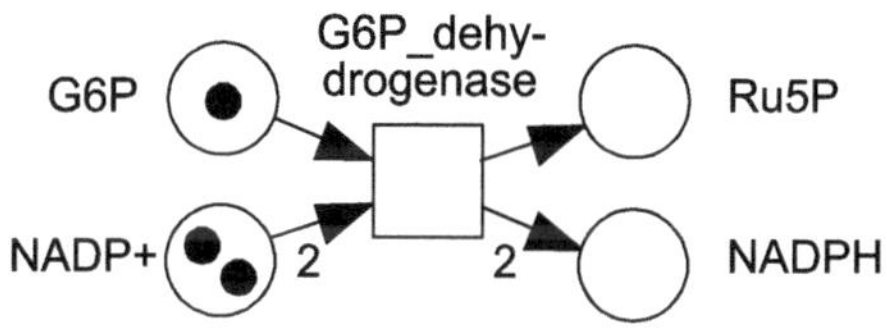

(3.7)

(3.7) shows a state where the reaction may occur. Its occurrence results in

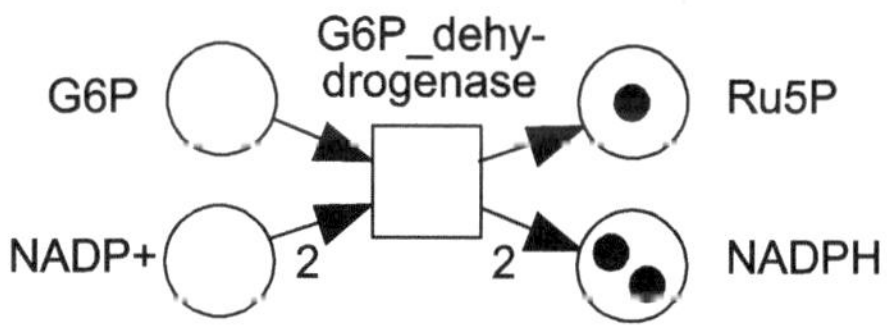

Intuitively formulated, an arrow is inscribed by the number of tokens "passing through the arrow" upon the corresponding reaction's occurrence.

3.2.5 *Alternative Composition*

On our way to integrate, all so far considered components into a description of a part of the glycolysis and the pentose phosphate pathway, we observe that (3.5) and (3.7) both require G6P as a resource for the occurrence of the corresponding reaction. The composition of (3.5) and (3.7) results in

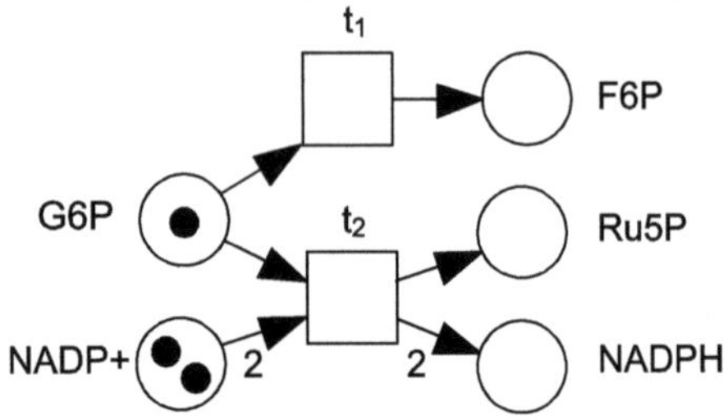

(3.8)

In the depicted state, the token at G6P is a scarce resource, that can be consumed *either* by Phosphoglucose_ isomerase, that is, t_1, *or* by G6P_ dehydrogenase, t_2. Consequently, only one of these two reactions will occur, but not both of them. Which one will eventually be chosen, is not represented in (3.8). The model describes *nondeterministic choice* in this respect.

3.2.6 Modeling the Interface

As we chose to model only a part of a pathway process, the model necessarily exhibits an *interface* to the rest of the biological system. The interface consists of chemical reactions that produce substances "from nowhere" and that dispose substances without leaving mark in the model. The Petri net model of interface reactions is

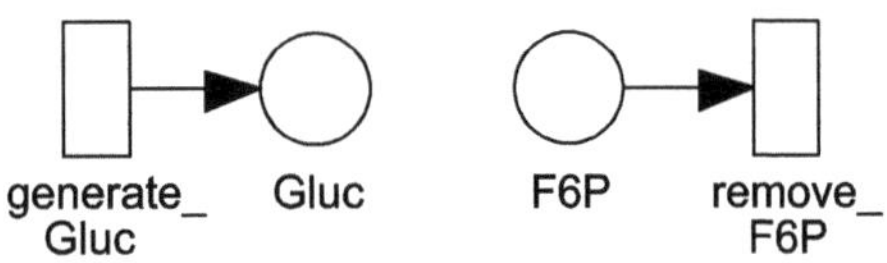

respectively.

3.2.7 Putting it All Together

We have now seen all means necessary to understand an interesting part of the glycolysis and the pentose phosphate pathway in erythrocytes (red blood cells), as depicted in the net N_1 of Fig. 3.1. As a new phenomenon, we see in N_1 that substances [such as $NADP^+$, NADPH, GSSG (oxidized gluathione) and GSH (reduced glutathione)] may follow a *cycle*, that is, are repeatedly consumed and produced by iterated occurrences of reactions.

3.3 The Static Structure of Petri Nets

Section 3.2 has already shown all types of static components, as they can occur in Petri nets. In this section, we just systematically repeat and generalize what they are intended to be used in modeling concrete systems. Dynamic aspects will be detailed in Sect. 3.4.

3.3.1 *Places*

A Petri net includes a set of *places*. Each place is depicted as a *circle*. Frequently, a place is *named*, with different places carrying different names. In a Petri net model, a place *always* represents a *passive* system component. Typically, a place is to store items, to make items visible, or to represent a potentially reachable (local) state. The example of Fig. 3.1 employs places to model (the presence of) *signals*. As a signal is usually bound to a chemical substance, the structural similarity of the substance/signal interpretation of places is very convenient.

3.3.2 *Transitions*

A Petri net includes a set of *transitions*. Each transition is depicted as a *box*. Frequently, a transition is named, with different places and transitions carrying different

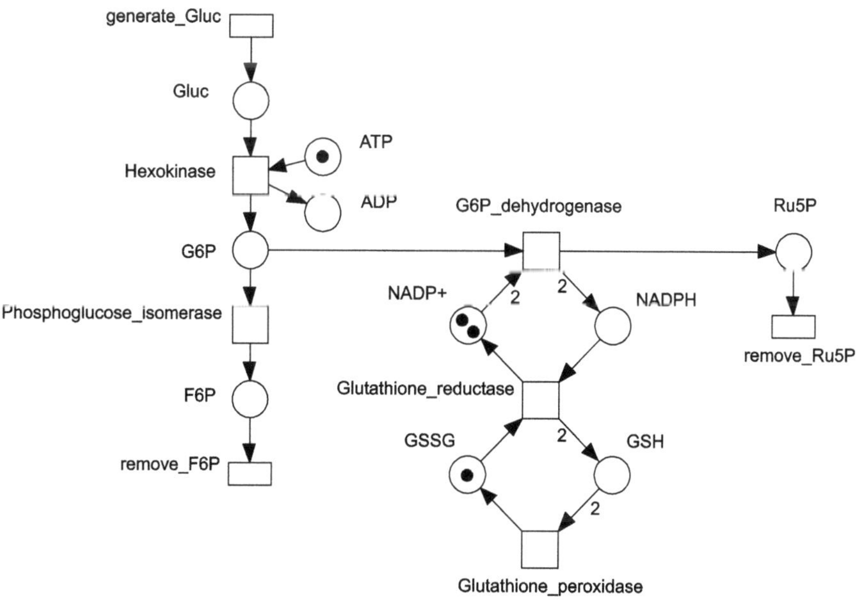

Fig. 3.1 Petri net N_1, modeling a part of the glycolysis and the pentose phosphate pathway in erythrocytes

names. In a Petri net model, a transition *always* represents an *active* system component. A transition typically produces, consumes, transports, or recombines passive items. The example of Fig. 3.1 employs transitions to model chemical reactions. Likewise useful are transitions that model the generation, combination and distribution of *signals*.

3.3.3 Arcs

Places and transitions of a Petri net are linked by *directed arcs*. Graphically, an arc is depicted as an *arrow*. An arc *never* models a system component, but always an abstract *relationship* between components. This relationship is frequently only in the eye of the beholder. Typical examples include causal relationship, local vicinity, immediate link. In the example of Fig. 3.1, an arc never links two places or two transitions; in fact, an arc either starts at a place and ends at a transition, or vice versa starts at a transition and ends at a place. This is neither coincidental nor coerced: It is a structural regularity that inevitably arises whenever the modeling technique of Petri nets is applied accordingly, that is, if passive and active components are separated in a reasonable manner.

An arc may be given a *weight*, that is, an integer value. The role of arc weights will become obvious later on.

3.3.4 Markings

A Petri net models the behavior of a system, as it proceeds stepwise from state to state. A state of a system modeled as a Petri net N is a distribution of *tokens* on the set P of places of the net, that is, a mapping

$$m : P \to \mathbb{N}$$

assigning each place p a number $m(p)$ of *tokens*. In a graphical representation of a Petri net, one usually depicts an *initial* state m_0, drawing a black dot for each token. As an example, the net N in Fig. 3.1 depicts an initial state m_0 with

$$m_0(\mathsf{NADP}^+) = 2$$
$$m_0(\mathsf{GSSG}) = 1$$
$$m_0(p) = 0 \quad \text{for all other places } p$$

3.3.5 Static Net Structures

Summing up, a Petri net N can be written as

$$N = (P, T, F, W, m_0)$$

with

- two finite sets P and T ("*places*" and "*transitions*", respectively), constituting the *elements* of N, that is, each element is either a place or a transition;
- a relation $F \subseteq (P \times T) \cup (T \times P)$, the "flow relation" of N, including *arcs*, shaped (p, t) or (t, p), (where p is a place and t is a transition);
- a mapping $W : F \to \mathbb{N}$, assigning each arc (x, y) an integer, its *weight* $W(x, y)$;
- the *initial marking* $m_0 : P \to \mathbb{N}$, assigning each place p its initial token load, $m_0(p)$.

The graphical representation of N has been discussed in Sect. 3.2 already: Places, transitions and arcs are depicted as circles, boxes, and arrows, respectively. Each arrow is inscribed by its weight (with weight 1 usually not written explicitly). Each place p is inscribed by $m_0(p)$ dots ("tokens").

This completes the *static* structure of Petri net models. *Dynamic behavior* comes on top of the static structure, and will be discussed in the sequel.

3.4 Dynamic Behavior of Petri Nets

A biochemical system evolves in a sequence of *steps*. A step converts a given state into a new state upon the occurrence of a (chemical) reaction. A corresponding Petri net models a state as a marking, and a step as an occurrence of a transition. *Sequences* of steps describe the dynamic behavior of nets.

3.4.1 Enabled Transitions

A transition may be *enabled* in a given marking. At a marking m, in general more than one transition t may exhibit the potential to occur. In technical terms, m *enables* t iff each arc (p, t) from a place p to a transition t is weighted with a number $W(p, t)$ that does not exceed $m(p)$. Shortly:

$$W(p, t) \leq m(p) \tag{3.9}$$

for each arc (p, t). As an example, the marking as given in (3.8) enables both transitions t_1 and t_2. In the net N_0 of Fig. 3.1, the indicated marking only enables $\mathsf{t}_0 = \mathsf{generate\_Gluc}$: This is due to the extreme case that no place p with an arc (p, t_0) exists. So, (3.9) is trivially fulfilled. Consequently, each marking enables t_0.

3.4.2 Steps

To define steps, we start with a technicality: The arc weight W is extended to *all* pairs (d, e) of elements of N, by

$$\overline{W}(d, e) = \begin{cases} W(d, e), & \text{iff } (d, e) \in F \\ 0, & \text{otherwise} \end{cases} \tag{3.10}$$

Thus, $\overline{W}(d,e)=0$ in case there exists no arc from d to e. As an example, in Fig. 3.1, $\overline{W}$(Gluc, generate_Gluc) $= 0$. The idea of $\overline{W}(p,t)$ is to describe the amount of tokens to be removed from a place p upon occurrence of transition t. Correspondingly, $\overline{W}(t,p)$ is the number of tokens added to p upon t's occurrence.

At a marking m of a net N, occurrence of an enabled transition t yields the marking m', defined for each place p of N by

$$m'(p) = m(p) - \overline{W}(p,t) + \overline{W}(t,p) \tag{3.11}$$

As an example, occurrence of the (enabled) transition of (3.7) yields

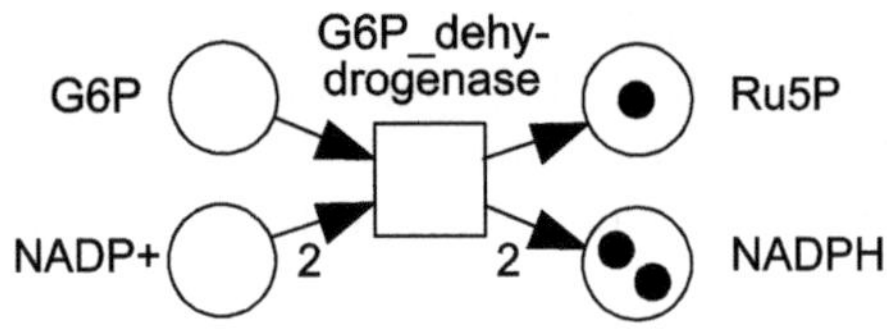

Occurrence of generate_Gluc of Fig. 3.1 produces an additional token at Gluc.

For two markings m, m' and a transition t of a net N, if m enables t and occurrence of t at m yields m' according to (3.11), the triple (m,t,m'), usually written

$$m \xrightarrow{t} m'$$

is a *step of* N. As an example, the marking as indicated in (3.8) enables the transitions t_1 and t_2. Occurrence of t_1 yields a token on Ru5p, as well as two tokens at NADPH.

3.4.3 Step Sequences and Reachable Markings

Starting at the initial marking m_0 of a Petri net N, it is worthwhile to consider steps $m_{i-1} \xrightarrow{t_i} m_i$ $(i = 1, 2, \ldots)$ of N and to *compose* them to step sequences

$$m_0 \xrightarrow{t_1} m_1 \xrightarrow{t_2} m_2 \xrightarrow{t_3} \cdots \xrightarrow{t_n} m_n.$$

A marking m is *reachable* from the initial marking m_0 in N, if such a step sequence exists, with $m = m_n$.

Step sequences are not the only way to represent behavior of a Petri net: Frequently, it is useful to explicitly represent the *independence* of transition occurrences. Details will be provided in Sect. 3.5.9.

3.4.4 The Role of Infinity

Notice that the set of reachable markings of a net N is in general infinite, though N has only finitely many places. As an example, the net in Fig. 3.1 has infinitely many reachable markings. This happens whenever the net has at least one place, p, where the number of tokens may grow without bound. In Fig. 3.1, this is the place Gluc. This in turn is due to the transition generate_Gluc. As discussed above, this transition is enabled at each marking. Its occurrence adds a further token to Gluc.

The amount of substance or of signals in a component of a real biological system has an upper bound, of course. So, one may expect its counterpart in a model, that is, a place p, to correspondingly exhibit an upper bound for tokens. This contradicts the above mentioned properties of Gluc. The reason for this mismatch is the artificial interface given by the transition generate_Gluc. In the model N_1 of Fig. 3.1, this transition comes without any ingoing arcs, and thus may occur in every marking. In the real world, the system is embedded into a larger environment, that would prevent the corresponding chemical reaction to occur too often. This explains why the net N_1 provides a faithful model, in deed.

3.5 Analysis Techniques for Petri Nets

3.5.1 Important Properties

Models are made to support better understanding of a system. To this end, we compute properties and draw consequences in the model, that would reflect interesting properties of the system. Typical such properties of a Petri net model N concerns questions such as

- Is the overall amount of tokens on N always the same?
- Can the initial marking be re-gained after some steps?
- Is it guaranteed that the initial marking will eventually be reached again?
- Is the token load on place p limited by a number, n, for each reachable marking?
- Is there always at least one token on one of the places $p_1, \ldots, p_n$?
- Does each reachable marking enable at least one transition?
- Is it possible for each reachable marking to continue such that eventually transition t occurs?

Many of those questions can be solved, at least in part (i.e., with necessary or sufficient conditions to answer them), by help of linear algebra: Each Petri net N is assigned a matrix $\underline{N}$, with numerical entries. This matrix gives rise to equational systems

$$\underline{N} \cdot x = 0 \quad \text{and} \quad \underline{N^T} \cdot x = 0$$

Here, $\underline{N^T}$ denotes the transposition of the matrix N. (Literature frequently employs "C" instead of "$\underline{N}$".) Solutions of those equational systems provide nontrivial insight into the behavior of N, and help answer the above question.

In order to construct the matrix $\underline{N}$ of a Petri net N, we have to assume an *order* on the set of places, and on the set of transitions. The order is naturally given in case the places are indexed, that is, named $p_1, \ldots, p_n$, or are named $a, b, c, \ldots$. If no such order is visible (such as in N_1), it is usually given by the order of writing the elements down. Choice of the order is entirely irrelevant. But once fixed, it should be retained. In chemistry and biochemistry, this incidence matrix is called "stoichiometry matrix". The equation systems describe flux and substance balance in the steady state.

3.5.2 The Linear Algebra of Steps

Now assume a Petri net N with places $P = \{p_1, \dots, p_k\}$, ordered along the indices $1, \dots, k$. Then each marking $M : P \to \mathbb{N}$ can be written as the column vector

$$\underline{M} = \begin{pmatrix} a_1 \\ \vdots \\ a_k \end{pmatrix}$$

with $a_i = M(p_i)$ for $i = 1, \dots, k$. Hence, the ith entry is the number of tokens on the ith place ($i = 1, \dots, k$).

Correspondingly, we assign each transition t a column vector, $\underline{t}$. From (3.11), we can easily deduce that for each step $m \xrightarrow{t} m'$, and each place p, the amount of tokens added to p or removed from p, is $\overline{W}(p,t) - \overline{W}(t,p)$, hence is independent of m: The update at p is constant for all occurrences of t. So we can assign each transition t and each place p the *effect* of t's occurrence upon p; given by

$$\underline{t}(p) =_{\text{def}} \overline{W}(t,p) - \overline{W}(p,t)$$

for all places p. This defines the *update vector* $\underline{t}$ of t, given by

$$\underline{t} = \begin{pmatrix} \underline{t}(p_1) \\ \vdots \\ \underline{t}(p_k) \end{pmatrix}$$

Applying the usual addition of vectors, for a step $M \xrightarrow{t} M'$ apparently holds

$$\underline{M'} = \underline{M} + \underline{t}$$

3.5.3 The Matrix of a Petri Net

Now assume also the transitions $t_1, \dots, t_l$ of a net N are ordered, for example, according to their index. Together with the ordered places $p_1, \dots, p_k$ (as discussed

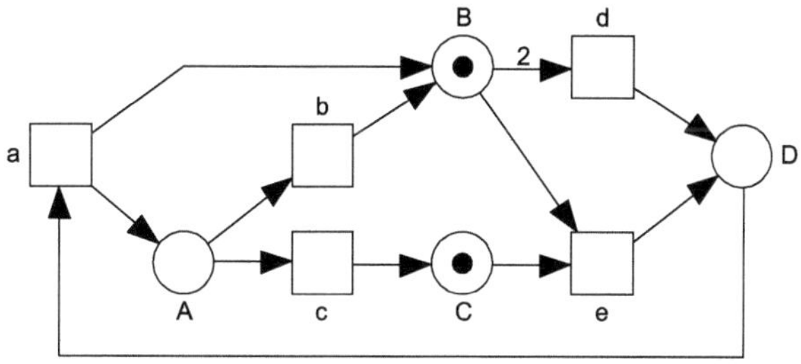

Fig. 3.2 A technical example of a Petri net, N_2

Fig. 3.3 Matrix $\underline{N_2}$ and initial marking m_0 of N_2 in Fig. 3.2

$\underline{N_2}$	a	b	c	d	e	m_0
A	1	-1	-1			
B	1	1		2	-1	1
C			1		-1	1
D	-1			1	1	

in 3.5.2), the vectors $\underline{t_i}$ define the matrix $\underline{N}$ of N, defined by

$$\underline{N} =_{def} (\underline{t_1}, \ldots, \underline{t_l}) = \begin{pmatrix} z_{11} & \cdots & z_{l1} \\ \vdots & & \vdots \\ z_{1k} & \cdots & z_{lk} \end{pmatrix}$$

with

$$z_{ij} = \overline{W}(t_j, p_i) - \overline{W}(p_i, t_j)$$

(Literature frequently writes "C" instead of "$\underline{N}$".) By construction of this matrix, for each step $M \xrightarrow{t_j} M'$ and each place p_i holds:

$$\underline{M'}(p_i) = \underline{M}(p_i) + \underline{N}(i, j)$$

Figure 3.2 shows a small example of a Petri net, N_1, based on the equation system. Figure 3.3 shows its matrix $\underline{N_1}$, and its initial marking, m_0. Entries with value zero are skipped.

3.5.4 *Place Invariants*

We are now interested in solutions of the homogeneous linear equational system

$$\underline{N^T} \cdot x = 0 \tag{3.12}$$

where $\underline{N^T}$ denotes the transposed matrix of $\underline{N}$. A solution $\underline{n} = (n_1, \ldots, n_k)$ of (3.12) returns a number n_i for each place p_i $(i = 1, \ldots, k)$. For example, Fig. 3.4 shows a solution i of (3.12) for the net N_2 in Fig. 3.2. For reasons to become obvious later, a solution of (3.12) is called a *place invariant* of N.

Fig. 3.4 Solution i of $x \cdot \underline{N_2} = 0$, and solutions j_1, j_2 of $\underline{N_2} \cdot x = 0$

$\underline{N_2}$	a	b	c	d	e	i
A	1	-1	-1			1
B	1	1		2	-1	1
C			1		-1	1
D	-1			1	1	2
j_1	1	1		1		
j_2	1		1		1	

Place invariants (P-invariants) $\underline{n}$ of a net N give rise to deep-rooted properties of N. Those properties are based on the *constant of* n, defined as the number

$$n_1 \cdot m_0(p_1) + \cdots + n_k \cdot m_0(p_k)$$

that is, the sum of tokens on all places in the initial marking m_0, where the tokens on place p_i are "weighted" by n_i. For the example N_2 of Fig. 3.2 with its initial marking m_0 and its place invariant i of Fig. 3.3, we get

$$\begin{aligned} & i_A \cdot m_0(A) + i_B \cdot m_0(B) + i_C \cdot m_0(C) + i_D \cdot m_0(D) \\ & \quad = 1 \cdot 0 + 1 \cdot 1 + 1 \cdot 1 + 2 \cdot 0 = 2 \end{aligned} \tag{3.13}$$

Hence, the constant of i is 2.

The decisive aspect of a place invariant i is the observation that its constant remains when in (3.13) the initial marking m_0 is replaced by *any* reachable marking. For example, in N_2 of Fig. 3.2, the marking m with $m(D) = 2$ and $m(p) = 0$ for all other places p is reachable. In fact, upon replacing m_0 in (3.13) by m, we gain $1 \cdot 0 + 1 \cdot 0 + 1 \cdot 0 + 2 \cdot 1 = 2$.

The above observations are compiled in the following theorem.

Theorem 3.1 *Let N be a Petri net with places $p_1, \ldots, p_k$, and let $n = (n_1, \ldots, n_k)$ be a solution of $x \cdot \underline{N} = 0$. Furthermore, let $c =_{def} n_1 \cdot m_0(p_1) + \cdots + n_k \cdot m_0(p_k)$. Then for each reachable marking m of N holds: $n_1 \cdot m(p_1) + \cdots + n_k \cdot m(p_k) = c$.*

The reader may convince himself that the net N_1 in Fig. 3.1 has three place invariants. According to Theorem 3.1, they imply for each reachable marking m:

$$m(\mathsf{ATP}) + m(\mathsf{ADP}) = 1$$

$$m(\mathsf{NADP}^+) + m(\mathsf{NADPH}) = 2$$

$$2 \cdot m(\mathsf{GSSG}) + m(\mathsf{GSH}) = 2$$

In the biological interpretation, P-invariants correspond to substance conservation. For closed systems, a biochemical Petri net should be CPI.

3.5.5 Transition Invariants

As a symmetrical counterpart to (3.12), we now consider solutions of the equational system

$$\underline{N} \cdot x = 0 \tag{3.14}$$

A solution $m = (m_1, \dots, m_l)$ of (3.14) returns a number m_i for each transition t_i $(i = 1, \dots, k)$. For example, Fig. 3.4 shows two solutions, j_1 and j_2 of (3.14) for the net N_2 in Fig. 3.2. For reasons to become obvious later, a solution of (3.14) is called a *transition invariant* of N.

Transition invariants (T-invariants) give rise to behavioral properties. They are based on the *counting vector* of step sequences

$$\sigma: \quad m_0 \xrightarrow{u_1} m_1 \xrightarrow{u_2} m_2 \xrightarrow{u_3} \dots \xrightarrow{u_n} m_n \tag{3.15}$$

of a net N. Each u_i is one of the transitions $t_1, \dots, t_l$. For example, with m_0 the initial marking of N in Fig. 3.2, and $m_1, \dots, m_6$ obvious from context,

$$\sigma: \quad m_0 \xrightarrow{\mathsf{e}} m_1 \xrightarrow{\mathsf{a}} m_2 \xrightarrow{\mathsf{b}} m_3 \xrightarrow{\mathsf{d}} m_4 \xrightarrow{\mathsf{a}} m_5 \xrightarrow{\mathsf{c}} m_6 \tag{3.16}$$

is a step sequence of N. With $t_1, \dots, t_l$ the transition of a net N, the *counting vector* $c = (c_1, \dots, c_l)$ of an occurrence sequence σ of N returns the number c_i of occurrences of t_i in σ. For instance,

$$c = (2, 1, 1, 1, 1) \tag{3.17}$$

is the counting vector of (3.16) (with the canonical order a, ..., e on the transitions).

An occurrence sequence such as (3.15) *reproduces the initial marking*, if $m_n = m_0$. For example, (3.16) reproduces the initial marking. The following theorem relates occurrence sequences that reproduce their initial marking with solutions of $\underline{N} \cdot x = 0$:

Theorem 3.2 *Let N be a Petri net and let m be a solution of $\underline{N} \cdot x = 0$. If m is the counting vector of an occurrence sequence σ of N, then σ reproduces its initial marking.*

As an example, (3.17) solves $\underline{N_2} \cdot x = 0$: (3.17) is the sum $j_1 + j_2$ of the two transition invariants j_1 and j_2 in Fig. 3.4. The net N_1 in Fig. 3.1 has no transition invariant: Occurrence of transition Hexokinase moves the token from ATP to ADP. There is no way to move it back. Let N_1' be derived from N_1 by deleting the two places ATP and ADP. The net N_1' has two transition invariants, of which all other transition invariants can be gained by linear combination. We leave details as an exercise to the reader.

T-invariants are important for the analysis of biochemical Petri nets. First, the Petri net should be CTI because a transition which is not member of at least one

T-invariant does not contribute to the net behavior, and can, thus, be removed. Second, T-invariants can be interpreted as basic pathways and should, therefore, have a biological meaning. Otherwise, in most cases a modeling error could be the reason for a *wrong* pathway, or, in rare cases, a new pathway was detected. Third, the transitions of a T-invariant and the places in between form a subnet exhibiting a specific biological function. Thus, the computation of T-invariants provides an automatic network decomposition. More details are described in Chap. 4. All these invariants can be used to validate the model.

3.5.6 Traps

A *trap* of a Petri net N is a subset Q of the places of N, which is embedded into N in a regular way: Each transition that removes tokens from some places of Q, also contributes a token to Q. For example, the two places B and D of the net N_2 of Fig. 3.2 form a trap: Transitions a, d and e remove tokens from B or D, but each of them also contributes tokens to B or D. This structural regularity implies the preservation of at least one token: If at least one place of a trap Q carries initially a token, then at each reachable marking at least one place of Q has a token.

3.5.7 Syphons

Arguments on backward/forward symmetry imply the definition of *syphons*: A syphon of a Petri net N is a subset Q of its places, where each transition that contributes a token to Q also removes a token from Q. For example, the three places A, B and D of Fig. 3.2 form a syphon: Occurrence of one of the transitions a, b, d and e contributes a token to A, B or D. Each of the transition coincidently removes a token from one of the places A, B or D. This structural regularity implies the preservation of emptiness: Once empty, a syphon never again gains a token. The term "syphon" refers to the French version of a bottle with gas under pressure.

3.5.8 The Marking Graph

The marking graph of a Petri net N has the reachable markings of N as its vertices, and the (reachable) steps of N as its edges. Figure 3.5 shows the marking graph of the net N_2 in Fig. 3.2. The marking graph of N is infinite in case infinitely many markings are reachable in N. In finite case, the graph can grow rather large, in fact more than exponentially in the size of the net. A lot of properties of a Petri net N can be decided by help of its marking graph, G:

- N *terminates*, that is, each step sequence eventually reaches a marking that enables no transition: G is finite and cycle-free.

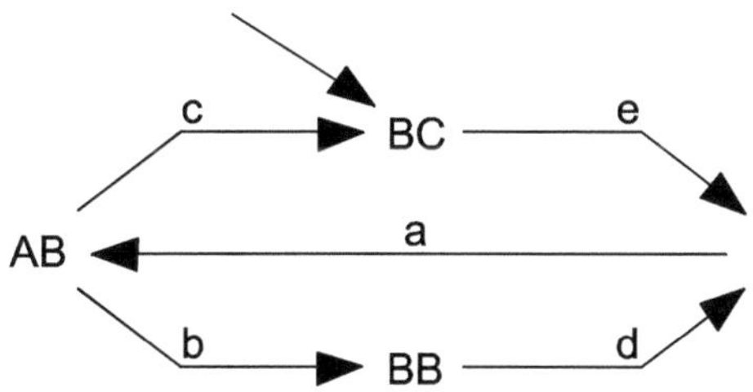

Fig. 3.5 The marking graph of the net N_2 in Fig. 3.2

- N *diverges*, that is, each reachable marking enables at least one transition: At each vertex of G starts an edge.
- N is *live*, that is, for each reachable marking m and each transition t, a marking m' is reachable from m that enables t: At each node of G, for each transition t starts a path with a t-inscribed edge.
- N is *weakly live*, that is, to each transition t there exists a reachable marking m that enables t: Each t is the inscription of at least one edge.
- N is *bounded*, that is, for some number b, no reachable marking has more than b tokens at any place: G is finite.
- N is *reversible*, that is, the initial marking m_0 is reachable from each reachable marking m: G is strongly connected (i.e., each node is connected to each other node along some path).

3.5.9 Concurrent Runs

As mentioned in Sect. 3.4.3 already, the step sequences and their representation in the marking graph of a net N are complemented by means to explicitly represent concurrent (i.e., mutually independent) occurrences of transitions. As an example, the net N_3 in Fig. 3.6 extends N_2 of Fig. 3.2. it is easy to see that a marking M is reachable in N_3 with tokens on both places A and E. Hence, the transitions b, c and f are enabled. Observe that b and c *compete* for the token at A: Only one of b or c will occur. However, f will occur in any case. This gives rise to two different *concurrent runs*, as depicted in Fig. 3.7. We refrain from formal arguments here, and appeal to the reader's intuition.

3.6 Petri Nets as a Modeling Technique for Systems Biology

Here, we provide evidence why Petri nets are a particularly good choice as a modeling technique for systems biology.

Dynamic systems can be modeled in many different ways. For example, physics models dynamic systems behavior as functions in n-dimensional spaces over real numbers. One axis denotes the flow of time, all other axes represent the values of variables, changing during the flow of time. Petri nets follow a fundamentally different approach, emphasizing fundamentally different aspects.

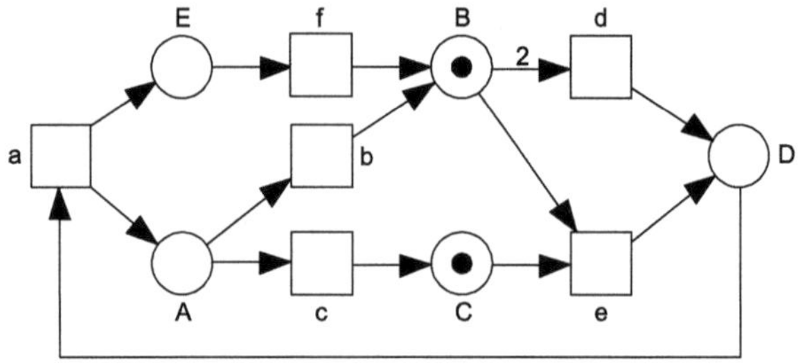

Fig. 3.6 Net N_3, an extension of N_2 in Fig. 3.2

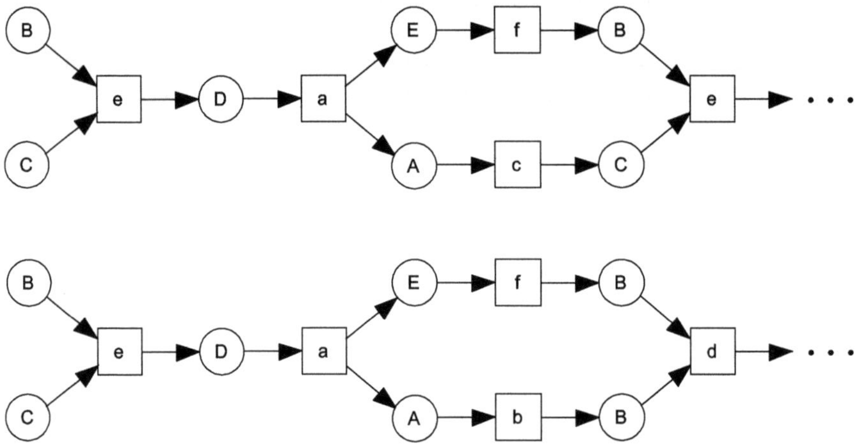

Fig. 3.7 Two concurrent runs of N_3 (Fig. 3.6)

3.6.1 Discrete Steps

First of all, Petri nets describe dynamic behavior in *discrete steps.* We will not launch into the discussion whether or not the "real" world proceeds continuously or in discrete steps. A model is a matter of utility, not of truth. And it has become fairly clear that the assumption of discrete steps decisively supports the understanding and intuition of systems biology. Stoichiometric equations are the most successful description technique for processes of system biology. The level of abstraction taken by Petri net models exactly corresponds with the level of abstractions of stoichiometric equations. In a nutshell, Petri nets allow to formulate causal relationships among stoichiometric equations.

3.6.2 Local Cause and Effect

A discrete step of a system, and in particular a biological system, never affects the entire system. It is rather the presence of some locally bounded objects (such as chemical substances) and conditions (such as temperature and pressure) that cause a step. The effect of a step, mostly a relocation and composition of substances, is likewise locally bounded. Petri nets emphasize this aspect, limiting the cause and effect of a transition to its surrounding places.

Even more important, the scope of a place p is strictly limited to the transitions in its vicinity, that is, those that deliver new tokens to p and those that remove given tokens from p.

3.6.3 Invariance of Substance

A (bio-) chemical step relocates and recombines a given amount of substances. It never "creates" material from nowhere, nor does it dispose of material. Place invariants allow to trace the flow of substances. Arc weights balance the amount of substance as represented by a token. A place p not belonging to any place invariant is a strong indication that the model may be wrong (or the place has links to the system's environment, i.e., the system is open.): The amount of substance as modeled by tokens on p may increase or decrease without bounds. In this context, it is interesting to observe that Petri nets are *reversible*: Given a marking m' that has been reached by occurrence of transition t, it is easy to recompute the previous marking m, that is, the entire step $m \xrightarrow{t} m'$. This means that substances neither arise from "nowhere" nor disappear to "nowhere".

3.6.4 Other Models of Dynamic Systems

Any other model of dynamic behavior fails to properly address at least one of the above-mentioned aspects. As mentioned in the introduction text of this section, differential equations model continuous behavior, but not discrete steps. Any formalism that resembles programming languages and employs assignment to program variables such as

$$x := f(y)$$

does not limit the scope of variables (that would correspond to places) at all: A variable can be addressed everywhere in the model. Furthermore, such formalisms lack of any means that correspond to the "preservation of amount" as available by place invariants. An assignment statement is in general not reversible: The old value of x is lost after executing, for example, the assignment statement $x := 1$.

3.7 References and Tools

Literature on Petri nets has reached formidable numbers. This applies, to some extent, also to applications of Petri nets in systems biology. We recommend the interested reader the Petri net portal

http://www.informatik.uni-hamburg.de/TGI/GI-Fachgruppe0.0.1

This portal offers good facilities to trace specific aspects. Furthermore, it offers up-to-date information on recent developments of software tools that support any effort of modeling and analyzing systems by means of Petri nets.

3.8 Problems

3.1 Let N_3 be the net of Fig. 3.6. Construct

1. the matrix $\underline{N_3}$,
2. some place invariants,
3. some transition invariants,
4. a trap,
5. a siphon, and
6. the marking graph

of N_3.

Part II
Modeling Techniques

Chapter 4
Discrete Modeling

Andrea Sackmann

Abstract A discrete Petri net (PN) considers only discrete objects, that is, its marking is specified by an integer number of tokens distributed over its places. This chapter deals with model development and qualitative analysis of discrete PNs. Latter lays emphasis on the net's invariants. Caused by the special usage of P-invariants representing an anticoincidence, read arcs appear in the net, here given as loops. In order to nevertheless base a sound consideration on the set of T-invariants, they are processed to built feasible T-invariants. Their examination aims at a validation of the net's structure. For this purpose, initially MCT-sets are built on the set of feasible T-invariants. Subsequently, they may be clustered in T-clusters aiming at gaining knowledge about the involved components and their relationships. In-/dependency of involved processes can be verified. The sets of T-invariants may additionally serve as basis of theoretical knockout analyses as given for instance through the Mauritius maps.

4.1 Modeling Concepts

As given above by definitions in Chap. 3, Petri nets (PNs) are directed bipartite graphs with two types of nodes, namely places and transitions. A net is called *discrete* PN, if it deals only with discrete objects, that is, its tokens can be specified by integer values. Representing biological components by places and biological (re-) actions by transitions, the underlying directed graph of a discrete PN descriptively identifies the interactions between the components. Here, one important advantage of PNs takes effect, namely the inherent possibility of incorporating different levels of abstraction within one net. Consequently, biological processes can be modeled

A. Sackmann (✉)
Institute of Computing Science, Poznań University of Technology, Piotrowo 2, 60-965 Poznań, Poland
e-mail: asackmann@cs.put.poznan.pl

I. Koch et al. (eds.), *Modeling in Systems Biology*, Computational Biology 16,
DOI 10.1007/978-1-84996-474-6_4,

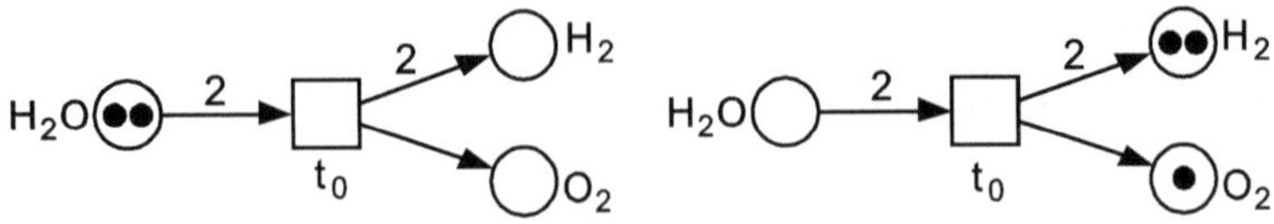

Fig. 4.1 An example network modeling electrolysis of water, that is, $2\ H_2O \rightarrow 2\ H_2 + O_2$. The net shown on the *left* (on the *right*) illustrates a marking before (after) firing of transition t_0

even if the mechanisms of all contained subprocesses are not completely understood. The resulting net may benefit the user comprehension as there are a lot of tools for visual representation as well as for simulation of PNs. An animation of the token flow may provide an intuitive understanding of the net's dynamic behavior. Furthermore, the firm mathematical foundation of a PN finally enables a well-structured net analysis. In the following, a brief introduction of concepts required for the net analysis is given.

In discrete modeling, two main branches may be distinguished, namely a qualitative and a quantitative approach. The qualitative discrete PN is constituted by the underlying graph which describes the components' interactions as mentioned above. Including information which quantifies these interactions, for instance, by specifying reaction rates, leads to a quantitative PN. Figure 4.1 depicts a small PN as an illustrating example how a metabolic reaction can be directly translated into a PN. Here, the reactants of a considered stoichiometric equation are the pre-places of a transition representing the chemical reaction and its post-places give the reaction products.

As outlined in Fig. 4.1, the stoichiometric coefficients define the arc weights and therefore, the tokens represent the quantity of the biological species (e.g., in mole). Figure 4.1 can be read as: two moles of water (H_2O) are decomposed into two moles of hydrogen (H_2) and one mole of oxygen (O_2), that is, $2\ H_2O \rightarrow 2\ H_2 + O_2$, representing the electrolysis of water. In the case that the necessary stoichiometric data is available, whole biological networks may be translated into PNs. The interested reader is referred to a corresponding example [211] introducing the sucrose breakdown pathway in the potato tuber as a PN model. In the following, a more general case is considered by discussing purely qualitative nets. In biological context, a huge amount of qualitative data is available for example, due to high-throughput techniques. Thus, this data may serve as a basis for modeling corresponding PNs even if not all biological mechanisms are understood in detail. After validating the consistency of the qualitative PN model, the PN can be extended by further available information of the biological system. Including for example kinetic data would lead to a hybrid or continuous PN, see [129] and Chaps. 6, 8, 13. Additionally taking into account temporal information or probabilities leads to time PN or stochastic PN, respectively (cf. related work below and Chap. 7 of this book).

In order to translate the information about components' interactions into subnets, they may be considered as basic logical statements, that is, the components are connected by logical operators such as implication, conjunction, disjunction, and negation (cf. [327]). A transition's pre-places represent preconditions whose fulfillment

Table 4.1 Basic logical operators symbols

Symbol	Meaning	Symbol	Meaning
$\Rightarrow$	Implication (if...then)	$\vee$	Inclusive disjunction (OR)
$\neg$	Negation (NOT)	$\dot{\vee}$	Exclusive disjunction (XOR)
$\wedge$	Conjunction (AND)		

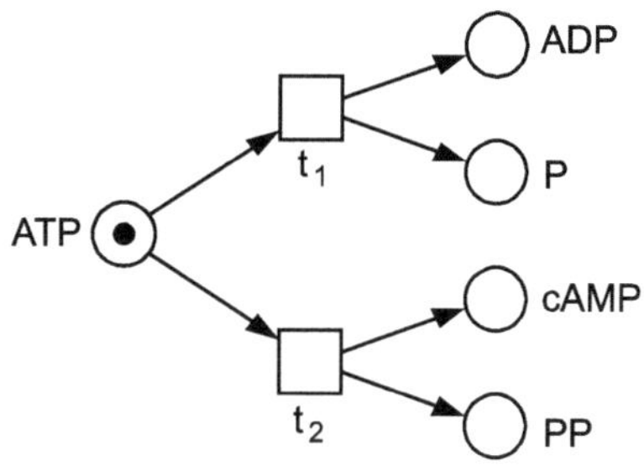

Fig. 4.2 An example network representing a disjunction coupled implication. Transitions t_1 and t_2 are in conflict, in the considered marking only one of them can actually fire, that is, $ATP \Rightarrow (ADP \wedge P) \dot{\vee} (cAMP \wedge PP)$

causes the transition's firing. This process leads to the fulfillment of the transition's post-conditions. Following this modeling methodology, different substructures are distinguished. Table 4.1 recaps the logic symbols which are used in the following. Transition t_0 shown in Fig. 4.1 represents a conjunction coupled implication, that is, $2\ H_2O \Rightarrow (2\ H_2 \wedge O_2)$. Note that several pre- as well as several post-places of a transition are linked among each other with a logical AND (because of the transition firing rule). In contrast, several post-transitions of a considered place are in static conflict and hence, represent a disjunction (i.e., logical OR). If they are enabled under the same condition and only one of them can actually fire, they are in dynamic conflict.

In fact, transitions t_1 and t_2 shown in Fig. 4.2 are in dynamic conflict since place *ATP* is their common preplace and marked with only one token. The transitions represent different possibilities of ATP utilization. While t_1 represents a hydrolytic dephosphorylation splitting off one phosphate group of ATP, t_2 stands for the synthesis of cAMP from ATP by adenylyl cyclase based splitting of two phosphate groups. The marking shown in Fig. 4.2 results in a nondeterministic behavior of the net since the transitions represent an exclusive disjunction, that is, $ATP \Rightarrow (ADP \wedge P) \dot{\vee} (cAMP \wedge PP)$.

A logical negation is represented by $\neg p_0 = p_1$ and vice versa $p_0 = \neg p_1$. Extending this concept to more than two conditions, that is, places, the corresponding subnet represents an anti-coincidence since the coincident marking of these places is excluded, see places p_0, p_1 and p_2 in Fig. 4.3 for which it holds $p_0 \dot{\vee} p_1 \dot{\vee} p_2$. This anti-coincidence represents a form of mutual exclusion [271]. Depending on firing of transitions t_0, t_1 and t_2, the depicted token circulates between the considered places. Since the weighted sum of tokens over these places is constant (here equal to 1), p_0, p_1 and p_2 form a P-invariant (cf. (4.1) below). Note that a P-invariant whose weighted sum of tokens is greater than 1 may contain subsets of

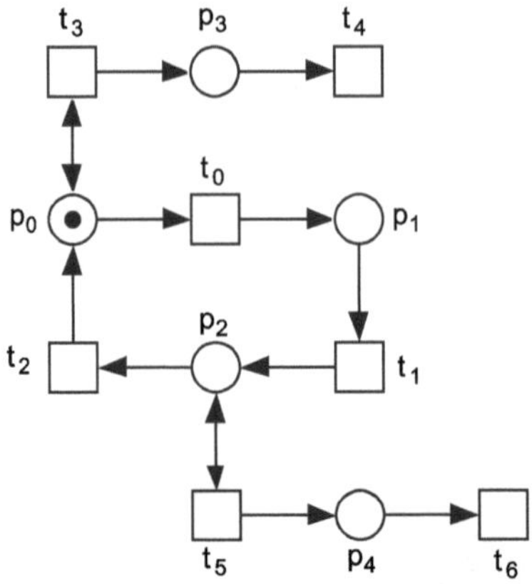

$$C = \begin{pmatrix} -1 & 0 & 1 & 0 & 0 & 0 & 0 \\ 1 & -1 & 0 & 0 & 0 & 0 & 0 \\ 0 & 1 & -1 & 0 & 0 & 0 & 0 \\ 0 & 0 & 0 & 1 & -1 & 0 & 0 \\ 0 & 0 & 0 & 0 & 0 & 1 & -1 \end{pmatrix}$$

Fig. 4.3 Subnet representing an anticoincidence of places p_0, p_1 and p_2. The token shown at p_0 circulates only between these places by firing of transitions t_0, t_1 and t_2. Matrix C is the incidence matrix of the net

places being marked simultaneously. Thus, not each P-invariant involves mutual exclusion among all its places. From a technical point of view, the places forming an anti-coincidence can be considered as a kind of switcher whose state is indicated by the current location of the token. In the context of a biological application, the corresponding places may represent different possible states of a considered component (e.g., a protein in different modifications each given by one place). To preserve the P-invariant structure in Fig. 4.3, adjacent transitions $t3$ and $t5$ are connected via read arcs, represented here by a loop, that is, by a bidirectional arrow. In this way, places p_0 and p_2 constitute preconditions of transitions $t3$ and $t5$ but their firing does not remove the token from p_0 and p_2 (firing of a transition is considered to be a timeless process). With respect to the read arcs, the disjunction springing from p_0 (and p_2, respectively) is not uniquely exclusive or inclusive. Namely, while $p_0 \Rightarrow p_1$ removes the token from p_0 by firing of t_0 and thereby is exclusive, t_0 remains enabled subsequent to $p_0 \Rightarrow p_3$, that is, the firing of t_3. Taking this into account, such a kind of disjunction is denoted by $p_0 \Rightarrow (p_1 \dot{\vee} (p_3 \vee p_1))$ and analogously, $p_2 \Rightarrow (p_0 \dot{\vee} (p_4 \vee p_0))$. Note that $\vee$ represents an inclusive disjunction, that is, may be realized either as OR or as AND (cf. Table 4.1).

Beyond translating basic logical statements into subnets, there are other approaches bringing together mathematical logic and PN structures, see, for instance, the PN model construction based on Boolean networks as introduced in [374] and described in Chap. 5.

4.2 Qualitative Analysis

The sections below give an overview about qualitative analysis approaches. Chapter 14 points out a further approach concerning discrete models by dealing with the network structure. Generally, qualitative analysis provides knowledge on the involved components and may expose novel information about their relationships. Novel information that may be utilized in view of a prediction of the network behavior may be rather obtained out of a quantitative analysis. A qualitative one may serve as a basis of a model validation by checking the consistency of a nets structure.

Here, the emphasis is put on the analysis of the net's invariants. Hence, the analysis is mainly based on the topology of the PN, in particular given in the form of the *incidence matrix* C, that is, the $(k \times l)$-matrix giving at entry c_{ij} the token change at place p_i by firing of transition t_j where $k = |P|$ and $l = |T|$, see Chap. 3. However, as specified below, in special cases, additionally the marking of the net has to be taken into consideration. The approach presented here does not take into account quantitative information (often the considered qualitative PNs are ordinary ones, that is, all arcs are weighted with one). Therefore, the analysis is based on the T-invariants' support instead of examining the firing frequency of their involved transitions (see Sect. 4.4 below). As it will be explained in detail below, for an analysis approach based on T-invariants the considered PN should be transition-bordered, that is, in case that any place of the net has no pre- or post-transition, these should be included representing the interface of the modeled system to its surroundings.

The decision about dynamic net properties is usually based on its reachability graph (i.e., the graph containing as nodes all markings which are reachable from the initial marking by firing of transitions). Thus, it is computationally expensive to decide about them. Nevertheless, it should be mentioned that generally, liveness and reversibility are dynamic properties (defined in Chap. 3) which are considered to be important in the context of biological models [211]. A net is *live* for an initial marking m_0 if no transition is or will become (after any firing sequence) permanently unfirable (a *dead* transition). A net is *reversible* if the initial marking is reachable (via a firing sequence) from each reachable marking.

4.3 P-Invariants

As introduced in [219] a P-invariant is defined as a vector $y \in \mathbb{N}^k$ (where $k = |P|$) satisfying the equation

$$C^T \cdot y = 0 \tag{4.1}$$

where C denotes the incidence matrix. A P-invariant represents a set of places over which the weighted sum of token is constant. Hence, in metabolic networks they are commonly used to model substrate conservations, see, for instance, [407]. As discussed above, they may also be utilized in their function representing an anticoincidence. In any case, a requirement for a P-invariant to contribute to the net's behavior, is that it is marked in the initial marking. Concerning biological systems, these tokens are typically placed in the P-invariants in a way that the initial marking represents an inactive state [327], or a state considered to be the physiologically normal one [328], respectively.

A PN in which all places are included in a P-invariant is called to be *covered by P-invariants* (CPI), that is, it holds:

$$\forall \text{ places } p_j \in P \; \exists \text{ P-invariant } y\colon (y)_j \neq 0. \tag{4.2}$$

According to the P-invariant's definition, the weighted sum of tokens in a PN holding the CPI property is constant.

A *trap* is a set of places, whose all output transitions are also input transitions of that set. Thus, a trap will not be empty of tokens if it contains tokens in the initial marking. The other way around, a nonempty set of places, whose all input transitions are also output transitions of that set, is called a *siphon.*[1] A siphon can not be marked with tokens again once there are no tokens in that set. A trap is called *maximal* if it is not a proper subnet of any other trap and a siphon is called *minimal* if it does not properly contain any other siphon. In case that the maximal trap in each minimal deadlock is sufficiently marked, that is, it contains a place which carries sufficiently many tokens so that all its post-transitions are enabled, then no dead states are reachable in the model [370]. A *dead state* is a state in which no transition can fire anymore. Assuming that the only structural traps and siphons of a given net are its minimal P-invariants (which should be sufficiently marked in the initial marking as discussed above), then the net won't reach a dead state since at least the transitions of the P-invariant may fire.

Considering again the PN shown in Fig. 4.3, as mentioned above, places p_0, p_1 and p_2 form P-invariant

$$y_1 = (1 \quad 1 \quad 1 \quad 0 \quad 0).$$

In view of initial marking $m_0 = (1\,0\,0\,0\,0)$, y_1 contains a token (initially placed at p_0) which may circulate among the places of the P-invariant. This net does not hold the CPI property and contains no other trap or siphon than y_1.

4.4 T-invariants

A T-invariant is defined as a vector $x \in \mathrm{N}^l$ (where $l = |T|$) satisfying the equation

$$C \cdot x = 0 \tag{4.3}$$

with C denoting the incidence matrix (definition introduced in [219]). A T-invariant reproduces a given marking by firing of all its contained transitions (each the required number of times). Thus, the minimal T-invariants are considered to represent the basic behavior of a PN [152, 153]. Returning again to the PN shown in Fig. 4.3, its three minimal T-invariants are given as x_1, x_2 and x_3 with:

$$x_1 = (1 \quad 1 \quad 1 \quad 0 \quad 0 \quad 0 \quad 0)^T,$$
$$x_2 = (0 \quad 0 \quad 0 \quad 1 \quad 1 \quad 0 \quad 0)^T,$$
$$x_3 = (0 \quad 0 \quad 0 \quad 0 \quad 0 \quad 1 \quad 1)^T.$$

[1] Note that some authors term a siphon as a *deadlock*, cf. [370].

All of these T-invariants are *nontrivial*. A *trivial* T-invariant comprises two transitions which represent a reversible reaction in the net (split into forward and backward reaction each given by one of these transitions), cf. [211]. In the considered example, all transitions are contained in an invariant. Generally, this property of a PN is called *covered by T-invariants* (CTI). The *support* of an invariant x is given as the set of transitions being included in the invariant, that is, $supp(x) = \{t_i \in T : (x)_i \neq 0\}$. Therefore and analogous to (4.2), a net is CTI , iff:

$$\forall \text{ transitions } t_i \in T \ \exists \text{ T-invariant } x : (x)_i \neq 0$$
$$\iff \quad \forall \text{ transitions } t_i \in T \ \exists \text{ T-invariant } x : t_i \in supp(x). \qquad (4.4)$$

Holding the CTI property ensures that each process represented by a particular transition appears as a part of the basic behavior of the net (given in the form of T-invariants). Thus, CTI is considered to be an important property in context of biological PN application and analysis (cf. Chap. 10 of this book). In case that a PN turns out to not fulfill the CTI property, the transitions not included in invariants should be examined to check if there are, for example, unwanted token accumulations in the net. Avoiding net behavior like that, the net architecture may be rearranged,[2] so that a net is built which is covered by T-invariants [208, 327].

4.4.1 Feasible T-invariants

Note that the connections via read arcs given by loops are not considered in the incidence matrix and thus, they are not considered in calculating the invariants (cf. definition in 4.3 of Chap. 3). There are approaches transforming PNs including loops into pure ones but precondition for this transformation is an unequal weight of arcs forming a loop [220], which does not hold for nets containing read arcs as the example net presented in Fig. 4.3. Here, transitions t_3 and t_5 are consequently represented as input transitions in the incidence matrix and the precondition given by places p_0 and p_2 are not taken into account, compare incidence matrix C as given in Fig. 4.3. With regard to the calculation of T-invariants in PNs with read arcs, the concept of *feasible T-invariants* is introduced [327]. Accordingly, a T-invariant is nonfeasible iff one of its transitions is connected via a read arc with a place neither being marked in the initial marking nor is marked by firing of another transition within the considered invariant. Thus, in the PN shown in Fig. 4.3, transition t_3 is not affected by this issue since p_0 is marked in the initial marking. Whereas minimal T-invariant x_3 is nonfeasible (as a firing sequence in respect of given marking m_0) since it contains t_5 without containing any transition providing empty p_2 with a token, namely t_1. In order to get feasible T-invariants, the set of minimal T-invariants is processed. According to this approach [327], a processing step creates new invariants by joining a

[2]For example, a place-bordered net can't be CTI. Therefore, it is demanded above that the net is transition-bordered.

nonfeasible one with all invariants containing transitions which provide token at the critical place.

To summarize that in a well-structured way, consider a PN with l transitions, i.e. $|T| = l$, with k places, that is, $|P| = k$, and with m minimal T-invariants x_i, that is, $|X| = m$. $(x_i)_j$ denotes the jth entry of x_i, that is, $(x_i)_j$ refers to transition t_j and $(m_0)_n$ refers to the initial marking of place p_n. Then it holds:

$$\begin{aligned}
&\text{a T-invariant } x_i \text{ is nonfeasible iff}\\
&\exists j \in \{1,\dots,l\}\colon (x_i)_j \neq 0 \text{ with}\\
&\exists n \in \{1,\dots,k\}\colon {}^\bullet t_j \ni p_n \in t_j^\bullet\colon \text{ with } (m_0)_n = 0 \text{ and}\\
&\forall o \in \{1,\dots,l\} \text{ with } {}^\bullet t_o \not\ni p_n \in t_o^\bullet : (x_i)_o = 0.
\end{aligned} \tag{4.5}$$

Here, transition t_j is connected via a read arc with place p_n from which t_o is a post-transition. Considered T-invariant x_i contains t_j but not t_o. In this case, add index i of the nonfeasible invariant of (4.5) in a (formerly empty) set X^-. Invariants x_i are combined with invariants x_q each containing a transition t_r providing a token at place p_n (see (4.5) and (4.6)). Count the number of newly generated feasible T-invariants by c

$$\begin{aligned}
&\text{for } q = 1,\dots,m\colon \text{if } \exists r \in \{1,\dots,l\} \text{ with } \left({}^\bullet t_r \not\ni p_n \in t_j^\bullet\right) \text{ and } \left((x_q)_r \neq 0\right)\colon\\
&c = c + 1\\
&x_{m+c} = x_i + x_q\\
&\text{end for.}
\end{aligned} \tag{4.6}$$

Let X' denote the resulting set of feasible T-invariants. X' is built by applying (4.6) to all i from (4.5), resulting in:

$$X' = \left\{x_i \in X : i \notin X^-\right\} \cup \{x_{m+d}\ \forall d = 1,\dots,c\}. \tag{4.7}$$

Paraphrasing, X' is the union of minimal (already feasible) T-invariants with newly formed, that is, joint, (formerly non-feasible) T-invariants.

In a PN utilizing substructures such as P-invariants representing anticoincidences as discussed above, the processing generating feasible T-invariants aims to bridge the read arc connection which is not realized in the incidence matrix (which the T-invariants' calculation is based on). Thus, discussing such a PN in a biological context, only the feasible T-invariants may represent pathways, or transduction paths from a signal to a cell response, or a regulatory loop, respectively.

Note that CTI is a property which ensures that for all nonfeasible T-invariants x_i there is an invariant x_q as given in (4.6) (which contains a transition providing a token at the critical place p_n), compare (4.4). In other words, the assumption that a net is CTI leads to the fact that the set of all minimal T-invariants can be transformed to a set of T-invariants which all are feasible (as denoted in the Equations above). Furthermore, CTI implies that the considered net also is covered by feasible invariants of set X' as specified in (4.7). As mentioned above, CTI is an important property for biological PN models. In case that it holds, each process represented by a transition takes place in a feasible T-invariant.

According to the processing given in (4.5), (4.6), (4.7), the minimal T-invariants of the example net in Fig. 4.3 lead to the following feasible invariants:

$$x_1 = (1 \quad 1 \quad 1 \quad 0 \quad 0 \quad 0 \quad 0)^T,$$
$$x_2 = (0 \quad 0 \quad 0 \quad 1 \quad 1 \quad 0 \quad 0)^T,$$
$$x_4 = (1 \quad 1 \quad 1 \quad 0 \quad 0 \quad 1 \quad 1)^T.$$

Because of their combinatorial generation, the number of feasible T-invariants may increase exponentially (since a nonfeasible one generates a new one with each invariant providing a token at the critical place). One possibility to avoid redundant information within the set of feasible T-invariants is the formulation of biologically motivated restrictions. For example, in a PN which models the iron homeostasis process [328, 329] it makes sense to delete feasible invariants containing more than one (of four possible) ways of iron uptake.

4.4.2 *MCT-sets*

One important step of the net validation of a model is the examination of the invariants' set [152], that is, first ensuring that the PN is CTI, and afterwards checking the biological plausibility of its invariants. However, the set of T-invariants which have to be evaluated may be very big, determined by the size and density of the considered PN. In case that the considered set is too big, manual inspection of each T-invariant would be indefensible. Thus, there are concepts facilitating this investigation. As a first step, *maximal common transition sets* (*MCT-sets*) [327] pool transitions which occur always together with each other in the considered set of T-invariants. Let X denote this set of minimal (or feasible [328]) T-invariants x. A support-oriented grouping leads to sets ϑ of transitions holding

$$\forall x \subset X: \quad \vartheta \subseteq supp(x) \;\dot{\vee}\; \vartheta \cap supp(x) = \emptyset. \tag{4.8}$$

Maximizing these sets $\max_t \vartheta \subseteq T$ gives the MCT-sets. The set of resulting MCT-sets constitutes a partition of transition set T where each non-trivial MCT-set (means each with a cardinality >1) represents a functional unit as a kind of building block of the net [327]. Note that the MCT-sets are not necessarily connected, that is, may contain transitions which are not adjacent via places. Depending on the biological context, the transitions which are contained in a MCT-set may stand for reactions which show a similar regulation pattern. Their adjacent places may show an expression behavior similar to each other.

Returning to the example PN shown in Fig. 4.3, the MCT-sets built based on its feasible T-invariants x_1, x_2 and x_4 are listed in Table 4.2.

Table 4.2 The MCT-sets of the PN shown in Fig. 4.3

MCT-set	Contained transitions
1	t_0, t_1, t_2
2	t_3, t_4
3	t_5, t_6

4.4.3 *T-clusters*

A further approach facilitating the investigation of the T-invariants is their classification into functionally distinct clusters, *T-clusters* [143]. In order to structure the invariants in a clear way, first a distance matrix D is calculated. The *distance measure* d_{ij} for two T-invariants x_i and x_j is defined as $d_{ij} = 1 - s(x_i, x_j)$ where the Tanimoto coefficient [19] serves as similarity measure $s(x_i, x_j)$, that is, two T-invariants x_i, x_j have the distance measure:

$$d_{ij} = 1 - s(x_i, x_j) = 1 - \frac{|supp(x_i) \cap supp(x_j)|}{|supp(x_i) \cup supp(x_j)|}. \tag{4.9}$$

Obviously, the resulting distance matrix is symmetric. Considering again the PN shown in Fig. 4.3 whose feasible T-invariants x_1, x_2, and x_4 are given above. According to (4.9), the distances between those invariants are:

$$d_{12} = 1 - \frac{0}{5} = 1, \qquad d_{14} = 1 - \frac{3}{5} = 0.4, \qquad d_{24} = 1 - \frac{0}{7} = 1.$$

Based on the resulting distance matrix, it is possible to cluster the feasible T-invariants of a considered PN. The agglomerative, hierarchical clustering algorithms used for this purpose, such as UPGMA, merge in each iteration the two most similar objects (i.e., T-invariants, or already built clusters, respectively). In order to determine the distance between two objects, several methods may be applied. According to the UPGMA method the distance Δ_{kl} between two clusters C_k and C_l is given by the average distance of all possible pairs of their contained invariants, that is, by the equation

$$\Delta_{kl} = \frac{1}{|C_k| \cdot |C_l|} \cdot \sum_{x_i \in C_k, x_j \in C_l} d_{ij}. \tag{4.10}$$

Generally, the algorithm terminates when all T-clusters are joined in one cluster. This way of binary clustering results in an ordered sequence of partitions, that is, in a dendrogram. Here, the T-invariants, are located at the leaves of the dendrogram. The more similar two of them are (means the more transitions are contained in both of them), the more proximal their branches are located to each other. The dendrogram can be cut at any level to yield different clustering, that is, partitions, of the data. Different cluster validity measures are discussed in [143] to assess the quality of a clustering partition. The measure showing the best results is the Silhouette Width. Note that in the same work also different similarity measures and different clustering algorithms are compared.

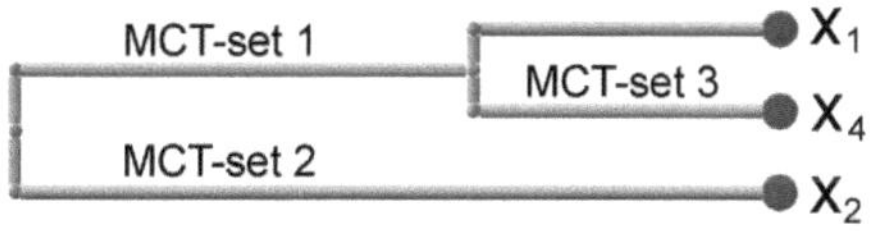

Fig. 4.4 Dendrogram of the three feasible T-invariants of the PN shown in Fig. 4.3

Table 4.3 Three example PNs. Their transitions are grouped into MCT-sets according to (4.8). The numbers of MCT-sets given here refer to real partitions of T, that is, includes all trivial MCT-sets. The nets' T-invariants are clustered according to (4.9) and (4.10)

Number of			Discussed T-clusters	Presented in
Transitions	MCT-sets	T-invariants		
57	24	85	10	[328]
42	22	102	15	[329]
88	40	107	34	[147]

Since the Tanimoto coefficient and the UPGMA method provide the best results in many cases, they are introduced here. A dendrogram of the of the PN shown in Fig. 4.3 is given in Fig. 4.4. In a next step, invariants x_1 and x_4 would build one cluster (with $\leq$60% accordance within this cluster since their distance is $d_{14} = 0.4$ as calculated above). Obviously, applying this method to a PN as small as the considered example PN (with three T-invariants) is not needful. Several applications to greater PNs with more transitions and more invariants can be found in the literature, see Table 4.3 for a comparison of three PNs. These nets should illustrate how the approach facilitates the analysis. For instance, instead of examining 85 objects (invariants) characterized by 57 elements (transitions), after applying the approach only 10 objects (T-clusters) characterized by 24 elements (MCT-sets) are considered. How many nontrivial MCT-sets are built depends on the density of the net. The number of considered T-clusters is determined by the cut of the dendrogram (see above). This cut can be chosen by the user since there is no exact method to measure/get a "best" cut. Often several cuts should be examined.

On closer examination of the T-clusters, dependency and independency of the different processes can be verified. For instance, for the PN shown in Fig. 4.3, the T-clusters as given in Fig. 4.4 prove that the processes given by MCT-set 3 depend on those of MCT-set 1. Processes given by MCT-set 2 are independent of both, compare Table 4.2. Obviously, in case of such a small net these conclusions can be drawn directly from the net. In greater models, the manual inspection is not feasible anymore, compare the PNs instanced in Table 4.3.

4.4.4 Mauritius Maps

Furthermore, in silico knockout analyses are possible based on the T-invariants by excluding several transitions and investigating the remaining invariants [327].

For this purpose, the concept of Mauritius maps is introduced [147]. From a biological viewpoint, it is interesting to know which parts of a considered system will be affected by the knockout of a certain piece, for example, a reaction. Or, vice versa it can be examined which reactions should be knocked out to achieve a desired system behavior. Such kind of analysis can be worked out by an theoretical approach in order to save the corresponding biological experiments or in order to point out future experiments. Here, the usage of a Mauritius maps representation facilitates an exhaustive knockout analysis. This approach visualizes the dependencies of T-invariants in terms of a binary tree. By knocking out one transition, it is directly visible which part of the tree is affected and which part works independently of this knockout. A formal definition of Mauritius maps as well as the application to the case study (i.e., the gene regulation in Duchenne Muscular Dystrophy) can be found in [207].

4.5 Related Work

Besides the approaches based on PNs, dedicated methods have been developed in order to model and analyze biochemical pathways. An example for a well established concept are the elementary (flux) modes [348] being defined as vectors x satisfying the equation $C \cdot x = 0$ with the incidence matrix C of the underlying directed graph. This approach is predominantly (but not only, cf. [32]) applied for analyzing metabolic networks considered to be in steady state in order to analyze them using methods derived from convex algebra [302]. Closely related to the concept of elementary modes are the extreme pathways introduced in [342]. Considering the solution space of the given system of linear inequalities, that is, the systems $C \cdot x = 0$ and $x \geq 0$, the solution space is given by the corresponding convex polyhedron. While the extreme pathways are the edges of this cone, all elementary modes lay on the surface of the cone, including all its edges [291]. Thus, the extreme pathways are a subset of the elementary modes. The concept of elementary modes corresponds to the one of minimal T-invariants [427]. Also in the analysis of elementary modes, classification methods are introduced. However, contrary to the T-clusters, the resulting classes do not necessarily represent partitions, that is, one elementary mode may occur in several classes [298]. The knockout experiments mentioned above in the context of Mauritius maps are closely related to concepts as minimal cut sets (MCS) [196] being defined as a minimal set of structural interventions repressing a certain functionality specified by a deletion task (in form of elementary modes) [194]. Further remarks to elementary mode analysis can be found in Chap. 10; a current review to the topic is given, for instance, by [397]. The concept of MCT-sets is analog to partially coupled metabolic fluxes as introduced in [53]. The analysis about in-/dependencies of several processes based on MCT-sets and T-clusters can be compared with the flux coupling analysis, see, for example, [217]. Later one takes into account the coefficient of the corresponding process, that is, in terms of Petri nets it is not based on the support vector but on the Parikh vector dealing with the firing number of each considered transition.

The above discussed types of Petri nets are all considered to include no temporal information and as mentioned, firing of transitions is considered to be a timeless process. Please note that there are several different Petri net types containing time concepts which are associated either with places or with transitions in form of time durations or time intervals. General surveys may be found in [88, 409], and biological applications, for instance, in [39, 308].

The analysis approach described above is mainly based on the nets' invariants. Please note that there are different approaches developed in order to decrease the computational complexity of the invariants' calculation, see, for instance, decompositions of PNs [423, 424].

4.6 Software

There are a lot of different tools for editing, animating and analyzing PNs. Reviews can be found, for instance, in [63, 276, 295]. The PNs presented here were edited by *Snoopy*, analyzed by *Charlie* and the clustering of their invariants performed by the *Petri net Invariant Analyser (PInA)*, all three tools freely available via http://www-dssz.informatik.tu-cottbus.de/. The visualization of the dendrogram is done by *WiDa—Wilmascope for Distance Analysis* being freely available via http://nwg.bic-gh.de/wida.

4.7 Problems

4.1 Build a PN including the following information:

1. p_0 is marked with one token
2. $p_0 \Rightarrow p_1 \dot{\vee} (p_2 \vee p_1)$
3. $p_1 \Rightarrow p_0 \dot{\vee} ((p_3 \wedge p_4) \vee p_0)$
4. $p_3 \Rightarrow p_2$

Include output transitions for p_2 and p_4.

4.2 Calculate the minimal P-invariants for the PN built in Exercise 1 and decide about mutual exclusion. Calculate the minimal and the feasible T-invariants. What are the resulting MCT-sets?

4.3 Consider the sub-nets each formed by a feasible T-invariant calculated in Exercise 2. They are given by the contained transitions together with their adjacent places. Compare the resulting sets of places (each denoted e.g., by a binary vector $z_i \in \{0, 1\}^5$ indicating per entry the presence or absence of the corresponding place) with the four statements listed in Exercise 1.

4.4 Build a PN including the following information:

1. $p_0 \Rightarrow p_1 \dot{\vee} p_2$
2. $p_1 \Rightarrow p_0 \dot{\vee} (p_3 \wedge p_4)$
3. $p_3 \Rightarrow p_2$

Include output transitions for p_2 and p_4 and input transition for p_0.

4.5 For the PN built in Exercise 4: Calculate minimal P-invariants as well as minimal and the feasible T-invariants. Calculate the sets of places affected by respective transitions of the feasible T-invariants analog to Exercise 3. Compare the two built PNs based on the analysis results.

Chapter 5
Modeling Genetic Regulatory Networks

Richard Banks, Victor Khomenko, and L. Jason Steggles

Abstract Cellular systems are regulated by complex genetic control structures known as *genetic regulatory networks* (*GRNs*). In this chapter, we present a range of practical techniques for qualitatively modeling and analyzing GRNs using Petri nets. Our starting point is the well-known Boolean network approach, where regulatory entities (i.e., genes, proteins and environmental signals) are viewed abstractly as binary switches. We present an approach for translating synchronous Boolean networks into Petri net models and introduce the support tool GNAPN which automates model construction. We illustrate our techniques by modeling the GRN for carbon stress response in *Escherichia coli* and, in particular, consider how existing Petri net techniques and tools can be used to understand and analyze such a GRN model. While asynchronous GRN models are considered more realistic than their synchronous counterparts, they often suffer from the problem of capturing too much behavior. We investigate how techniques from asynchronous electronic circuit design based on *Signal Transition Graphs* (*STGs*) and *Speed-Independent circuits* can be used to address this, by identifying and refining conflicting behavioral choices within a model. We illustrate these techniques by developing an asynchronous model for the lysis–lysogeny switch in phage λ.

5.1 Introduction

The development and function of cellular systems is controlled by complex networks of interacting genes, proteins and metabolites referred to as *genetic regula-*

R. Banks · V. Khomenko · L.J. Steggles (✉)
School of Computing Science, Newcastle University, Newcastle upon Tyne, UK
e-mail: L.J.Steggles@ncl.ac.uk

R. Banks
e-mail: Richard.Banks@ncl.ac.uk

V. Khomenko
e-mail: Victor.Khomenko@ncl.ac.uk

I. Koch et al. (eds.), *Modeling in Systems Biology*, Computational Biology 16,
DOI 10.1007/978-1-84996-474-6_5, © Springer-Verlag London Limited 2011

tory networks (*GRNs*) [43]. In order to be able to understand and investigate the complex behavior of GRNs, various formal modeling techniques have been proposed, ranging from simple qualitative approaches, such as Boolean networks, to detailed quantitative approaches based on differential equations or stochastic techniques. Some of these techniques are described in this book; see also [175, 182, 343] for an overview. While quantitative models can provide detailed results about the behavior of a GRN, the current lack of quantitative data concerning reaction rates, and the noise associated with such data, means that the application of these methods is restricted in practice. For this reason, qualitative modeling techniques have emerged as an important first approach to documenting and understanding GRNs.

In this chapter, we present a range of practical techniques for qualitatively modeling and analyzing GRNs using Petri nets. We take as our starting point the synchronous Boolean network approach [4, 43], where regulatory entities (i.e., genes, proteins, and environmental signals) are viewed abstractly as binary switches which update their state together. We describe a systematic process for constructing Petri net models from such synchronous Boolean networks based on using *logic minimization* [48] to extract concise logical descriptions of the fundamental relationships between regulatory entities in a GRN. Often the information about a GRN can be incomplete, and we explain how our modeling approach is able to cope with this by using the nondeterminism associated with Petri nets. Our Petri net modeling approach is fully supported by a purpose built software tool GNAPN (Genetic Networks as Petri Nets) [24]. GNAPN fully automates the model construction process and allows the resulting models to be exported in a variety of Petri net file formats, making them amenable to the wide range of available Petri net analysis tools [301].

We illustrate our modeling approach by presenting a detailed case study in which a GRN for the carbon starvation stress response in the bacterium *E. coli* [159, 323] is modeled and analyzed. Using the detailed data provided in [323], we construct a qualitative Petri net model capturing the synchronous behavior of the given GRN. This Petri net is then validated and analyzed using PEP [296], a standard Petri net tool. In particular, we explain how well-known Petri net analysis techniques, such as *model checking* [79], can be used in a biological context, and consider performing biologically inspired mutant analysis to gain insights into our model.

Historically, the synchronous update semantics for Boolean networks has been favored in the literature, since the resulting models have deterministic behavior and are therefore easier to analyze and understand. However, the assumption of synchronous updates can be argued to be biologically unrealistic [123, 390], which leads to reservations about the results obtained from such models. Hence, the asynchronous semantics seems to be more appropriate. However, asynchronous models tend to have too rich behavior, not all of which is realizable in practice. This behavior also tends to be highly nondeterministic, that is, (nonconverging) *choices* are common when choosing the next state.

It turns out that many such choices can be resolved either by assuming that the environment of the biological system is slow (i.e., the system always has enough time to react to its changes), or by using knowledge of the relative speeds of chemical reactions. Therefore, in practice the behavior has far less nondeterminism than such asynchronous models suggest. (This may explain why synchronous Boolean networks, which are always deterministic, were often favored over asynchronous ones, in spite of synchronous updates being biologically unrealistic.)

We consider how Petri net techniques from *asynchronous circuit design* [84] can be used to develop realistic asynchronous models of GRNs. We introduce the Petri net modeling formalism of *Signal Transition Graphs* (STGs) [75, 84, 325], and describe associated techniques for developing *Speed-Independent* (SI) circuits [84, 268]. In particular, we see how identifying choices within a model that violate the SI property highlight candidate points for refinement. We explain how these choices can be refined by adding information about the environment and relative reaction rates to the model using a firing order enforcement transformation.

We illustrate our STG modeling approach by considering a case study in which we develop an asynchronous model of the GRN controlling the switch between the lysogeny and lysis cycles in phage λ [289, 311]. We begin by constructing an STG model based on the Boolean network presented in [390]. We then refine it by finding the points where this STG violates the SI property and appropriately resolving the corresponding conflicts by making assumptions about the relative rates of reactions. We also see how some violations of SI highlight the key stochastic choice between lysogeny and lysis modes; this choice is not resolved and remains in the final model. The case study makes use of the STG support tool PETRIFY [84] and demonstrates its practical role in asynchronous model development.

The chapter is organized as follows. In Sect. 5.2, we introduce an approach for systematically developing qualitative Petri net models of GRNs from synchronous Boolean networks. We illustrate this approach with a case study which highlights how Petri net techniques can be used to understand and analyze a synchronous GRN model of the carbon stress response in *E. coli*. In Sect. 5.3, we consider how techniques from asynchronous circuit design can be applied to develop realistic asynchronous models of GRNs. We illustrate these techniques by developing an asynchronous GRN model of the lysogeny–lysis switch in phage λ. In Sect. 5.4, we briefly consider related work and give pointers for further reading. Then, in Sect. 5.5 we summarize the techniques and results presented in this chapter. Finally, in Sect. 5.6 we provide some instructive exercises for the interested reader.

5.2 Synchronous Models of GRNs

In this section, we present an approach for constructing qualitative Petri net models of GRNs (see [373, 374]) and introduce the associated support tool GNAPN [24, 374]. We take as our starting point a *synchronous Boolean network* (see [4, 43]) description of a GRN, and then extract from it a compact logical description using *logic minimization* [48]. We then directly translate this logical description into appropriate Petri net control structures which capture the GRN's synchronous behavior. We illustrate the above approach by modeling and analyzing the GRN underlying the carbon stress response in *E. coli* [159, 323].

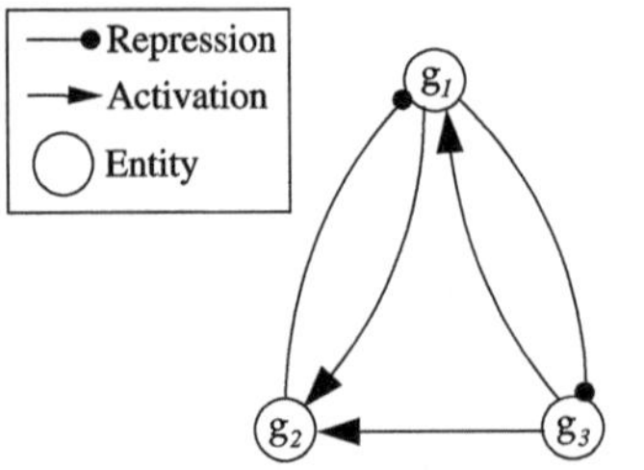

g_2	g_3	$[g_1]$
0	0	1
0	1	1
1	0	0
1	1	1

g_1	g_3	$[g_2]$
0	0	0
0	1	0
1	0	0
1	1	1

g_1	$[g_3]$
0	1
1	0

Fig. 5.1 A simple example of a Boolean network with defining truth tables for the next-state functions of the entities, where $[g_i]$ represents the next state of entity g_i

Recall that a *Boolean network* [4, 43] consists of a set of regulatory entities $\{g_1, \ldots, g_n\}$ which can be in one of two possible states, either 1 representing the entity is active (e.g., a gene is expressed or a protein is present) or 0 representing the entity is inactive (e.g., a gene is not expressed or a protein is absent). The behavior of each entity g_i is described by a Boolean *next-state function,* which, given the current states of the entities that affect it (referred to as its *neighborhood states*), returns the next state for g_i.

As an example, consider the Boolean network in Fig. 5.1 which contains three entities, g_1, g_2 and g_3. The next state $[g_i]$ of each entity g_i is defined by the corresponding truth table.

A Boolean network can be semantically interpreted in two distinct ways [43]: either *synchronously,* where all entities update their states together; or *asynchronously*, where entities update their states independently. In this section, we focus on the synchronous semantics and return to the asynchronous semantics in Sect. 5.3.

5.2.1 Constructing a Qualitative Petri Net Model

Given a set of truth tables defining the Boolean behavior of all the entities in a GRN, we extract a compact representation of the regulatory relationships between entities using well-known techniques from Boolean logic [48, 146]. The idea is to consider the truth table for each entity and to extract two Boolean expressions, one that specifies when the entity becomes active and one which specifies when it becomes inactive.

As an example, consider the truth table for entity g_1 given in Fig. 5.1. We can see that the neighborhood states 00, 01, and 11 result in g_1 being 1 in its next state (where xy denotes the state $g_2 = x$, and $g_3 = y$). We can represent each state as a Boolean term [146], using conjunction (logical *and*), where the variable g_i represents that an entity g_i is in state 1, and the negated variable $\overline{g_i}$ represents that an entity g_i is 0. So the neighborhood state 01 for g_1 is represented by the term $\overline{g_2}g_3$. Applying this approach and then summing the derived terms using disjunction (logical *or*) allows us to derive a Boolean expression in *disjunctive normal form*

(*DNF*) [146] that defines precisely the conditions under which the entity is active. Continuing with our example, we derive the following Boolean expression for g_1:

$$\overline{g_2}\,\overline{g_3} + \overline{g_2}g_3 + g_2g_3$$

(where the notation $\overline{x}$, $x + y$ and xy is used to represent the Boolean operators *not*, *or* and *and*, respectively). In a similar way, we can derive an expression in DNF that defines the conditions under which an entity is inactive (i.e., we group together all those neighborhood states which result in the entity being inactive in the next state). By applying this approach to each entity, we are able to construct a logical description that completely specifies the functional behavior of a GRN. In our example, we derive the following expressions defining the behavior of g_1, g_2 and g_3, which we present as logical equations:

$$\begin{aligned}
&[g_1] = \overline{g_2}\,\overline{g_3} + \overline{g_2}g_3 + g_2g_3, && [\overline{g_1}] = g_2\overline{g_3},\\
&[g_2] = g_1g_3, && [\overline{g_2}] = \overline{g_1}\,\overline{g_3} + \overline{g_1}g_3 + g_2\overline{g_3},\\
&[g_3] = \overline{g_1}, && [\overline{g_3}] = g_1.
\end{aligned}$$

The Boolean expressions derived above are often unnecessarily complex and can often be simplified using *logic minimization* [48, 146]. From a biological point of view, this simplification process is important, as it helps to identify the underlying regulatory relationships that exist between entities in a GRN. For brevity, we omit the full details of Boolean logic minimization here (we refer the interested reader to [48]) and simply present the final minimized logical equations for our running example:

$$\begin{aligned}
&[g_1] = \overline{g_2} + g_3, && [\overline{g_1}] = g_2\overline{g_3},\\
&[g_2] = g_1g_3, && [\overline{g_2}] = \overline{g_1} + \overline{g_3},\\
&[g_3] = \overline{g_1}, && [\overline{g_3}] = g_1.
\end{aligned}$$

The Boolean expressions derived above compactly capture the behavior of a Boolean network and provide a good basis from which to start constructing a Petri net model of the underlying GRN. The approach we take is to represent the Boolean state of each entity g_i in a Petri net by the well-known technique (see, for example, [65, 318]) of using two complementary places p_i and $\overline{p_i}$, where a token on place p_i indicates the entity is active (i.e., $g_i = 1$) and a token on place $\overline{p_i}$ that it is inactive (i.e., $g_i = 0$). Note that the total number of tokens on places p_i and $\overline{p_i}$ will therefore always be equal to 1. We then translate the logical equations derived for the GRN into appropriate Petri net structures to implement the specified logical behavior.

In order to model the synchronous update semantics of a Boolean network [125] within the asynchronous Petri net framework, we make use of a two phase commit protocol to synchronize updates in the Petri net model. In phase one, each entity decides what its next state will be and records this decision. When all the entities have made a decision about their next states, the second phase of the protocol begins, and the state of each entity is updated according to the recorded decisions. The synchronous Petri net model construction is defined in Fig. 5.2.

Synchronous Petri Net Model Construction:

1. Assume we have n entities in our model $g_1, \ldots, g_n$. We add a place $p_{0/\mathrm{done}}$ to our Petri net model, and for $i = 1, \ldots, n$, we also add places p_i, $\overline{p_i}$, $p_{i/\mathrm{on}}$, $p_{i/\mathrm{off}}$, $p_{i/\mathrm{start}}$, $p_{i/\mathrm{syn}}$ and $p_{i/\mathrm{done}}$. Note that in the initial marking, either p_i or $\overline{p_i}$ (but not both) must contain a token, and $p_{i/\mathrm{start}}$ will contain a token, for $i = 1, \ldots, n$. All other places must be unmarked.
2. **Phase One**:
   Consider each logical equation $[g_i] = T_1 + \cdots + T_m$ which defines when g_i becomes active. For each product term T_j, $1 \leq j \leq m$, add a transition $t_{i/j}$ to the Petri net model such that:
   - place $p_{i/\mathrm{start}}$ is an input place and places $p_{i/\mathrm{on}}$ and $p_{i/\mathrm{syn}}$ are output places of $t_{i/j}$;
   - for each variable g_k (respectively, $\overline{g_k}$) in T_j, add a read arc connecting place p_k (respectively, $\overline{p_k}$) to $t_{i/j}$.

   We now use the same approach to model the logical equation $[\overline{g_i}] = T_1 + \cdots + T_m$ which defines when g_i becomes inactive. We add a transition $\overline{t}_{i/j}$ to model each product term T_j, $1 \leq j \leq m$, using the same scheme as detailed above, but with $p_{i/\mathrm{on}}$ replaced by $p_{i/\mathrm{off}}$ as an output place for the transition.
3. **Phase Two**:
   - We add a transition to initiate the update process which has input places $p_{i/\mathrm{syn}}$, for $i = 1, \ldots, n$, and one output place $p_{0/\mathrm{done}}$ [see Fig. 5.4(a)].
   - For each entity g_i, we add four transitions to update the state of g_i. These transitions are shown in Fig. 5.4(b) and represent the four possible update scenarios:
     - move token from place $\overline{p}_i$ to p_i;
     - leave token on p_i;
     - move token from place p_i to $\overline{p}_i$;
     - and leave token on $\overline{p}_i$.
   - We add a transition to reset the control places which has input place $p_{n/\mathrm{done}}$ and output places $p_{i/\mathrm{start}}$, for $i = 1, \ldots, n$ [see Fig. 5.4(c)].

Fig. 5.2 Petri net construction method for synchronous Boolean networks

5.2.1.1 Phase One: Next State Decision

In the first phase of the update protocol, each entity g_i in the model decides whether it should be active or not in the next state. This decision is recorded using two places, $p_{i/\mathrm{on}}$ and $p_{i/\mathrm{off}}$, where a token on $p_{i/\mathrm{on}}$ indicates g_i is active in the next state, and a token on $p_{i/\mathrm{off}}$ that it is not. We model this decision process by using the derived Boolean expressions (in DNF) that compactly capture the conditions under which the entity becomes active or inactive. We directly translate each of these expressions into a set of transitions in our Petri net model, using one transition to represent each product term they contain. For each entity g_i, we use an additional place $p_{i/\mathrm{start}}$ to indicate when a decision about the next state of g_i is required, and a place $p_{i/\mathrm{syn}}$ to indicate when an update decision has been made. Clearly, when all the synchronization places $p_{i/\mathrm{syn}}$ have been marked, we know phase one of the protocol is complete.

As an example, consider the Petri net fragment presented in Fig. 5.3, which models the decision process of entity g_1 in our running example. Transitions $t_{1/1}$ and $t_{1/2}$ record that entity g_1 will become active in the next state if g_2 is inactive or g_3 is active, respectively. Note if both conditions are true then either transition can fire,

$$[g_1] = \overline{g_2} + g_3$$
$$[\overline{g_1}] = g_2\,\overline{g_3}$$

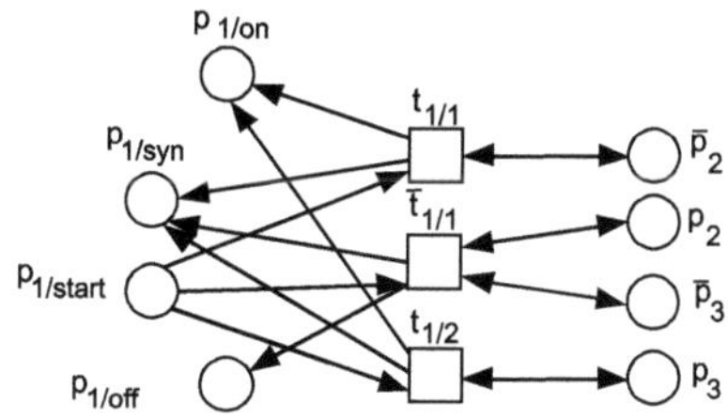

Fig. 5.3 Transitions modeling the decision process for entity g_1 (see Fig. 5.1)

leading to the same marking. Transition $\bar{t}_{1/1}$ records that g_1 will become inactive whenever g_2 is active and g_3 is inactive.

5.2.1.2 Phase Two: Synchronous State Update

When all the entities have made a decision about their next state, the second phase of the protocol begins, and the state of each entity is updated according to the recorded decisions. The Petri net structures needed to control this synchronous update are depicted in Fig. 5.4. The idea is to use a place $p_{i-1/\text{done}}$ to indicate when entity g_i should be updated; after g_i has been updated, place $p_{i/\text{done}}$ is marked, allowing the sequential update of all entities. To initiate the process there is a transition which fires when phase one is complete and places a token on $p_{0/\text{done}}$, see Fig. 5.4(a). This phase performs a synchronized update step, in which the state of each entity g_i is updated in turn by placing a token on p_i if place $p_{i/\text{on}}$ is marked, or on $\overline{p}_i$ if place $p_{i/\text{off}}$ is marked. The Petri net structure used for this update consists of four transitions representing the four possible update situations that can occur for an entity, as shown in Fig. 5.4(b). Once the state of entity g_i has been updated, a token is placed on place $p_{i/\text{done}}$ to indicate that the next entity can be updated. When the last entity g_n has been updated, place $p_{n/\text{done}}$ will be marked, and the control transition depicted in Fig. 5.4(c) initiates a reset step which remarks all the start places, allowing the whole update protocol to begin again.

The modeling approach presented above has assumed that we start with a set of complete and consistent truth tables which correctly describe the qualitative behavior of the GRN in question. However, in practice it is rarely the case that a GRN is fully understood, and indeed, this is one important reason for modeling such a system. We therefore often find that the description provided is *incomplete*, in the sense that information is missing about what happens in certain states, or *inconsistent*, in that we have conflicting information. The result is that under certain conditions the behavior of an entity may be unknown.

Such incomplete and/or inconsistent behavioral information can not be modeled using standard synchronous Boolean network techniques. However, Petri nets are a nondeterministic modeling formalism [318] able to represent unknown behavior by incorporating all possible choices. The idea is to identify for each entity all the problematic states in which the next state is unknown, and then to include these states when deriving Boolean expressions using logic minimization for both the active and

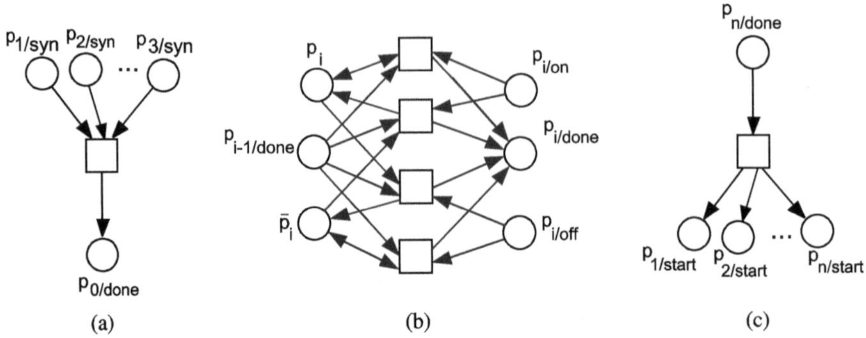

Fig. 5.4 Petri net structures for synchronous state update: initiating synchronous update (**a**); updating state of entity g_i (**b**); and reset step (**c**)

inactive next state behavior. The result is a Petri net model in which any unknown next state for an entity is modeled using two conflicting transitions representing a choice between an active and inactive next state for the entity. We therefore allow the model to exhibit both possible behaviors and rely on the nondeterministic choice mechanisms of the Petri net model. Such nondeterministic models can still be analyzed to provide meaningful results, since the Petri net tools are designed to cope with nondeterministic choices. As more data becomes available for a GRN, the Petri net model can be refined to reduce the amount of nondeterminism it contains, and so Petri nets provide an interesting means of documenting the development of knowledge about a genetic network. For a more detailed explanation and example of this approach, we refer the interested reader to [374].

5.2.2 Tool Support

The Petri net modeling approach presented above is supported by a software tool GNAPN (Genetic Networks as Petri Nets) [24], which completely automates the model construction process. GNAPN was implemented using the Java programming language and makes use of an auxiliary open source tool MVSIS [273] to perform logic minimization. GNAPN is freely available for academic use, and can be obtained from the project's website http://bioinf.ncl.ac.uk/gnapn.

GNAPN takes as input a series of truth tables describing the logical behavior of a GRN; these can be either directly supplied in an appropriate file format or created using a provided GUI utility which allows the entities, network structure and logical interactions in a GRN to be specified. Note that the tool is able to cope with partial GRN descriptions following the approach described in Sect. 5.2.1. The tool automatically constructs a Petri net model from the given logical description based on either the synchronous or asynchronous (see Sect. 5.3) update semantics. The resulting Petri net can be exported in a number of formats, including PNML [144], PEP [296] and ASTG [83]. This enables the model to be investigated and

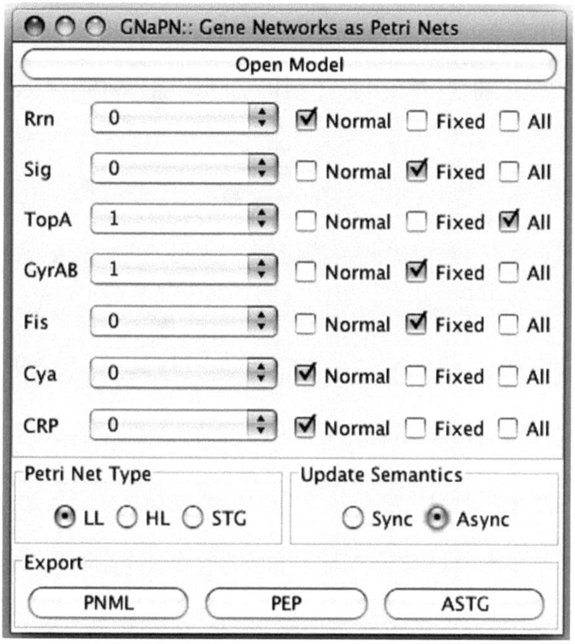

Fig. 5.5 The GUI provided by GNAPN to allow the user to set key properties of the Petri net model

analyzed using the wide range of tool support available for Petri nets (see, e.g., [301]). To aid the analysis process, GNAPN provides a GUI (see Fig. 5.5) which allows the user to set the initial state of entities and facilitates the creation of mutant models by allowing the state of one or more entities to be fixed (e.g., see the case study given in the next section).

For formal analysis of the constructed Petri nets we used the model checking capabilities of the PEP tool [296]. It accomplishes verification by constructing a condensed representation of the set of all reachable markings of the Petri net.

5.2.3 *Case Study: Nutritional Stress Response in E. coli*

To demonstrate the modeling techniques, we have so far introduced, and to illustrate the practical application of Petri net analysis techniques, we now present a case study. We consider modeling and analyzing a simplified synchronous version of the GRN responsible for the carbon starvation nutritional stress response in *E. coli*. This case study is based on the comprehensive data collated in [323] and was first presented in [373].

5.2.3.1 Constructing the Petri Net Model

The bacterium *E. coli* under normal environmental conditions, when nutrients are freely available, is able to grow rapidly, entering an *exponential phase* of

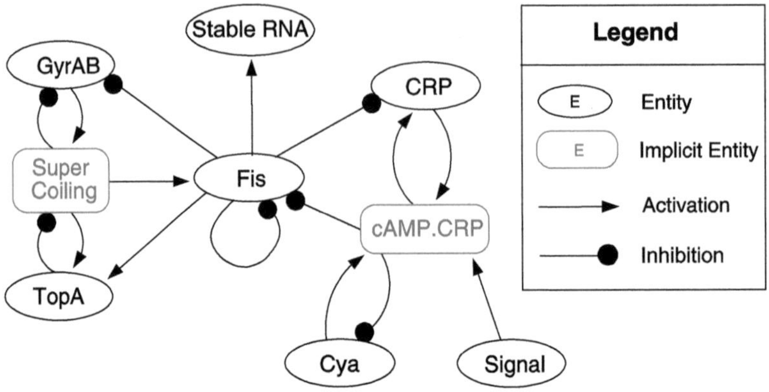

CRP	Cya	Signal	[Cya]
0	0	0	1
0	0	1	1
0	1	0	1
0	1	1	1
1	0	0	1
1	0	1	1
1	1	0	1
1	1	1	0

GyrAB	TopA	Fis	[GyrAB]
0	0	0	1
0	0	1	0
0	1	0	1
0	1	1	0
1	0	0	0
1	0	1	0
1	1	0	1
1	1	1	0

GyrAB	TopA	Fis	[TopA]
0	0	0	0
0	0	1	0
0	1	0	0
0	1	1	0
1	0	0	0
1	0	1	1
1	1	0	0
1	1	1	0

Fig. 5.6 The GRN for carbon starvation stress response in *E. coli*: the top-level view (*top*); and the truth tables defining the Boolean behavior of Cya, GyrAB and TopA (*bottom*)

growth [159]. However, as important nutrients become depleted and scarce, the bacteria experiences nutritional stress and responds by slowing down growth, eventually resulting in a *stationary phase* of growth. This nutritional stress response mechanism is reversible, and *E. coli* resumes the exponential phase of growth as soon as nutrients are available again.

The GRN underlying the stress response in *E. coli* to carbon starvation is shown abstractly in Fig. 5.6 (adapted from [323], where they refer to entities which will not be directly modeled as *implicit entities*). The network has a single input signal indicating the presence or absence of carbon starvation, and uses the level of stable RNA (ribosomal RNA and transfer RNA) as indicative of the current phase of *E. coli*, that is, during the exponential phase the level of stable RNA is high to support rapid growth, while under the stationary phase the level drops, since only a maintenance metabolism is required [323]. The carbon starvation signal is transduced by the activation of adenylate cyclase (Cya), an enzyme which results in the production of the metabolite cAMP. This metabolite immediately binds with and activates the global regulator protein CRP, and the resulting cAMP.CRPcomplex is responsible for controlling the expression of key global regulators including Fis and CRP itself. The global regulatory protein Fis is central to the stress response, and is responsible for promoting the expression of stable RNA from the *rrn operon* [159, 323]. Thus, during the exponential phase high levels of Fis are normally observed, and the mutual repression that occurs between Fis and cAMP.CRP is thought to play a key role in the regulatory network [323]. The expression of Fis is also promoted by high levels of *negative supercoiling* being present in the DNA. The level of DNA supercoiling is tightly regulated by two topoisomerases [159, 323]: GyrAB (composed of the products of genes gyrA and gyrB) which promotes supercoiling, and TopA which removes supercoils. An increase in DNA supercoiling results in increased expression of TopA, which prevents excessive supercoiling. A decrease in supercoiling results in increased expression of gyrA and gyrB, and the resulting high level of GyrAB acts to increase supercoiling.

Using the data provided in [323], we were able to derive truth tables defining the Boolean behavior of each regulatory entity in the stress response GRN for carbon starvation. Following the approach in [323], the level of cAMP.CRP and DNA supercoiling are not explicitly modeled as entities, and they are therefore referred to as *implicit entities*. As an example, the truth tables defining the behavior of entities Cya, GyrAB and TopA are shown in Fig. 5.6.

The next step is to apply logic minimization to the truth tables we have derived to extract Boolean expressions which compactly define the qualitative behavior of each regulatory entity. This process is automated by the support tool GNAPN, and the result is the set of logical equations presented below

$$[\mathsf{Cya}] = \overline{\mathsf{Signal}} + \overline{\mathsf{Cya}} + \overline{\mathsf{CRP}}, \qquad [\mathsf{CRP}] = \overline{\mathsf{Fis}},$$
$$\left[\overline{\mathsf{Cya}}\right] = \mathsf{Signal\ Cya\ CRP}, \qquad \left[\overline{\mathsf{CRP}}\right] = \mathsf{Fis},$$

$$[\mathsf{GyrAB}] = \overline{\mathsf{GyrAB}}\ \overline{\mathsf{Fis}} + \mathsf{TopA}\ \overline{\mathsf{Fis}},$$
$$\left[\overline{\mathsf{GyrAB}}\right] = \mathsf{GyrAB}\ \overline{\mathsf{TopA}} + \mathsf{Fis},$$

$$[\mathsf{TopA}] = \mathsf{GyrAB}\ \overline{\mathsf{TopA}}\ \mathsf{Fis}, \qquad [\mathsf{SRNA}] = \mathsf{Fis},$$
$$\left[\overline{\mathsf{TopA}}\right] = \overline{\mathsf{GyrAB}} + \mathsf{TopA} + \overline{\mathsf{Fis}}, \qquad \left[\overline{\mathsf{SRNA}}\right] = \overline{\mathsf{Fis}},$$

$$[\mathsf{Fis}] = \overline{\mathsf{Fis}}\ \overline{\mathsf{Signal}}\ \mathsf{GyrAB}\ \overline{\mathsf{TopA}} + \overline{\mathsf{Fis}}\ \overline{\mathsf{Cya}}\ \mathsf{GyrAB}\ \overline{\mathsf{TopA}} + \overline{\mathsf{Fis}}\ \overline{\mathsf{CRP}}\ \mathsf{GyrAB}\ \overline{\mathsf{TopA}},$$
$$\left[\overline{\mathsf{Fis}}\right] = \mathsf{CRP\ Cya\ Signal} + \mathsf{Fis} + \overline{\mathsf{GyrAB}} + \mathsf{TopA}.$$

These equations can then be used to construct a Petri net model of the nutritional stress response GRN for carbon starvation. We used GNAPN to automate the model construction, and the result is a Petri net model that contains 45 places and 49 transitions (unfortunately, this model is too large to be drawn here).

5.2.3.2 Analyzing the Petri Net Model

Once we have constructed a Petri net model of a GRN, we may then validate and analyze it using the wide range of tools available for Petri nets (see, for example, [301]). In this case study, we consider using PEP [296], a general purpose Petri net tool, and in particular, make use of model checking techniques [79]. Our aim is to illustrate the range of analysis possible using available tools, from simple validation tests to more in-depth gene knockout and overexpression analysis.

Table 5.1 Simulation test results showing the model correctly switches from exponential to stationary growth phase (*left*) and vice a versa (*right*)

Signal	CRP	Cya	GyrAB	TopA	Fis	SRNA	Signal	CRP	Cya	GyrAB	TopA	Fis	SRNA
1	0	1	1	0	1	1	0	1	0	0	0	0	0
1	0	1	0	1	0	1	0	1	1	1	0	0	0
1	1	1	1	0	0	0	0	1	1	0	0	1	0
1	1	0	0	0	0	0	0	0	1	0	0	0	1
1	1	1	1	0	0	0	0	1	1	1	0	0	0

Model Validation We begin our analysis by validating that our Petri net model is a reasonable representation of the GRN in question. The idea is to perform a range of simple simulation tests on the model to ensure that it satisfies the basic behavioral properties indicated by the literature. In the case of the GRN for the carbon stress response in *E. coli*, this involves validating that the switch between the exponential and stationary phases of growth [323] occurs correctly.

The first simulation test ensures the model correctly switches from the exponential to the stationary phase of growth when carbon stress is present. We initialize the Petri net to a state representing the exponential growth phase (see [323]) and activate Signal to indicate carbon stress. The resulting simulation results are presented in Table 5.1 (left), and show that the model correctly switches to the stationary phase by entering an attractor cycle of period two in which entity SRNA remains inactive. Similarly, we can check that the model correctly switches back from the stationary to the exponential growth phase when carbon stress is removed. The corresponding simulation results, presented in Table 5.1 (right), show that the model does correctly return to the exponential growth phase.

Property Analysis After performing a range of validation tests to gain confidence in the correctness of the model, we can now consider analyzing the behavior and properties of the model in more detail. To investigate the behavior of the model, we make use of the extended reachability analysis provided by the model checking tools of PEP [144, 189]. For example, it appears from the literature that the entities GyrAB and TopA should be mutually exclusive, that is, whenever GyrAB is significantly expressed then TopA should not be, and vice versa. We formulate the following constraints on places:

$$GyrAB + TopA > 1, \qquad GyrAB/done = 1,$$

which characterize a state in which the mutual exclusion property does not hold (the condition $GyrAB/done = 1$ is needed to ensure that only the states reached after a complete pass of the two phase commit protocol are considered). The model checking tool is able to confirm that no state satisfying these constraints is reachable from any reasonable initial state, which proves that GyrAB and TopA are mutually exclusive.

We can attempt to prove a similar mutual exclusion property for CRP and Fis using the same approach. However, this time the model checking tool confirms that

Table 5.2 Results of mutant analysis on the model

Entity	Knock out	Overexpressed	Knock out(s)	Overexpressed(s)
CRP	Yes	Yes	Yes	Yes
Cya	Yes	Yes	Yes	Yes
GyrAB	No	Yes	No	Yes
TopA	Yes	No	Yes	No

it is able to reach a state satisfying the constraint, proving that CRP and Fis are not mutually exclusive in our model. In fact, the tool returns a witness firing sequence leading to such a state, which can be simulated to gain important insight into why this behavior can occur.

Mutant Analysis We complete our analysis by investigating how "fixing" the state of a single entity affects the normal function of the model. This corresponds to the experimental approach of creating mutants, in which genes are knocked out or overexpressed, and is a useful analysis technique that can provide important insights into a GRN. In order to create a mutant model, we fix the state of an entity by treating it as an input to the model (e.g., as we do for Signal), and simply ignore its corresponding behavioral definition when the model is constructed. The tool GNAPN (see Sect. 5.2.2) provides a simple interface which allows mutant models to be automatically constructed.

For the given GRN, we are interested in identifying situations in which SRNA can be prevented from being active in the absence of carbon stress, and where SRNA can become active in the presence of carbon stress. We therefore perform a series of tests by first setting Signal, Fis and SRNA to be inactive, and then systematically knocking out and overexpressing the remaining entities in turn. We can then repeat this series of tests with Signal set to be active. The observed results of this analysis are presented in Table 5.2, where 'Yes' indicates that SRNA was able to become active, 'No' that this did not occur, and an appended '(s)' indicates the presence of carbon stress.

We see that when CRP or Cya are knocked out or overexpressed, SRNA is able to become active regardless of the presence or absence of carbon stress. However, when we knock out GyrAB with no carbon stress, SRNA does not become active. This can be explained by noting that since GyrAB indirectly activates Fis via supercoiling, TopA is allowed to reduce the amount of supercoiling without competition, and so reduces the level of Fis. Meanwhile, the cAMP.CRP complex represses Fis, and so overall SRNA is repressed.

Another interesting case is when carbon stress is present and we overexpress GyrAB. One should note that under normal conditions when Signal is active, SRNA should never become active. However, with GyrAB overexpressed, the level of supercoiling can increase without being affected by the inhibitory effect of TopA, and so Fis increases. This increased level of Fis also reduces the amount of the cAMP.CRP complex, which in turn reduces the repression of Fis. As a result, SRNA is activated under these abnormal conditions, which appears to be consistent with [323].

5.3 Asynchronous Models of GRNs

In the previous section, we considered using Petri nets to model and analyze synchronous Boolean network models of GRNs. Historically, the synchronous Boolean models have been much favored by the biological community, since they result in deterministic behavior which is simpler to understand and analyze. However, the assumption that all entities update their state synchronously can be argued to be biologically unrealistic [390], and this leads to reservations about the reliability of results obtained from such models.

The asynchronous update semantics therefore appears to provide a more realistic modeling approach. However, asynchronous models suffer from the problem of capturing too much behavior, not all of which is realizable in practice. This behavior can often be highly nondeterministic, resulting in many (nonconverging) *choices* when selecting the next state. Thus, the asynchronous approach can result in a complex model which is difficult to understand and analyze.

In fact, many of the choices present in an asynchronous model can be resolved either by assuming that the environment of the biological system is slow (i.e., the system always has enough time to react to its changes), or by using knowledge about the relative speeds of chemical reactions. It therefore turns out that the behavior of a biological system has much less nondeterminism than such asynchronous models suggest. (This may explain why synchronous models were often favored over asynchronous ones by the biological modeling community.) Therefore, the problem in practice is to be able to identify the critical choices that arise in a model and to have the appropriate kinetic knowledge to resolve them.

These considerations motivate us to use techniques from *speed-independent* (*SI*) circuits [84] when asynchronously modeling GRNs. SI circuits are a subclass of asynchronous circuits that work correctly (i.e., according to their specification) regardless of the delays associated with logic gates. We follow the classical Muller's approach [268] which regards each logic gate as an atomic evaluator of a Boolean function, with a delay element associated with its output (the wires are assumed to have negligible delays). In the SI framework, no assumptions are made about the gate delays (except that they are positive), that is, individual gates can be arbitrarily slow/fast and even have variable unbounded delays. SI circuits tend to be deterministic, though they can handle certain kinds of non-determinism using *arbiters* [84]—special devices deciding which of two inputs arrives first (this proves to be important from a biological perspective as illustrated in Sect. 5.3.3).

We therefore make the following important methodological assumption:

GRNs can be modeled by speed-independent circuits.

That is, if a GRN can not be qualitatively modeled by an SI circuit then either its behavior is inherently nondigital or its corresponding model is incorrect and/or misses some important information.[1] Thus, SI techniques can provide important

[1] In particular, such systems *in principle* can not be modeled by synchronous Boolean networks, which are deterministic by definition.

insights into a model's correctness and highlight key areas that need refining. We will discuss this issue in detail later in this section.

It turns out that whether a circuit is SI or not almost always depends on its *environment*, that is, a circuit can be SI in one environment and non-SI in another one. That is, *whether the circuit is SI or not can not be deduced solely from the structure of the circuit*! This suggests that *traditional asynchronous Boolean networks lack some important information* (*viz. the behavior of the environment*). In Sect. 5.3.2, we explain this phenomenon in more detail; there we demonstrate that a C-element circuit is not SI in the most general environment that can at any time change the value of any input, but becomes SI in a restricted environment.

In this section, we make a case for using another formalism, viz. *Signal Transition Graphs* (*STGs*) [75, 325], which allows one to capture in a natural way the behavior of both the circuit and its environment. STGs are Petri nets in which transitions are labeled with the rising and falling edges of circuit signals. They have been used extensively for the design of asynchronous control circuits.

We investigate how the sufficient conditions ensuring that an STG can be implemented by an SI circuit [84] can be interpreted in the context of GRNs. We observe that violations of these properties provide important insights into a model and highlight areas which need to be refined. In particular, the violation of the *output-persistency* (*OP*) condition [84] indicates the presence of choices that either require further information to resolve or indicate some stochastic effects in the system that have to be carefully documented. STGs provide a formal means of documenting and refining this information, and thus provide a well-supported formal framework for GRNs that allows realistic models to be incrementally developed and analyzed.

We illustrate our proposed approach by considering a case study in which we develop and analyze an SI STG model of the GRN controlling the switch between the lysogeny and lysis cycles in phage λ [289]. The case study makes use of the STG support tool PETRIFY [84] and demonstrates its practical role in model development.

5.3.1 Signal Transition Graphs and Speed-Independent Circuits

Signal Transition Graphs (*STGs*) [84] are a particular type of labeled Petri nets developed specifically for modeling asynchronous digital circuits. The idea is to associate a set of Boolean variables, referred to as *signals*, with a Petri net to represent the state of the actual digital signals (i.e., wires within a circuit or entities within a GRN). The Petri net's transitions are then labeled to represent changes in the state of these signals; a transition label either has the form a^+ to indicate a signal a goes from 0 to 1, or a^- to indicate the signal goes from 1 to 0. Thus, the underlying Petri net specifies the causal relationship between signal changes and is intended to capture the behavior of a system. Clearly, for an STG to correctly represent a circuit one has to ensure that the labels a^+ and a^- are correctly alternated between for each signal (the so called *consistency condition* [84]). In general, several transitions can have the same label, for example, a^+; in such a case, these transitions are named $a^+, a^+/1, a^+/2$, etc.

Since the behavior of an STG is based on its underlying Petri net behavior, the concepts of enabling and firing of transitions still hold. STGs are therefore amenable to general Petri net analysis tools, but are also supported by a range of specific tools, such as PETRIFY [83, 84]. These are able to analyze and optimize STGs, as well as synthesize digital circuits from them. An STG can be represented graphically as a labeled Petri net. However, a short-hand notation is often used, in which transitions are simply represented by their labels, and nonmarked places with only one input and one output transition are contracted (see Fig. 5.7(c) for an example).

The signals (i.e., entities) of an STG are partitioned into *input*, *output* and *internal* signals; the output and internal signals are collectively referred to as *local* signals. The inputs are controlled by the environment of the STG (in the context of biological systems, this could be either the actual environment of the organism, or the other systems within the organism, whose outputs affect the behavior of the system), and the outputs are controlled by the system itself and are observable by the environment (e.g., they can be inputs of other systems within the organism). Internal signals represent some auxiliary entities needed to produce outputs; like outputs, they are controlled by the system, but they are not observable by the environment. The partitioning of signals is an important part of the modeling process and represents key design decisions when developing an STG.

From a biological point of view, we can interpret STGs as follows. Signals are used to represent the states of biological entities, and transitions are used to capture changes in these states, for example, through chemical reactions or fluctuations in protein concentration. In particular, the input signals from the environment can be used to represent external factors such as temperature, or simply the output from other subsystems in the organism, and the internal signals can be used to represent any auxiliary biological entities that are required for the system to function correctly. The output signals can therefore be interpreted as the result of these environmental factors and internal mechanisms, such as a changes in the concentration of key proteins.

Intuitively, an STG represents a contract between the system and its environment, and is interpreted in the following way. If an input signal transition is enabled, then the environment is allowed (but is not obliged) to send this input, and vice versa, the environment is not allowed to send inputs which are not enabled. If a local transition is enabled, then the system is obliged eventually to produce this signal (or it is eventually disabled by another transition), and vice versa, it is not allowed to produce outputs which are not enabled. That is, an STG specifies the behavior of a system in the sense that the system must provide *all and only* the specified outputs, and that it must allow *at least* the specified inputs (in fact, it could optionally allow more inputs, which means that it could work in a more demanding environment).

As an example, consider Fig. 5.7(a) which shows a commonly used component in asynchronous circuit design called a *C-element* [84]. The C-element takes two inputs a and b from the environment and produces a single output c back into the environment, as defined by the truth table in Fig. 5.7(b). The C-element's behavior

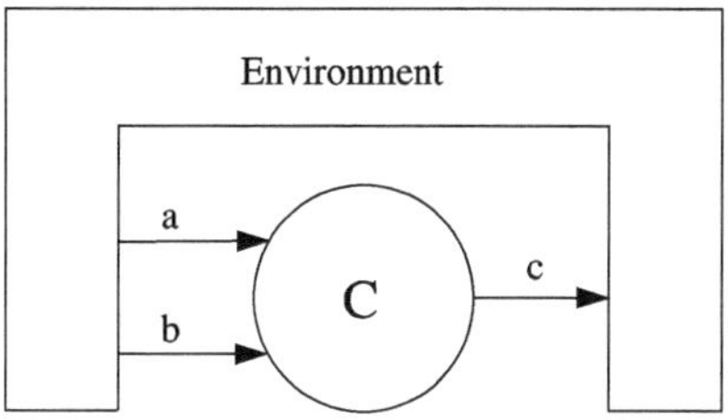

(a)

a	b	c	$[c]$
0	0	0	0
0	0	1	0
0	1	0	0
0	1	1	1
1	0	0	0
1	0	1	1
1	1	0	1
1	1	1	1

(b)

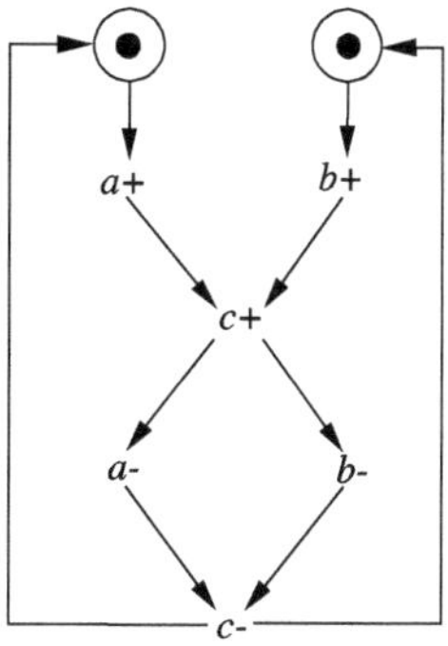

(c)

Fig. 5.7 C-element (**a**); truth table for C-element (**b**); and STG representing the required behavior of the C-element (**c**)

is specified by the STG shown in Fig. 5.7(c), in which the initial value of each signal is set to 0. It specifies that the system waits until the environment raises (in any order) the inputs a and b (transitions a^+ and b^+), and then raises the output c (transition c^+). (Observe that the environment is assumed not to reset the raised inputs until c^+ fires.) Then the environment resets (in any order) the inputs a and b (transitions a^- and b^-), and in response the system resets its output c (transition c^-). (Again, the environment is assumed not to raise the reset inputs until c^- fires.)

In order for an STG to represent an SI circuit, it needs to satisfy the following properties:

Boundedness: An STG has finitely many reachable states iff it is *bounded,* that is, the number of tokens in each place can never exceed some bound k. Since a digital circuit (or a Boolean network) can have only finitely many reachable states, boundedness is taken as an implementability requirement.

Consistency: *Consistency* is a basic well-formedness property, stating that the reachable signal values must be binary. That is, in every trace of the STG the transition labels for each signal a must alternate between a^+ and a^-, always beginning with the same sign.

Output-persistency: *Output-persistency* (*OP*) property requires that if some local signal becomes enabled, it can not be disabled by firing some other transition, that is, there should be no choices involving local transitions. The rationale for this is that once a signal becomes enabled, its voltage starts, for example, to rise from 0 to 1. If the signal is disabled during this process, the voltage is pulled down, resulting in a glitch. This glitch can be interpreted in different ways by the logic gates listening to this signal, depending on whether the voltage has crossed the threshold between 0 and 1 or not. Hence, the behavior of the circuit becomes nondeterministic. Such a situation can be interpreted in biological terms as well, with the voltage replaced by, for example, the concentration of some protein.
Visually, if OP is violated then there are two transitions with different labels in the STG with at least one of them marked by a local signal, which share some

preplaces and can be enabled simultaneously (unless both transitions are connected to these shared preplaces by read arcs).

Identifying and understanding the choices within a model can provide important insights into its behavior and correctness. Below, we consider several kinds of choices that can arise and their implications for OP.

- A choice involving only inputs is not regarded as a violation of OP, and simply models a non-deterministic choice in the environment.[2] (For example, the environment might nondeterministically decide either to rise the temperature above normal, or to reduce it below normal.) That is, *this choice does not have to be implemented by the system itself.*
- A choice between a local signal and an input constitutes a serious problem in the model. Intuitively, the decision to fire the transitions involved in the choice has to be made independently by the system and the environment, and so *both* transitions can actually start firing, and only later the system senses the change of the input and aborts the firing of a local transition. Such choices are the points where the circuit's behavior becomes nondigital,[3] and hence the STG ceases to be an adequate model of the circuit. In practice, such a choice often indicates an implicit assumption about the slowness of the environment, and the input transition will never fire in this state. Alternatively, the model should be reconstructed to allow for parallel firing of these transitions.
- A choice involving only local transitions can still be implemented in a speed-independent way (in spite of the violation of OP) using an *arbiter* [191]—a special component that can handle the meta-stable behavior associated with such a choice. In such a case, the behavior of the circuit becomes nondeterministic. When modeling a biological system, the violations of OP that correspond to true (e.g., stochastic) choices between local transitions, that can not be deterministically resolved by the reaction rates, should be left in the model; however, any such violation should be carefully investigated by the model designer and clearly documented.

Note that arbitration should be used only for representing truly stochastic phenomena, like the choice between lysogeny and lysis modes in phage λ (see Sect. 5.3.3). Other violations of OP indicate that some important information is missing in the model, for example, some assumptions about the environment's behavior should be made, or the reaction rates can be used to resolve the choice. Methodologically, violations of OP are detected automatically, and if there are any, the user should either document the associated stochastic choice or refine the model, as we illustrate by an example in Sect. 5.3.3.

Complete State Coding (*CSC*): If the STG has two reachable states in which the values of all the signals coincide but the values of the next-state function for some

---

[2]Only an *abstraction* of the environment can be included into the model; such abstractions are often nondeterministic even if the environment itself is deterministic.

[3]Note that even though the circuit is constructed of logic gates only, such gates can exhibit a nondigital behavior under certain circumstances.

local signal are different, then these two states are said to be in a *Complete State Coding* (*CSC*) conflict. The STG satisfies the *CSC property* if no two of its reachable states are in a CSC conflict.
An STG not satisfying the CSC property can not directly represent an SI circuit. Intuitively, during its execution the system can 'see' only the values of its signals, but not the marking of the STG. Hence, if two semantically different reachable states with the same values of all the signals exist, the system can not distinguish between them, and so can not know what to do next.
At the circuit level, CSC conflicts are resolved by inserting new internal signals helping to distinguish between the conflicting states, in such a way that its 'external' behavior does not change. (One has to take care to preserve the consistency and other SI properties when inserting new signals.) Intuitively, insertion of a signal introduces additional memory into the circuit, helping it to trace the current state.
In an STG modeling a biological system, CSC conflicts can be interpreted as a lack of information about the internal workings of the system. That is, they indicate the presence of some auxiliary internal entities (e.g., proteins) which are not visible to the environment but help the system to accomplish its function. An STG with CSC conflicts might be useful in some cases as a high-level view of the system (in such a case, all the internal signals can be abstracted away to simplify the model), but if a detailed description of the system is needed, the STG should satisfy the CSC property.

Checking the properties discussed above is automated by the STG support tool PETRIFY [84], and in the next section we show how to apply the developed theory to modeling GRNs.

5.3.2 *Refining Asynchronous Models of GRNs*

In this section, we consider applying the techniques and tools from SI circuit design to developing realistic asynchronous models of GRNs. We describe how a Boolean network model of a GRN can be translated into an STG which satisfies the consistency, boundedness and CSC properties. We then consider how such an STG can be semi-automatically refined by identifying and removing OP violations.

To gain an initial insight into the proposed refinement methods, we consider a small example based on the C-element. The behavior described by the STG in Fig. 5.7(c) can be modeled by a Boolean network (i.e., circuit) whose behavior is defined by the logical equation $[c] = ab + c(a + b)$ (or equivalently, the truth table in Fig. 5.7(b)). This model is SI in the intended environment, as specified in Fig. 5.7(c). However, *just by looking at the logical equation* (*i.e., the Boolean network*) *it is impossible to say what were the assumptions about the environment*; in particular, there are environments where the behavior of this circuit becomes non-SI, for example, if the environment, after raising a and b, resets either of them before

The circuit-STG construction:

- Each signal (i.e., regulatory entity) g_i is represented by two places, g_i and $\overline{g_i}$, indicating whether the entity is active or inactive, respectively. Exactly one of these places is marked at any time.
- Since we do not have any information about the environment's behavior, it is taken to be the most general (i.e., it can always change the value of any input). This is modeled for each input signal g_i by adding transitions g_i^+ (consuming a token from $\overline{g_i}$ and depositing a token to g_i) and g_i^- (consuming a token from g_i and depositing a token to $\overline{g_i}$).
- For each local signal g_i, the circuit computes the next-state value $[g_i]$ of g_i using the given Boolean equation $[g_i] = E_i$ (see Sect. 5.2.1). For each product term T_j in the minimized DNF of $E_i|_{g_i=0}$ (where $E_i|_{g_i=b}$ denotes the Boolean expression resulting from substituting g_i by $b \in \{0, 1\}$ in E_i), we add a transition g_i^+/j which switches g_i on. We add an arc from place $\overline{g_i}$ to g_i^+/j and an arc from g_i^+/j to place g_i. For each g_k (resp. $\overline{g_k}$) occurring in T_j, we connect g_i^+/j to the place g_k (resp. $\overline{g_k}$) by a read arc. We use a similar process to define the transitions g_i^-/j which reset g_i based on $\overline{E_i|_{g_i=1}}$.

Fig. 5.8 Petri net construction method for asynchronous Boolean networks

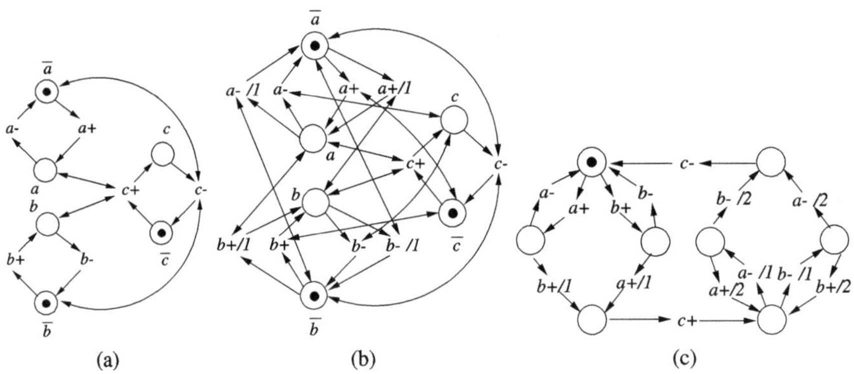

Fig. 5.9 The circuit-STG for the circuit $[c] = ab + c(a + b)$ (**a**); a way to resolve choices in it by assuming a slow environment (**b**); and the STG simplified using PETRIFY (**c**)

c^+ fires. This illustrates that *having an STG can be much more useful for analyzing the system than simply having a Boolean network definition.*

Any asynchronous Boolean network (i.e., circuit) can be converted into an STG using the approach outlined in Fig. 5.8 [23, 65, 318]. The idea is that we define a signal for each regulatory entity and then label the transitions in the model with appropriate signal labels (i.e., g_i^+ if the transition activates entity g_i and g_i^- if it de-activates it). Since the Boolean network does not provide any information about the environment's behavior, it is taken to be the most general (i.e., it can always change the value of any input). Note that any STG that results from the above construction always satisfies the properties of consistency, boundedness and CSC [23]. The above model construction process is fully automated by the support tool GNAPN.

Figure 5.9(a) illustrates this construction process for the C-element defined by the logical equation $[c] = ab + c(a + b)$. Note that the behavior of the resulting

STG strictly includes the behavior of the initial model Fig. 5.7(c), since the information about the behavior of the environment could not be retrieved from the logical equation, and the most general environment was modeled. Using PETRIFY (an STG tool developed specifically for asynchronous circuit design [83, 84]), we can automatically detect that the resulting STG is not SI in this environment, as an output c^+ can be disabled by a^- or b^-, and similarly, c^- can be disabled by a^+ or b^+.

If the circuit $[c] = ab + c(a + b)$ was used to model a system that is perceived to be deterministic, then some of this STG's behavior is not realizable in practice. Hence, the STG should be refined, so that it captures only the realistic behavior. The candidate points for refinement are where the OP property is violated, for example, due to conflicting choices involving a local transition. Such choices (unless they represent some truly stochastic phenomenon) have to be resolved either by making assumptions about the environment, or by looking at reaction rates. Methodologically, the points where the speed-independence is violated can be found automatically, but the resolution of choices requires interaction with the user.

Formally, we denote an OP violation by $t \rightharpoonup t'$, meaning that a transition t can be disabled by firing a transition t', where t and t' have different labels and t is labeled by a local signal. One can see that for the STG in Fig. 5.9(a), we have OP violations $c^+ \rightharpoonup a^-$, $c^+ \rightharpoonup b^-$, $c^- \rightharpoonup a^+$ and $c^- \rightharpoonup b^+$. This information is given to the user, who now can suggest a way to resolve these violations. In this particular case, the user might know that the environment is relatively slow, that is, if, say, a^- and c^+ are enabled simultaneously then c^+ will fire first. Alternatively, the relative rates of chemical reactions might determine which transition fires first. Of course, such rates must be provided by the user, since there is no way a tool can work them out from the STG or Boolean network. In practice, measuring reaction rates is a very effort-consuming task, but our method addresses this problem (i) by giving information about what rates have to be measured (in practice, few rates affect the qualitative behavior of the GRN), and (ii) by requiring only relative rates (i.e., it is enough to know that one reaction is faster than the other, rather than the absolute rates).

We use the following notation for the user-provided assumptions: we write $t \mapsto t'$ to denote that whenever transitions t and t' are enabled simultaneously then priority is given to t. (We assume that t and t' have different labels, at least one of these transitions is labeled by a local signal, t and t' share some preplaces, and not all of these shared places are connected to t and t' by read arcs.) In our example, the slowness of the environment can be expressed as $c^+ \mapsto a^-$, $c^+ \mapsto b^-$, $c^- \mapsto a^+$, $c^- \mapsto b^+$.

Such priority assumptions $t \mapsto t'$ can be applied to the STG, resulting in a transformed model which captures this information. The idea is to replicate the transition with lower priority t' to capture each situation in which t is not enabled and t' can safely fire. This *firing order enforcement* (*FOE*) transformation is formally defined in Fig. 5.10.[4]

[4]One can show that safeness and consistency are preserved by the FOE transformation, as it can only reduce the set of reachable markings. Moreover, though it does not preserve the CSC property, it preserves the stronger USC property, see [23]. Since the STGs derived from Boolean networks

The firing order enforcement (FOE) transformation:
Suppose $t \mapsto t'$ has been assumed, and let $p_1, \ldots, p_k$ be the preplaces of t which are not preplaces of t'. If $k = 0$ then t is enabled whenever t' is, and so t' can be simply eliminated from the STG, together with all the incident arcs, as in such a case it can never fire due to the assumption $t \mapsto t'$. Otherwise, t' is replicated $k - 1$ times, so that there are k copies (denoted by $t'_1 = t', t'_2, \ldots, t'_k$) of t' altogether. All these replicas are labeled by the same signal as t', and have exactly the same connections. Furthermore, a read arc is added between t'_i and $\overline{p_i}$ for each $i = 1, \ldots, k$, where $\overline{p_i}$ is $\overline{g_j}$ if p_i corresponds to g_j, and g_j if p_i corresponds to $\overline{g_j}$.
The FOE transformation guarantees that: (i) if t is enabled by some marking M then none of $t'_1, \ldots, t'_k$ is enabled; and (ii) if t is not enabled by some marking M but t' is enabled by M in the original STG, then at least one of $t'_1, \ldots, t'_k$ is enabled in the modified STG. That is, the choice is resolved to favor t.

Fig. 5.10 A transformation for enforcing the priority assumption $t \mapsto t'$

Our method allows for automatic application of user-given assumptions about the environment and relative reaction rates to the STG, in order to refine its behavior. In particular, it transforms the STG in Fig. 5.9(a) into the one in part (b) of this figure, which, after simplification by PETRIFY, becomes the STG in part (c) of this figure. The latter STG has less behavior than the STG in Fig. 5.9(a), and is SI. Somewhat unexpectedly, it has more behavior than the initial model in Fig. 5.7(c). This is explained by the fact that it poses fewer constraints on the environment (i.e., the system can actually cope with a more demanding environment than the one it was intended for).

5.3.3 Case Study: Lysis–Lysogeny Switch in Phage λ

We now illustrate the introduced STG modeling techniques by developing an STG model of the GRN responsible for the lysogeny-lysis switch in λ phage [289, 311]. Using the Boolean model presented in [390] as a starting point, we construct and refine an STG model of this GRN, utilizing the support tools GNAPN [24] and PETRIFY [84]. The model is refined by finding the points where OP violations occur, and then applying appropriate assumptions about the environment's behavior and relative reaction rates to resolve the associated hazards. Since the lysis-lysogeny decision is a stochastic phenomenon, it is not resolved and remains in the final model, which still turns out to be SI.

5.3.3.1 Model Construction

The temperate bacteriophage λ [289, 311] is an extensively studied virus which infects the bacteria *E. coli*. After infection of the host cell, a stochastic decision

using the construction in Fig. 5.8 are always safe (i.e., every place can contain at most one token), consistent and have USC [23], these three properties can be ignored when refining such models. However, for the manually constructed STGs they have to be checked.

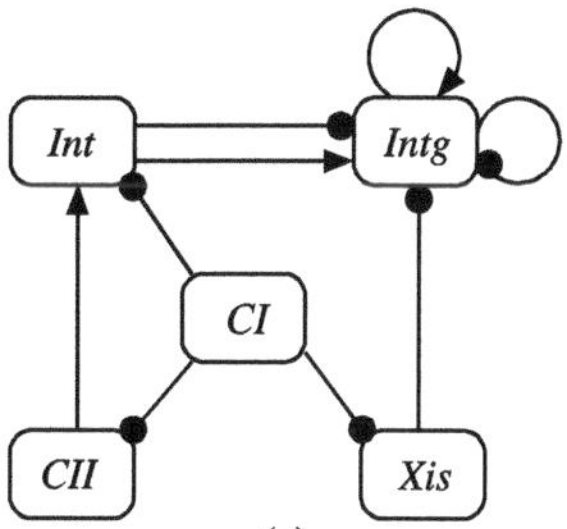

(a)

$$
\begin{aligned}
[\mathsf{CII}] &= \overline{\mathsf{CI}} \\
[\mathsf{Int}] &= \mathsf{CII} + \overline{\mathsf{CI}} \\
[\mathsf{Xis}] &= \overline{\mathsf{CI}} \\
[\mathsf{Intg}] &= \overline{\mathsf{Intg}}\,\mathsf{Int} + \mathsf{Intg}\,(\overline{\mathsf{Int}} + \overline{\mathsf{Xis}})
\end{aligned}
$$

(b)

Fig. 5.11 A high-level representation of the GRN of the phage λ switch (**a**); and the corresponding logical equations which define the GRN's behavior (**b**)

is made by phage λ, based on environmental factors, between two very different methods of reproduction, namely the *lytic* and *lysogenic* cycles [390]. In most cases, phage λ enters the lytic cycle in which it generates as many new viral particles as the host cell resources allow. It then uses an enzyme to breakdown and lyse the cell wall, releasing the new phage into the environment. Alternatively, phage λ may enter the lysogenic cycle where it integrates its DNA into the DNA of the host cell. In such a case, the genes expressed in the phage λ DNA, now a prophage, synthesize a repressor which blocks the expression of other phage genes including those involved in its own excision. As such, the host cell, now a lysogen, establishes an immunity to external infection from other phages, and the prophage is able to lie dormant, replicating with each subsequent cell division of the host.

A high-level pictorial representation of this GRN is presented in Fig. 5.11, along with the corresponding logical equations describing the qualitative behavior of each network entity [390]. Integration of the phage λ DNA into the host DNA requires the presence of the integrase Int. Furthermore, the phage λ DNA remains integrated unless the excisionase Xis is also present. Thus, integration and excision occurs in both directions when both Int and Xis are present, and so the stochastic lysis-lysogeny choice is qualitatively modeled as a nondeterministic one [390]. The signal Intg is used as an output to indicate the status of this process, taking the value 1 if the phage λ DNA is integrated and 0 if it is not integrated or has been excised. Both Int and Xis are repressed by the phage λ repressor CI, which we regard as an input since it is regulated outside the scope of this model. However, Int is also activated by CII, itself under negative control from CI. This additional control of Int therefore favors integration over excision [390].

From the Boolean network shown in Fig. 5.11, we are able to automatically construct an STG describing the behavior of the phage λ circuit using GNAPN. We define CI as an input signal from the environment, Intg as the output signal produced by the circuit, and CII, Int and Xis as internal signals which are invisible to the environment. (As discussed earlier, this partitioning of signals is a decision which

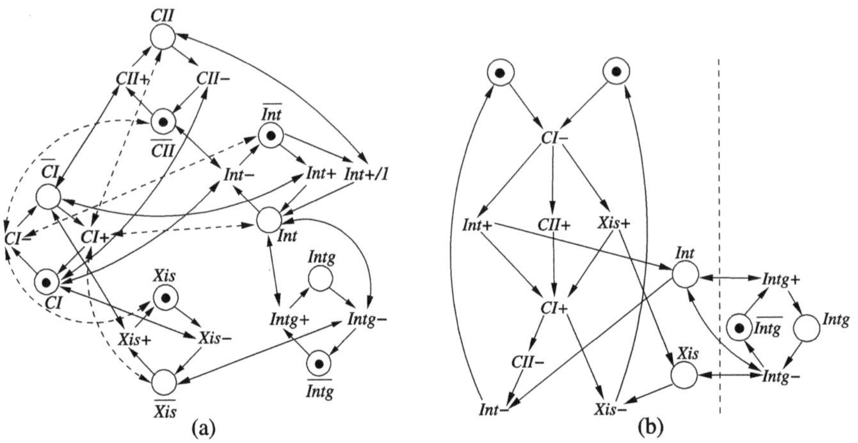

Fig. 5.12 STG model for phage λ, with the *dashed arcs* showing the FOE transformations that capture the assumption of a slow environment (**a**); and resulting STG after simplification by PETRIFY (**b**)

must be made by the modeler.) Furthermore, we choose the initial state[5] in which the values of all signals except CI are 0. Note that we allow CI to oscillate freely to represent the most general environment.

The resulting STG model is presented in Fig. 5.12(a). As explained in the previous section, STGs derived from circuits are bounded (in fact, safe), consistent and have CSC, and these properties are preserved by the subsequent transformations.

5.3.3.2 Model Analysis and Refinement

We can now check the basic properties of our STG model by running it through PETRIFY. It shows, as predicted by our theory, that the STG satisfies the boundedness, consistency and CSC properties. However, it highlights that there is a number of OP violations (resulting in nondeterministic behavior), which suggests the model may contain some behavior that is not realizable in practice:

(1) $\mathsf{Xis}^+ \rightharpoonup \mathsf{CI}^+$, (2) $\mathsf{Xis}^- \rightharpoonup \mathsf{CI}^-$, (3) $Int\mathsf{Int}^+ \rightharpoonup \mathsf{CI}^+$,
(4) $\mathsf{Int}^- \rightharpoonup \mathsf{CI}^-$, (5) $\mathsf{CII}^+ \rightharpoonup \mathsf{CI}^+$, (6) $\mathsf{CII}^- \rightharpoonup \mathsf{CI}^-$,
(7) $\mathsf{Intg}^- \rightharpoonup \mathsf{Int}^-$, (8) $\mathsf{Intg}^- \rightharpoonup \mathsf{Xis}^-$, (9) $\mathsf{Intg}^+ \rightharpoonup \mathsf{Int}^-$,
(10) $\mathsf{Int}^+/1 \rightharpoonup \mathsf{CII}^-$.

These violations of OP indicate the areas of the STG which are candidates for refinement by applying additional information about the environment's behavior or relative reaction rates. We proceed by considering OP violations (1)–(6) which

[5]Choosing a meaningful initial state is outside the scope of this paper; we just remark that typically a biological system has cyclic behavior, and that any state on this cycle can be taken.

involve conflicts between input and local transitions. Such conflicts can often be resolved by assuming that the environment is slow enough to allow the circuit to stabilize. We therefore apply the following FOE transformations to the model to resolve these violations:

$$\mathsf{Xis}^+ \mapsto \mathsf{CI}^+, \mathsf{Xis}^- \mapsto \mathsf{CI}^-, \mathsf{Int}^+ \mapsto \mathsf{CI}^+, \mathsf{Int}^- \mapsto \mathsf{CI}^-, \mathsf{CII}^+ \mapsto \mathsf{CI}^+, \mathsf{CII}^- \mapsto \mathsf{CI}^-,$$

which are also shown by dashed arcs in Fig. 5.12(a). Note that the support tool GNAPN provides a utility to allow such FOE transformations to be automatically applied to an STG model [24].

Interestingly, applying the above FOE transformations resolves also violation (10), leaving only violations (7)–(9) in the new model. Violations (7) and (8) show that excision (represented by the firing of Intg^-) when Int and Xis are 1 can be preempted if Int^- or Xis^- fires first, whilst violation (9) shows that integration (represented by the firing of transition Intg^+) can be preempted if Int^- fires first. These remaining OP violations are at the heart of the lysis-lysogeny switch in phage λ (which is a stochastic phenomenon in practice [390]), and so are not resolved. After applying the FOE transformations, we can simplify the model using PETRIFY and the result is the STG shown in Fig. 5.12(b).

The new STG in Fig. 5.12(b) is much less cluttered than the original one.[6] as the unrealizable behavior under the FOE transformations listed above has been stripped away, making it significantly simpler to interpret and analyze using for example, model checking [79]. Moreover, this simplified STG clearly separates into two components, which capture the crucial mechanisms governing the lysis-lysogeny switch:

- Component 1 (left) involves the input signal CI and the internal signals CII, Int and Xis. From the initial stable state, it waits for the environment to lower signal CI indicating the absence of immunity, after which CII^+, Int^+ and Xis^+ can fire in any order. This component then waits for the environment to raise signal CI, resulting in the firing of transitions Xis^- and CII^- (in any order), with the latter followed by Int^-, which returns the component to its initial state.
- Component 2 (right) is a simple flip-flop for signal Intg, which is controlled by the values of the signals Int and Xis in the first component. Note that the only connections between the two components are the read arcs between places of the former component and transitions of the latter one, that is, the latter component accesses the former one in the read-only fashion and hence does not affect its behavior.

After Component 1 has raised Int, transition Intg^+ is able to fire, representing the integration of the phage λ DNA into the host cell. Once Component 1 has raised both Int and Xis, Intg can freely oscillate, that is, there are no stable states in the absence of immunity [390]. Similarly, once the environment has raised CI, Component 1 executes Xis^- concurrently with CII^- followed by Int^-; the outcomes of the arbitrations between Intg^+ and Int^- and between Intg^- and Int^- or Xis^- determine

[6] This is very typical, as the original STG contained a lot of (rather random) behavior which is not realizable in practice.

the stable state of signal Intg in the presence of immunity. These arbitrations exactly correspond to the OP violations (7)–(9) still remaining in the STG in Fig. 5.12(b) and involving only local transitions.

Observe that CII^- 'delays' Int^-, modeling that the presence of CII causes phage λ to favor integration over excision; however, the latter is not a qualitative effect, and can not in fact be formally derived neither from this STG nor from the equations in Fig. 5.11(b) due to the arbitrary gate delays. In fact, one can see that CII can be removed from the model, without affecting its qualitative behavior; indeed, its only role is to change the probabilities involved in the stochastic choice made by phage λ, and so it is no longer required once this stochastic choice has been qualitatively modeled by a nondeterministic one.

5.4 Related Work

A number of formal techniques have been applied to modeling and analyzing GRNs. These range from qualitative approaches, like Boolean networks, to more detailed quantitative approaches based on differential equations or stochastic techniques (see, e.g., the survey papers [175, 182, 343]). Petri nets have emerged as a natural framework in which to model GRNs both qualitatively and quantitatively (e.g., see [63, 149]). In this section, we present a brief overview of the literature in this area, with the aim of providing a starting point for further reading.

One natural qualitative approach is to translate existing logical models for GRNs into Petri net models as was done in this chapter. (Further details and case studies illustrating our approach can be found in [23, 24, 373, 374].) Chaouiya et al. [65, 320] have developed a similar approach based on translating logical regulatory graphs (a logical model similar to asynchronous Boolean networks [390]) into Petri nets. This approach was used in [362] to develop an integrated model in which regulatory and biochemical pathways in the biosynthesis of tryptophan in *E. coli* were modeled and analyzed.

In [22, 66, 67], these approaches were generalized to the multi-value case, where the states of entities can take one of a discrete range of values [390]. In particular, the use of high-level Petri net models was considered in [22, 66]. An interesting approach is presented in [81], where a high-level Petri net model is used in conjunction with model checking to find GRN models that satisfy given behavioral constraints.

A range of quantitative modeling approaches have been considered for GRNs, for example see the early paper [294] on the use of stochastic Petri nets. The use of *hybrid Petri nets* (*HPN*) for modeling GRNs has been considered extensively by Matsuno et al. (e.g., see [248]). They have presented a refined class of HPNs for modeling biological systems, called *hybrid functional Petri nets* (*HFPNs*) [251, 255], which allow the firing rates of transitions to depend on the corresponding values held at input places.

5.5 Summary

In this chapter, we have presented a range of techniques based on Petri nets for qualitatively modeling and analyzing GRNs. We began by considering synchronous Boolean networks and presented a systematic approach for translating such models into Petri nets. This approach is supported by the tool GNAPN, which automates model construction and facilitates model analysis. We illustrated the practical application of our approach by modeling and analyzing the GRN underlying the carbon starvation stress response in *E. coli*. This case study demonstrated how Petri net tools can be used to validate and analyze GRN models. In particular, it detailed the use of model checking to investigate the behavior of the GRN and carry out a detailed mutant analysis by knocking out and overexpressing entities.

While asynchronous models appear to be more realistic than their synchronous counterparts, they suffer from the problem of capturing too much behavior, limiting their use in practice. We addressed this by applying techniques and tools from asynchronous circuit design to develop realistic asynchronous models. Key here was our methodological assumption that, given enough information, a GRN can be qualitatively modeled as an SI circuit. By using the STG modeling formalism to identify violations of speed-independence (in particular, output-persistency violations), one can highlight key areas of a model that either require refining or represent truly stochastic effects within a GRN. Thus, STGs can be seen as providing a well supported formal framework for GRNs, that allows realistic qualitative asynchronous models to be developed and analyzed. The practical application of these ideas was illustrated with a detailed case study, in which an SI STG model of the GRN for the lysis-lysogeny switch in phage λ was developed using the support tools GNAPN and PETRIFY.

Further work to extend the ideas presented in this chapter and strengthen the associated tool support is ongoing. One interesting area currently being investigated is the application of STG techniques to *synthetic biology* [35]. Given that STGs were developed to support the compositional construction of asynchronous circuits [84], they appear to be ideally suited to designing artificial genetic control systems.

The basic Boolean network model can be extended to *multi-valued networks* [390], where the Boolean state of entities is enlarged to a set of discrete values. Our STG refinement approach can be applied to multi-valued networks in a number of ways, such as using several Boolean variables to represent a signal's state or reformulating the consistency rule on signal labels. Work is currently underway to investigate these approaches (e.g., [22]).

5.6 Problems

The following simple exercises are aimed to help the reader familiarize themselves with the techniques we have introduced.

5.1 Consider the Boolean network presented below:

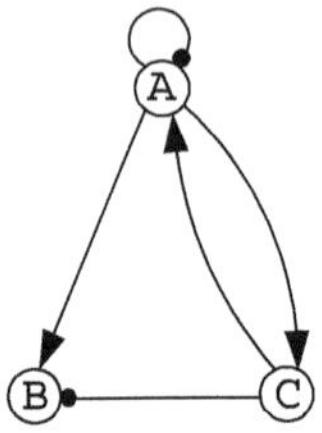

A	C	$[A]$
0	0	1
0	1	1
1	0	0
1	1	1

A	C	$[B]$
0	0	0
0	1	0
1	0	1
1	1	0

A	$[C]$
0	0
1	1

1. Following the approach detailed in Sect. 5.2.1, derive the logical equations that define when the entities are active or inactive.
2. Using GNAPN, construct a Petri net that models the synchronous behavior of the given Boolean network.
3. Using an appropriate Petri net tool (e.g., PEP), simulate the model from various initial states and observe its behavior.
4. Finally, use model checking to verify whether or not A and C are mutually exclusive (i.e., starting from any state in which they are not both active, they can not both simultaneously become active).

5.2 Consider an asynchronous Boolean network which has inputs A, B, and output C, where the behavior of C is defined by the logical equation $[C] = A + B$.

1. Using the circuit-STG construction presented in Fig. 5.8 construct by hand an STG for this asynchronous Boolean network.
2. Identify the four OP violations that exist in this STG.
3. Suppose we assume that the environment is faster than the system in this case. What effect will this have on the behavior of the STG?
4. Alternatively, suppose the environment is slower than the system (i.e., the system always has the chance to react to changes in its inputs). Draw the refined STG that will result from this assumption.

Acknowledgements We would like to thank the EPSRC for supporting R. Banks and the BBSRC for their support via the Centre for Integrated Systems Biology of Ageing and Nutrition (CISBAN). This research was also supported by the Royal Academy of Engineering/EPSRC post-doctoral research fellowship EP/C53400X/1 (DAVAC) and the Newcastle Systems Biology Resource Centre.

Chapter 6
Hybrid Functional Petri Net with Extension for Dynamic Pathway Modeling

Ayumu Saito, Masao Nagasaki, Hiroshi Matsuno, and Satoru Miyano

Abstract Using elementary Petri nets, it is difficult to mixturely model the discrete, continuous and other complicated events, for example, DNA, RNA and amino acid sequence events. Therefore, Hybrid Functional Petri Net with extension (HFPNe) was introduced to overcome this difficulty and also to allow representation of pathway models without loss of biological details. First, this book chapter explains how modeling with Petri net is done. After which, a full definition of HFPNe is given, along with its relation to hybrid Petri net and other related Petri net extension. Finally, to demonstrate the elegance of HFPNe architecture in handling complex pathway modeling, we chose a biological pathway that involves discrete, continuous and sequence events (gene regulatory network of the cell fate determination in *C. elegans*). We provide the biological mechanisms of this network and show how intuitively it could be modeled on HFPNe architecture using a software tool, Cell Illustrator.

6.1 Introduction

Reddy et al. [314] and Hofestädt [162] have shown that the concept of Petri net is very suited for modeling metabolic pathways. By following these pioneering works,

A. Saito (✉) · M. Nagasaki · S. Miyano
Human Genome Center, Institute of Medical Science, University of Tokyo, Tokyo 108-8639, Japan
e-mail: s-ayumu@ims.u-tokyo.ac.jp

M. Nagasaki
e-mail: masao@ims.u-tokyo.ac.jp

S. Miyano
e-mail: miyano@ims.u-tokyo.ac.jp

H. Matsuno
Graduate School of Science and Engineering, Yamaguchi University, Yamaguchi 753-8512, Japan
e-mail: matsuno@sci.yamaguchi-u.ac.jp

I. Koch et al. (eds.), *Modeling in Systems Biology*, Computational Biology 16, DOI 10.1007/978-1-84996-474-6_6, © Springer-Verlag London Limited 2011

Petri net and its extensions such as hybrid Petri net [10] have been applied for modeling and analyzing gene regulatory networks, metabolic pathways, and signaling pathways. The following are some of the reasons why such biological pathway modeling is gaining attention in biology: once an in silico model is developed, it can be easily updated with new biological discoveries and additional information. The consistency of the model with new experimental data can be checked by simulation. In silico mutant analysis can be also performed by modifying the wild-type model then via simulation.

In modeling biological pathways with the concept of Petri net, we correspond biological molecules in the cell to the places of Petri net such that the mark of a place represents the molecular density or the number of the molecules in the cell. Transitions of Petri net are used to describe the biochemical reactions among the molecules and the changes of the molecules in the cell, for example, enzyme reactions, protein phosphorylation, translocation of molecules in the cell. Dynamics is put into the transitions. The molecules and reactions in a pathway are connected by arcs which define the whole network structure of the pathway. For example, with hybrid Petri net, we can make a model that equips discrete genetic switches connected with a system of ordinary differential equations for a cascade of enzyme reactions in a metabolic pathway. Many studies have shown that most of the discrete and continuous events in pathways can be modeled with the concept of Petri net, and the pathways can be analyzed with some Petri net analysis methods [94–97, 140, 147, 163, 210, 211, 230, 231, 248, 250, 251, 255, 266, 338, 398, 407].

Most of the dynamics of logical and quantitative relations in pathways may be handled with hybrid Petri net and its extensions. However, less attention has been paid to one of the most important entities in cells, "sequence information", for example, DNA and RNA base sequence, amino-acid sequence of protein, etc., which are sources of molecules in pathways and whose changes may cause different system behaviors and structures in pathways. Although abstraction is a key to system modeling for simulation, it is also important for biology to organize the biological knowledge and facts into the models precisely as much as possible. For example, it is known that the gene *mdm2* inhibits the gene *p53*. The biological mechanism behind this abstractly described inhibition is that the complex of the p53 protein and the MDM2 protein accelerates the ubiquitination of the p53 protein for its destruction. A precise dynamic pathway model for this inhibition is constructed in [96] from the biological literature. Thus, for dynamic pathway modeling, we consider that it is crucial that the details of biological mechanisms can be described without loosing biological understandability and intuitions.

With this motivation, hybrid functional Petri net with extension (HFPNe) [274, 275, 277] was introduced to handle any kinds of objects based on the concept of Petri net. Objects include sequences in addition to discrete and continuous numbers. This architecture is implemented in a software tool namely, Cell Illustrator [59], and in practice, with this software tool we can build HFPNe pathway models by using its GUI and simulate them on it. One of the purpose of this book chapter is to give a full definition of HFPNe and explain its relation to hybrid Petri net and other related Petri net extensions. In Sect. 6.2, we give how the concept of Petri net is employed for biological pathway modeling. Section 6.3 gives the definition of HFPNe and its related notions.

Then, in Sect. 6.4, we show how HFPNe modeling is guided and done by using a concrete biological pathway. As the biological pathway involving discrete, continuous and sequence events, we consider a gene regulatory network of the cell fate determination of gustatory neurons in *C. elegans* that forms a double-negative feedback loop mediated by microRNAs (miRNAs). It is known that miRNAs play important roles in this complex gene regulatory network. We provide the biological mechanisms of this network and show how the HFPNe model of this network is developed.

6.2 Pathway Modeling with Concept of Petri Net

6.2.1 Petri Net (PN)

We briefly begin with an informal definition of the original Petri net (PN) [319]. PN is defined as a finite bipartite directed graph consisting of two kinds of nodes called *places* and *transitions*, where no edge is allowed between places or between transitions. Multiple edges are allowed between the same pair of place and transition. We denote the set of places by P and the set of transitions by T. Directed edges between places and transitions are called *arcs*. A place can hold a finite number of *tokens* and this nonnegative integer is called the *mark* of the place. A *marking* M of P is a mapping $M : P \rightarrow \mathbf{N}$ that assigns the *mark* (the number of tokens) to each place, where $\mathbf{N}$ is the set of nonnegative integers. An arc is labeled with a nonnegative integer called the *weight*. By definition, a transition has *input arcs* coming from some places (*input places*) and *output arcs* going out from the transition to some places (*output places*). A transition with these arcs defines a *firing condition* in terms of the marks of the places and the weights of the arcs. A typical firing condition is that the weights of input arcs are used as thresholds, that is, a transition can fire if, for each input arc, the mark of its input place is greater than or equal to the weight of the input arc. After firing the transition, the marks of the places are updated according to the weights assigned to the arcs. For example, for each input arc, the mark of its input place is decreased by the weight of the input arc while for each output arc, the mark of its output place is increased by the weight of the output arc.

6.2.2 Representation of Biological Pathway

Biological pathways consist of biological entities such as DNA, mRNA, protein in membrane, protein in nucleus, phosphorylated protein, protein complex, ligand, receptor, etc., and biological processes such as transcription, translation, phosphorylation, degradation, translocation, binding, dissociation, inhibition, etc. When we make a biological pathway model with PN, we associate a biological entity with a place and a biological process which is disabled. The connection structure of arcs describes how biological processes are running with biological entities.

Figure 6.1 shows an example of signaling pathway, Fas ligand induced apoptosis pathway. By receiving Fas ligands, this pathway sends out signals that ultimately cause the DNA fragmentation that means the apoptotic death. Starting from Fas ligand molecules at the left upper site, this pathway recruits several kinds of molecules such as Fas-associating protein with death domain (FADD), procaspase 8, etc., and it induces chains of biochemical reactions. These molecules and their complexes and modifications are represented by places to which some appropriate icons are attached. Biochemical reactions are represented by icons for intuitive understanding: (binding), (degradation), (activation), (translation), (cleavage), (translocation), (metabolic reaction), (DNA cleavage), (DNA damage), (DNA repair), (general molecular reaction), (unknown reaction), etc. Activated caspase 8 activates caspase 3 through two pathways. One is a pathway via mitochondria. Caspase 8 cleaves Bcl-2 interacting protein (Bid) and its COOH-terminal part translocates to mitochondria. Then it triggers releasing cytochrome c that binds to apoplectic protease protease activating factor-1 (Apaf-1) together with dATP and procaspase 9, and activates caspase 9. The activated caspase 9 cleaves procaspase 3 and activates caspase 3. The other pathway is a short cut from caspase 8 to procaspase 3 where caspase 8 cleaves procaspase 3 directly and caspase 3 is activated. Caspase 3 cleaves DNA fragmentation factor (DFF) in a heterodimeric factor of DFF40 and DFF45. Cleaved DFF45 dissociates from DFF40, inducing oligomerization of DFF40 that has DNase activity. The active DFF40 oligomer causes the internucleosomal DNA fragmentation. More details are found in [251], and the model can be downloaded from [61].

6.2.3 Timed Petri Net (TPN)

For dynamic biological pathway modeling, the notion of "time" should be inevitably included in Petri net. *Timed Petri net* (TPN) [319] is PN which counts time and allows delay in firing. Namely, in addition to firing condition, each transition has a nonnegative integer d called *delay*. The delay of a transition varies from transition to transition. Since TPN follows the time x, the marking is parameterized with x as $M(x)$. The marking $M(0)$ at time 0 is called the *initial marking*. For a place p, we denote the mark of p at time x by $M[p](x)$. Let t be a transition with delay d and let $a_{i_1}, \ldots, a_{i_k}$ be its input arcs with weights $w_{i_1}, \ldots, w_{i_k}$ and $p_{i_1}, \ldots, p_{i_k}$ be its

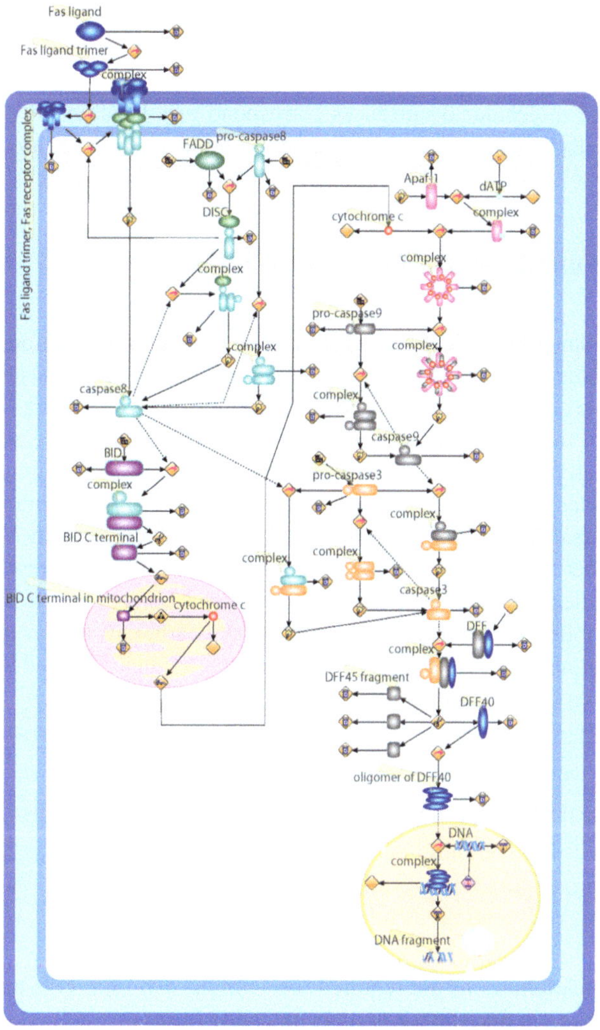

Fig. 6.1 Fas ligand induced apoptosis pathway represented by Cell Illustrator. By Fas ligands, the pathway sends out signals that ultimately cause the self DNA fragmentation that means the apoptotic death. Fas ligands, which usually exist as trimers, bind and activate their receptors by inducing receptor trimerization. Activated receptors recruit adaptor molecules such as Fas-associating protein with death domain (FADD), which recruit procaspase 8 to the receptor complex, where it undergoes autocatalytic activation. Activated caspase 8 activates caspase 3 through two pathways. The complex one is that caspase 8 cleaves Bcl-2 interacting protein (Bid) and its C-terminal part translocates to mitochondria where it triggers cytochrome c release. The released cytochrome c bind to apoplectic protease activating factor-1 (Apaf-1) together with dATP and procaspase 9 and activates caspase 9. The caspase 9 cleaves procaspase 3 and activates caspase 3. The another pathway is that caspase 8 cleaves procaspase 3 directly and activates it. The caspase 3 cleaves DNA fragmentation factor (DFF) 45 in a heterodimeric factor of DFF40 and DFF45. Cleaved DFF45 dissociates from DFF40, inducing oligomerization of DFF40 that has DNase activity. The active DFF40 oligomer causes the internucleosomal DNA fragmentation, which is an apoptotic hallmark indicative of chromatin condensation

input places. We say that transition t is *enabled* at time x if $M[p_{i_j}](x) \geq w_{i_j}$ for all $j = 1, \ldots, k$. Then, the transition t fires at $x + d$ time and the marking is updated.

For any kind of molecule in the cell, we can represent its occurrences in cell with a place so that molecular occurrences correspond to tokens on the place. TPN is useful to directly model molecular processes with places and transitions.

6.2.4 Continuous Timed Petri Net (CTPN)

On the other hand, biological pathways such as biochemical reactions are approximated as systems of ordinary differential equations instead of handling discrete molecular interactions. For this purpose, *continuous timed Petri net* (CTPN) [319] is a useful tool for modeling continuous features of biological processes. In order to make the differences between TPN and CTPN, we call place and transition of TPN as *discrete place* and *discrete transition* and those of CTPN *continuous place* and *continuous transition*. CTPN counts time and the continuous place holds a nonnegative real number as its token is called *continuous token* and can represent concentration of molecule. Different from TPN, the continuous transition does not have delay and it fires continuously and consumes/produces continuous tokens in places connected to the transition. Let t be a continuous transition in CTPN and let $a_{i_1}, \ldots, a_{i_k}$ be its input arcs connected to input places $p_{i_1}, \ldots, p_{i_k}$ and $a'_{i_1}, \ldots, a'_{i_{k'}}$ be its output arcs connected to output places $p'_{i_1}, \ldots, p'_{i_{k'}}$. Each input arc a_{i_j} has two nonnegative real numbers *weight* w_{i_j} and *input speed* s_{i_j}, and each output arc $a'_{i_{j'}}$ has a single nonnegative real number $s'_{i_{j'}}$ called *output speed*. The transition t fires continuously as long as its firing condition on weights and marks is satisfied as in the case of TPN. As firing, continuous token is removed from input place p_{i_j} with speed s_{i_j} through arc a_{i_j} and continuous token is added to output place $p'_{i_{j'}}$ with speed $s'_{i_{j'}}$ through arc $a_{i_{j'}}$. The marking $M(t)$ of CTPN is updated in this way.

More formal definitions are required for defining $M(t)$ for TPN and CTPN and they will be given in Section 6.3 in more extended form.

6.3 Hybrid Functional Petri Net with Extension

The purpose of this section is to define formally the notion of hybrid functional Petri net with extension (HFPNe) that is an extension of Petri net which can deal with not only discrete and continuous features but also any objects such as strings.

Genetic switches are easily modeled with discrete transitions while metabolic pathways comprising cascades of enzyme reactions are more suited for continuous transitions using Michaelis–Menten kinetics. If one wants to make a model consisting of genetic switches and their regulated metabolic pathways [95], the notion of hybrid system will be helpful.

We start with hybrid Petri net (HPN) [10]. HPN is defined as a hybrid system of TPN and CTPN with which we can make models having both discrete and continuous features. The original definition of HPN specifies that the input and output speeds of a transition must be the same. Thus, it gives the "firing speed" of the transition. By the definition of HPN and CTPN, the speeds assigned to arcs are constant numbers. These constraints may be rational when we analyze the systems behavior mathematically. On the other hand, due to a practical requirement, *hybrid dynamic net* (HDN) [99] removed the constant constraint on speeds so that the firing speed of a transition can depend on the marks of its input places, that is, the speed is defined by any function of the marks. But, the input and output speeds are kept the same. Some drawbacks of this equality constraint on speeds are discussed with biological examples in [251].

Hybrid functional Petri net (HFPN) [251] was introduced to resolve these inconveniences in biological pathway modeling. HFPN is defined by allowing any functions to be assigned to the input and output speeds, weights, and delays in the hybrid system of TPN and CTPN. Almost all biological pathways can be intuitively modeled with HFPN to some extent. In order to include more complex biological information such as sequence information of DNA, mRNA, and protein in biological pathway modeling, HFPN is further extended to *hybrid functional Petri net with extension* (HFPNe) [274, 275, 277]. HFPNe is a programming language with objects rather than a hybrid system of discrete and continuous systems. In order to handle "objects", we introduce *generic place* and *generic transition* for HFPNe.

The definition of HFPNe requires a series of formal notions. In HPN, two kinds of data types are used for places, that is, nonnegative integer (discrete) and nonnegative real number (continuous). For HFPNe, we use a more general framework of types for places and transitions. The definition here is slightly different from the originally defined one in [275], but the definition described below is given in a more general way.

We define the set $\mathscr{T}$ of *types* as follows:

$$\begin{aligned}\langle type\rangle ::= &\ \texttt{boolean} \parallel \texttt{integer} \parallel \texttt{integer+} \parallel \texttt{real} \parallel \texttt{real+} \parallel \\ &\ \texttt{string} \parallel \texttt{pair}(\langle type\rangle, \langle type\rangle) \parallel \texttt{list}\ \langle type\rangle \parallel \\ &\ \texttt{object}(\langle type\rangle, \ldots, \langle type\rangle).\end{aligned}$$

For each $\theta \in \mathscr{T}$, we associate it with its *domain* $D(\theta)$ as follows:

1. $D(\texttt{boolean}) = \{\texttt{true}, \texttt{false}\}$.
2. $D(\texttt{integer}) = \mathbf{Z}$ (the set of integers).
3. $D(\texttt{integer+}) = \mathbf{N}$ (the set of nonnegative integers).
4. $D(\texttt{real}) = \mathbf{R}$ (the set of real numbers).
5. $D(\texttt{real+}) = \mathbf{R}^{\geq 0}$ (the set of nonnegative real numbers).
6. $D(\texttt{string}) = \mathbf{S}$ (the set of strings over some alphabet).
7. $D(\texttt{pair}(\theta_1, \theta_2)) = D(\theta_1) \times D(\theta_2)$ (the pair of two domains).
8. $D(\texttt{list}\theta) = \bigcup_{k \geq 0} D(\theta)^k$ (the list with the same domain).
9. $D(\texttt{object}(\theta_1, \ldots, \theta_n)) = D(\theta_1) \times \cdots \times D(\theta_n)$ (the set of any domains).

We denote $D^* = \bigcup_{\theta \in \mathscr{T}} D(\theta)$.

For HFPNe, we newly introduce places and transitions which deal with D^* and these places and transitions are named "generic" while the names "discrete" and "continuous" are used for places and transitions of HFPN.

Let P be a finite set of places. Each place is labeled with either `discrete`, `continuous`, or `generic`. Place labeled with `discrete` (resp., `continuous`, `generic`) is called *discrete* (resp., *continuous*, *generic*). We assign a type $\tau(p)$ to each place p in P by a mapping called a *type function* $\tau : P \to \mathscr{T}$ which satisfies the following conditions:

1. If p is labeled with `discrete`, then $\tau(p) = \texttt{integer+}$.
2. If p is labeled with `continuous`, then $\tau(p) = \texttt{real+}$.
3. If p is labeled with `generic`, it can be any type in $\mathscr{T}$.

A *marking* of P is defined as a mapping $M : P \to D^*$ such that $M[p]$ is in $D(\tau(p))$ for $p \in P$. $M[p]$ is called the *mark* of p. We denote by $\mathscr{M}$ the set of all markings of P.

We define some notations which will be used for the definition of HFPNe. Let $\mathscr{D}_{\texttt{discrete}} = \{f \mid f : \mathscr{M} \to \mathbf{N}\}$, $\mathscr{D}_{\texttt{boolean}} = \{b \mid f : \mathscr{M} \to \{\texttt{true}, \texttt{false}\}$, and $\mathscr{D}_{\texttt{generic}} = \{f \mid f : \mathscr{M} \to D^*\}$.

Consider $f : \prod_{p \in P} D(\tau(p)) \to \mathbf{R}^{\geq 0}$. Note that the set $\mathscr{M}$ can be identified with the set $\prod_{p \in P} D(\tau(p))$. Let R be the set of all places labeled with `continuous` and let $Q = P - R$. For an element $v \in \prod_{p \in Q} D(\tau(p))$, let $f[Q = v] : \prod_{p \in R} D(\tau(p)) \to \mathbf{R}^{\geq 0}$ be the function defined by $f[Q = v](v') = f(v', v)$ for $v' \in \prod_{p \in R} D(\tau(p))$. We say that f is *continuous* if $f[Q = v]$ is continuous on $\prod_{p \in R} D(\tau(p))$ for any $v \in \prod_{p \in Q} D(\tau(p))$. Then, let $\mathscr{D}_{\texttt{continuous}} = \{f \mid f : \mathscr{M} \to \mathbf{R}^{\geq 0}$ is continuous$\}$.

Based on the above notations, we define HFPNe by introducing data types for places by giving functions which depend on markings to determine the weight, delay, and speed, etc. We embed the definition of HFPN into the definition of HFPNe as follows:

Definition 6.1 A *hybrid functional Petri net with extension* (HFPNe)

$$H = (P, T, A, \tau, w, u, d)$$

consists of the following:

1. P is a finite set of *places* and T is a finite set of *transitions*. We assume $P \cap T = \emptyset$. Each place is labeled with either `discrete`, `continuous`, or `generic`. Each transition is also labeled with `discrete`, `continuous`, or `generic`. The transition and place are called *discrete*, *continuous*, or *generic* according to its label.

   For each transition t in T, there are two sets $Input_t$ and $Output_t$ of arcs. Arc $a \in Input_t$ is an edge from input place p_a to the transition t called *input arc*. Arc $a' \in Onput_t$ is an edge from the transition t to output place $p_{a'}$ called *output arc*. Each arc is labeled with either `normal`, `test`, or `inhibitory`, and they are called `normal`, `test`, and `inhibitory`, respectively. We also say that

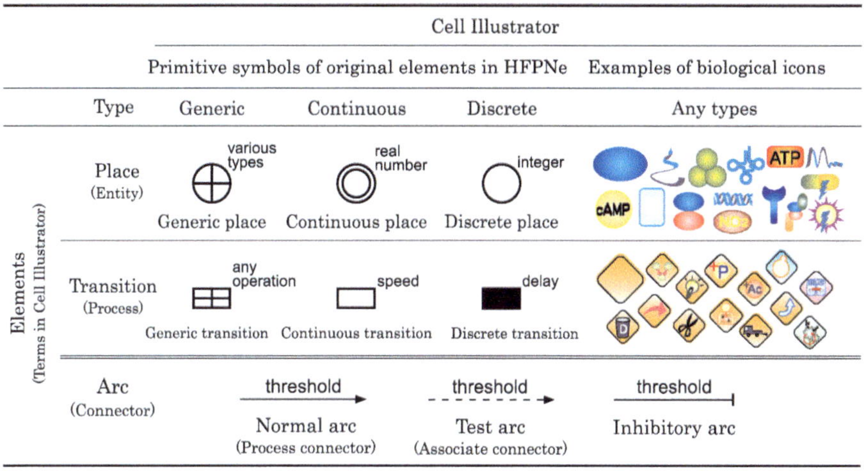

Fig. 6.2 All names in HFPNe elements and their graphical representation. Biological images are helpful for enhancing biological understanding of pathways modeled with HFPNe. These images are used in Cell Illustrator

arcs (a and a') are *discrete* (resp., *continuous*, *generic*) if transition t is discrete (resp., continuous, generic). For graphical representation, we use the symbols in Fig. 6.2. The labels of arcs, places and transitions satisfy the following rules:

a. The label of arc $a' \in Onput_t$ is `normal`.
b. For arc $a \in Input_t$ with input place p_a, the labels of arc a, place p_a, and transition t satisfies the connection rules in Table 6.1(a) and Fig. 6.3.
c. For arc $a' \in Onput_t$ with output place $p_{a'}$, the labels of arc a', place $p_{a'}$ and transition T satisfies the connection rules in Table 6.1(b) and Fig. 6.3.

We denote by PT and TP the set of input arcs and the set of output arcs of all transitions, respectively. We also denote arc a in PT as $a(p,t)$ by specifying input place and transition. In a similar way, arc a' in TP is denoted as $a'(t,p)$ by specifying transition and output place. The set A of arcs is given by $PT \cup TP$.

2. The types of places are given by a type function $\tau : P \rightarrow \mathcal{T}$.
3. For each input arc $a \in PT$, its *activity* $w(a)$ is given by an *activity function* $w : PT \rightarrow \mathcal{D}_{\mathtt{discrete}} \cup \mathcal{D}_{\mathtt{continuous}} \cup \mathcal{D}_{\mathtt{boolean}}$ which satisfies the following conditions:
   a. If a is discrete, $w(a)$ is in $\mathcal{D}_{\mathtt{discrete}}$.
   b. If $w(a)$ is in $\mathcal{D}_{\mathtt{continuous}}$.
   c. If a is generic, $w(a)$ is in $\mathcal{D}_{\mathtt{boolean}}$.

   For input arc $a(p,t)$, $w(a)$ is used as a function giving the threshold in discrete and continuous cases and the condition in generic case which is required for enabling the transition t.
4. For each arc c ($c = a(p,t) \in PT$ or $c = a'(t,p) \in TP$), the *update* $u(c)$ is given by an *update function* $u : A \rightarrow \mathcal{D}_{\mathtt{discrete}} \cup \mathcal{D}_{\mathtt{continuous}} \cup \mathcal{D}_{\mathtt{generic}}$ which satisfies the following conditions:
   a. If c is discrete, $u(c)$ is in $\mathcal{D}_{\mathtt{discrete}}$.

Table 6.1 (**a**) Connection rules from input places to transitions. √ means that the connection is allowed and – means that the connection is not allowed. (**b**) Connection rules from transitions to output places. Arc label is `normal` by definition. √ means that the connection is allowed and – means that the connection is not allowed. In Cell Illustrator, arc from discrete or continuous transition to generic place is allowed if the type of generic place is consistent with transition

Label of arc a		`Normal`		
Label of transition t		`Discrete`	`Continuous`	`Generic`
Label of place p_a	`Discrete`	√	–	√
	`Continuous`	√	√	√
	`Generic`	–	–	√
Label of arc a		`Test` or `inhibitory`		
Label transition t		`Discrete`	`Continuous`	`Generic`
Label of place p_a	`Discrete`	√	√	√
	`Continuous`	√	√	√
	`Generic`	√	√	√

(a) Input arc $p_a \xrightarrow{a} t$

Label of arc a'		`Normal`		
Label of transition t		`Discrete`	`Continuous`	`Generic`
Label of place $p_{a'}$	`Discrete`	√	–	√
	`Continuous`	√	√	√
	`Generic`	–	–	√

(b) Output arc $t \xrightarrow{a'} p_{a'}$

b. If c is continuous, $u(c)$ is in $\mathscr{D}_{\texttt{continuous}}$.
c. If c is generic, then $u(c)$ is in $\mathscr{D}_{\texttt{generic}}$ such that $u(c)[M]$ is in $D(\tau(p))$ for any marking $M \in \mathscr{M}$. $u(c)$ is used as a function which will update the mark of p.

5. For each discrete or generic transition t, the *delay* of t is given by a *delay function* $d : T_{\texttt{discrete}} \cup T_{\texttt{generic}} \to \mathscr{D}_{\texttt{continuous}}$, where $T_{\texttt{discrete}}$ (resp., $T_{\texttt{generic}}$) is the set of discrete transitions resp., generic transitions). The delay of firing on marking M is given by $d(t)[M]$.

We use the parameter $x \geq 0$ for the *time* in HFPNe. Do not confuse t for transition with x for time.

The *initial marking* I is a marking at time $x = 0$ and we denote the *marking at time* x by $M(x)$. The *reserved marking* $M_r(x)$ at time x represents the amount of "tokens" reserved for firing when firing conditions are satisfied. The mark of place p at time x is denoted by $M[p](x)$.

We call $M(x)$ $(x \geq 0)$ the *behavior* of H, that is, system state, starting at the initial marking $M(0) = I$. Given the initial marking of HFPNe, the marking $M(x)$ and the reserved marking $M_r(x)$ at time x are defined in the following way:

For time $x = 0$, $M(0) = I$ by definition. We define $M_r[p](0) = 0$ if p is discrete or continuous, and $M_r[p](0) = \texttt{null}$ (the empty list) if p is generic. By convention, let $\tilde{M}(x)$ be $\tilde{M}[p](x) = M[p](x) - M_r[p](x)$ if p is discrete or continuous and $\tilde{M}[p](x) = M[p](x)$ if p is generic.

Definition 6.2 For transition t at time x, we say that t is *enabled at time* x if the following conditions are satisfied. Otherwise the transition is said to be *disabled* at time x.

1. If t is discrete or continuous, then for all input arcs $a(p,t) \in PT$ the following conditions hold:
   a. $\tilde{M}[p](x) \geq w(a)[M(x)]$ if a is not labeled with `inhibitory`.
   b. $\tilde{M}[p](x) < w(a)[M(x)]$ if a is labeled with `inhibitory`.
2. If t is generic, then for all input arcs $a(p,t) \in PT$ the following conditions hold:
   a. $w(a)[\tilde{M}(x)] = \texttt{true}$ if a is not labeled with `inhibitory`.
   b. $w(a)[\tilde{M}(x)] = \texttt{false}$ if a is labeled with `inhibitory`.

Definition 6.3 If disabled transition t turns enabled at time x, we say that t is *triggered at time* x and x is called the *trigger time*. If enabled transition t turns disabled at time x, we say that t is *switched off at time* x and x is called the *switch-off time*.

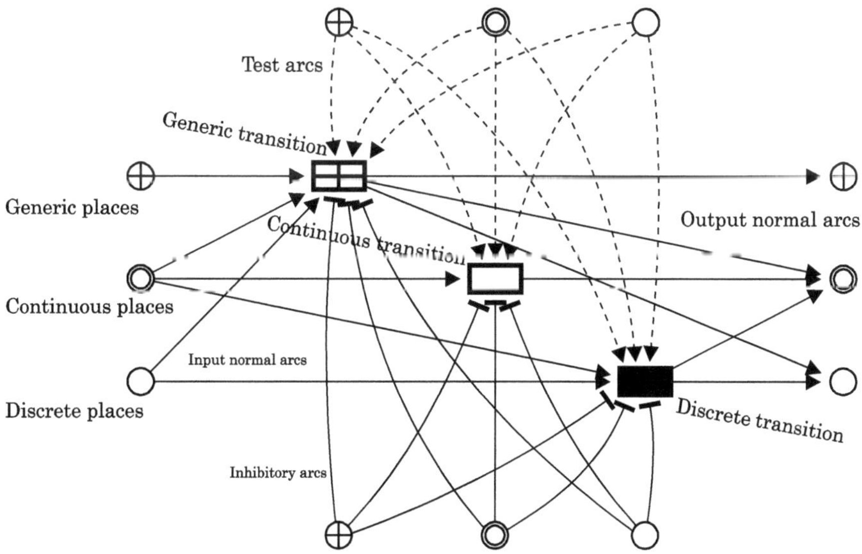

Fig. 6.3 This diagram shows all the connection rules among all the places and transitions with arcs. Both test and inhibitory arcs can be connected in all combinations of place type and transition type but normal arcs have several limitations

Definition 6.4 We define *firing* of discrete transition t. Assume that discrete transition t is triggered at time x. For each normal input arc $a(p,t)$, the place p must be discrete or continuous by definition. Then $M_r[p]$ reserves $\alpha \cdot u(a)[M(x)]$, that is, $\alpha \cdot u(a)[M(x)]$ is added to $M_r[p]$, for the time $y > x$ until $x + d(t)[M(x)]$, where $\alpha = \{0, 1\}$, if $\alpha = 0$, reserve is disabled; otherwise, token is reserved. If t is still enabled at $x + d(t)[M(x)]$, then at the same time $x + d(t)[M(x)]$, $M[p]$ is decreased by $u(a)[M(x)]$ and $M_r[p]$ releases $u(a)[M(x)]$, i.e., $u(a)[M(x)]$ is decreased from $M_r[p]$. Simultaneously, for each output normal arc $a'(t, p')$, $M[p']$ is increased by $u(a')[M(x)]$ at time $x + d(t)[M(x)]$ by arc $a'(t, p')$. The time $d(t)[M(x)]$ is called the *delay* that is determined by the function $d(t)$ of the mark $M(x)$ at time x.

As we will describe in Definitions 6.5 and 6.6 below, the reservation is not performed by generic or continuous transition. However, for the place p, there may be another discrete transitions $t_1, \ldots, t_\ell$ with input normal arcs $a_1(p,t_1), \ldots, a_m(p,t_\ell)$ which are triggered at time x. Then, each discrete transition t_i tries to reserve $u(a_i)[M(x)]$ from the same $M[p]$ at time x for $i = 0, \ldots, \ell$, where $a_0 = a(p,t)$ and $t_0 = t$. We say that there is a *conflict* with p at time x if $M[p](x) < \sum_{k=0}^{\ell} u(a_i)[M(x)]$. When a conflict occurs, some *conflict resolution* should be applied, for example, random selection of transitions, priorities on transitions, etc.

Even if some conflict resolution procedures among input transitions are applied to a selected transition t, the place p of $a(p,t)$ may be input places or output places of another discrete/continuous/generic transitions. By this, $M[p]$ and $M_r[p]$, and therefore $\tilde{M}[p]$, may be changed, the conditions of "enabled" are not be necessarily satisfied until the firing time $x + d(t)[M(x)]$. When t becomes disabled before $x + d(t)[M(x)]$, we say that a *system error* occurs with t.

Thus, triggered transition does not necessarily fire. If all of these actions succeed, we say that t *fires* at time $x + d(t)[M(x)]$. A example of discrete firing mechanism is shown in Fig. 6.4.

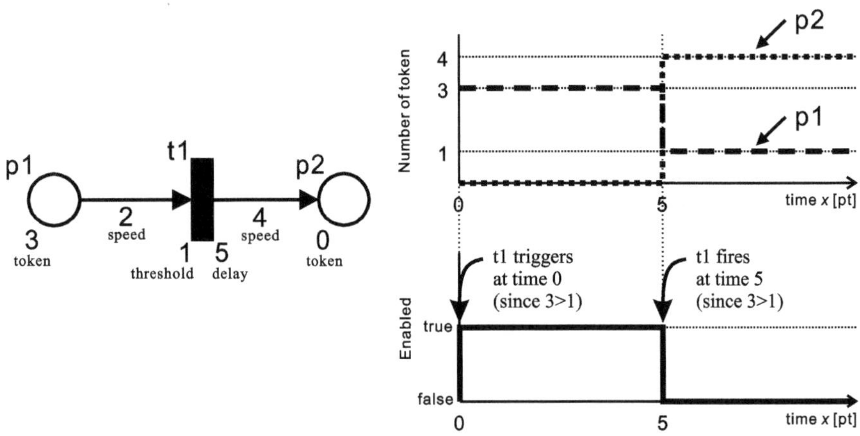

Fig. 6.4 A example of discrete firing

Definition 6.5 We define *firing* of generic transition t. Assume that generic transition t is triggered at time x. For each input normal arc $a(p,t)$, the place p can be discrete, continuous and generic. For each output normal $a'(t,p')$, p' can be also any kind of places. If t keeps enabled until time $x + d(t)[M(x)]$, then $M[p]$ at time $x + d(t)[M(x)]$ is updated to $u(a)[M(x)]$ and $M[p']$ is updated to $u(a')[M(x)]$ at time $x + d(t)[M(x)]$. We say that t *fires* at time $x + d(t)[M(x)]$ if this action succeeds. If p is generic, it is always that $M_r[p](x) = \texttt{null}$. No change is added to $M_r[p]$ by arc $a(p,t)$ if p is discrete or continuous.

In a similar way to discrete transition, if p is discrete or continuous, Mp and $M_r[p]$ have a possibility to be changed before $x + d(t)[M(x)]$ by another transitions. Therefore, $w(a)[\tilde{M}(y)] = \texttt{true}$ is not necessarily kept for $y \in (x, x + d(t)[M(x)])$. As in the case of discrete transition, it should be reported as system error. Since generic transition updates $M[p]$ and $M[p']$ at time $x + d(t)[M(x)]$, there is a possibility of conflict with another transitions which use p and p'. Thus, some conflict resolution should be applied or it should be reported as system error.

Definition 6.6 We define *firing* of continuous transition t. When continuous transition t is triggered, it starts firing and updates the marks of its connected places continuously with the speeds determined by the update function u and the marking M as long as it is enabled. Assume that continuous transition t is enabled at time x. For each input normal arc $a(p,t)$, the place p must be continuous by definition (see Fig. 6.3). Then, the mark $M[p]$ will be decreased through the arc $a(p,t)$ with the additional speed $u(a)[M(x)]$ at time x. No change is added to $M_r[p]$ by arc $a(p,t)$. For output normal arc $a'(t,p')$, the place p' must be continuous by definition. Then, the mark $M[p']$ will be increased through the arc $a'(t,p')$ with the additional speed $u(a')[M(x)]$ at time x. No change is added to $M_r[p']$ by arc $a'(t,p')$.

As we have investigated in Definitions 6.4 and 6.5, we consider continuous place p at time x. Continuous place p may be connected to some transitions with input/output normal arcs. Then, let $t_1, \dots, t_i$ be the continuous transitions which are enabled at time x and have p as input place with input normal arcs $a_1(p,t_1), \dots, a_i(p,t_i)$. Since continuous transition has no delay in firing, we consider continuous transitions $t'_1, \dots, t'_j$ which have p as output place with output normal arcs $a'(t'_1, p), \dots, a'(t'_j, p)$ and are enabled at time x. Then, the decreases and increases through these input and output normal arcs are summed up to define the derivative of $M[p]$ at time x:

$$\frac{dM[p](x)}{dx} = -\sum_{k=1}^{i} u(a_k)[M(x)] + \sum_{k=1}^{j} u(a'_k)[M(x)].$$

Starting from the time $x = 0$, we can solve the above equation (numerically) by using the initial mark $M(0)$ as the initial condition until one of the following events occur if no system error occurs:

E1: A new continuous transition is triggered or some already enabled continuous transition is switched off. This may change the definition of the equation.

E2: Some discrete or generic transition is triggered or fires. This may change the initial condition for solving the equation.

Thus, we can take a series of time points $x_0, x_1, \ldots$ such that E1 and/or E2 occur at time x_i and neither E1 nor E2 occurs in the time interval (left closed and right open) (x_i, x_{i+1}). Then, we can define the equation for the time interval $[x_i, x_{i+1})$ and solve it with $M(x_i)$ as the initial condition.

Cell Illustrator [59] has completely implemented this architecture for biological pathway modeling and simulation with biologically intuitive GUI. In this software, very common kinetics, for example, Mass Action, Hill, Michaelis–Mentens kinetics and kinetics with noise, can be easily assigned (for part of then please refer to Chap. 8.) The released version of Cell Illustrator does not support the complicated generic type, that is, `pair`, `list`, and `object`.

6.4 Modeling MicroRNAs Double-Negative Feedback Loop in Gustatory Neurons of *C. elegans*

This section discusses the importance of simulation and the necessity for modeling continuous, discrete and generic biological pathways. Here, we illustrate it by using the gene regulatory network of the cell fate determination in *C. elegans*.

ASE cells are a set of two ciliated neurons that are part of the amphid sensilla. *C. elegans* has these two gustatory neurons in the left and right sides of the body called ASEL (left) and ASER (right). ASE cells are differentiated from precursor of ASE cells and the cell fate determination is regulated by a complex double-negative feedback loop (DNFL) mediated by *lsy-6* and *mir-273* microRNA miRNAs [172, 174]. Figure 6.5 shows the mechanism of differentiation of ASE cells based on the DNFL as follows. In addition to these miRNAs, an NKx-type homeobox gene *cog-1* and a zinc-finger transcription factor *die-1* also play the key roles in this regulation [62, 172]. The gradually expressed gene *cog-1* promotes the differentiation into ASER while *die-1* promotes the differentiation into ASEL. For the differentiation into ASER, the *cog-1* and *mir-273* miRNAs regulate the mRNAs in the following manner. The *cog-1* mRNA contains an *lsy-6* complementary site in the 3′ untranslated region [172]. In contrast, the *die-1* miRNA contains two *mir-273* complementary sites in the 3′ untranslated region [62]. Thus, the actions of *cog-1* and *die-1* are inhibited under the abundance of *lsy-6* and *mir-273*, respectively, and differentiation into ASER and ASEL cells can not occur. In addition, *die-1* promotes the expression of *lsy-6* and *cog-1* promotes the expression of *mir-273*. Thus, the loop of *cog-1*, *mir-273*, *die-1*, and *lsy-6* forms the DNFL.

The *die-1* protein in the cytoplasm activates the expression of gene *lim-6* and then the *lim-6* protein migrates to the nucleus. After that, the *lim-6* protein activates the transcription of *lsy-6*. It also activates the expression of *flp-4* and *flp-20* and suppresses that of *gcy-5* and *gcy-22*. The *lim-6* protein in the nucleus activates the transcription of both *lsy-6* and *die-1*.

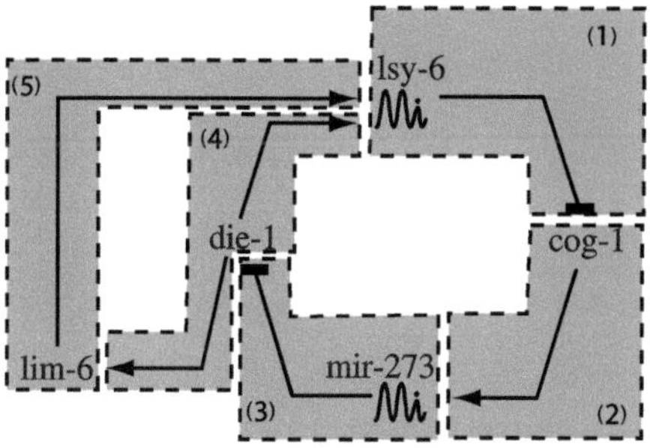

Fig. 6.5 Steps (1)–(4) forms the loop named double-negative feedback loop. The *arrow with triangle* as its head denotes activation to its target and *with rectangle* denotes suppression. The increased *die-1* leads to the activation of *lsy-6* (4) and the suppression of *cog-1* (1) which subsequently reduces *mir-273* (2). *Lsy-6* could also be activated via another path (5). On the other hand, the increased *cog-1* leads to the activation of *mir-273* (2) and the suppression of *die-1* (3) which subsequently reduces *lsy-6* (and *lim-6*) (4)

The ASEL cell expresses *gcy-6* and *gcy-7*, while the ASER cell expresses *gcy-5*, *gcy-22* and *hen-1*. In adult animals, *gcy-6* and *gcy-7* are suppressed by the *cog-1* protein in the cytoplasm and are expressed only in the ASEL cell, whereas *gcy-5*, *gcy-22*, and *hen-1* are expressed only in the ASER cell. Moreover, *flp-4* and *flp-20* are only expressed in ASEL cells [174].

If the expression level of *mir-273* is high, the reporter proteins of ASEL, that is, *flp-4*, *flp-20*, *gcy-6*, and *gcy-7*, are upregulated and the reporter proteins of ASER, that is, *gcy-5*, *gcy-22*, and *hen-1* are not observed. In contrast, if the level of *mir-273* is low, the results are completely reversed.

It is also known that *lsy-2* is another key regulator that activates the transcription of *lsy-6* by transporting it from the cytoplasm to the nucleus [173].

All these mechanisms are compiled into an HFPNe model with only continuous places and transitions as shown in Fig. 6.6. It involves the subcellular localization information of the proteins, mRNAs, etc. In [331], by controlling the initial concentration of *lsy-2*, we showed that the simulation results of this model and their mutants in silico are consistent with biological observations of the cell fate determination of gustatory neurons in vivo. The parameters of this HFPNe model and simulation results are available on the web site of Cell System Markup Language [61] and can be simulated using Cell Illustrator [59].

Such continuous models of biological pathways are useful for understanding the quantitative aspects of the mechanisms. Only logical regulatory relations given above are not enough to clearly elucidate the behavior of the differentiation that is affected by the amount of the initial expression of *lsy-2*. If the initial expression of *lsy-2* is lower, ASE is differentiated to ASER, and the expressions of *mir-273*, *gcy-5*, *gcy-22* and *hen-1* are increased. Otherwise, ASE is differentiated to ASEL, and the expressions of *lsy-6*, *gcy-6*, *gcy-7*, *flp-4*, and *flp-20* are increased. Thus, the continuous abstract of a biological pathway is helpful to understand the system.

On the other hand, with the recent advanced technology in biology, further details behind the mechanisms in a cell can be observed. Since the interests of biologists

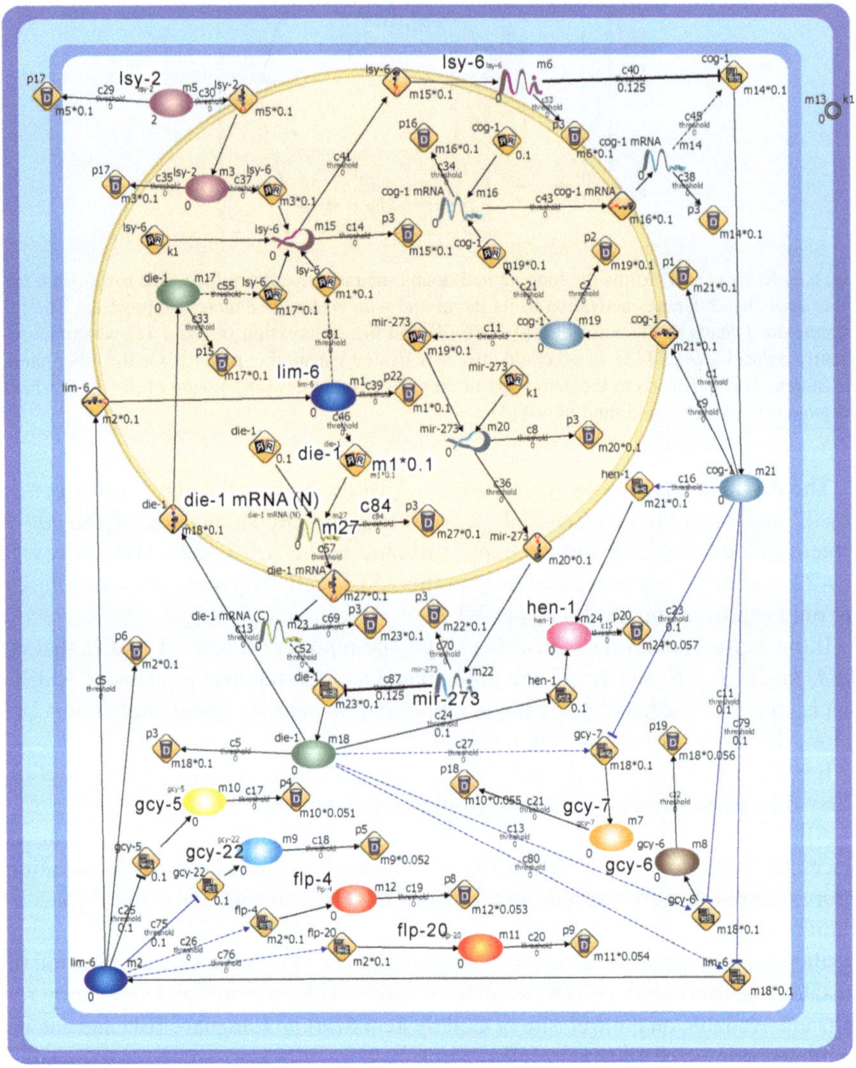

Fig. 6.6 Continuous model of cell fate determination in *C. elegans* build on Cell Illustrator. In Cell Illustrator, terms "entity", "process" and "connector" are used for "place", "transition" and "arc". The terms of components in this model are summarized in Fig. 6.2. In the above figure, for example, the place for *die-1* mRNA in the nucleus is attached its name "die-1 mRNA(N)" and its variable "m4" holding its mark. For a transition, for example, the transition from "lim-6" to "die-1 mRNA(N)" is attached its name "die-1" and its kinetics "m1*0.1". An arc is attached a name, for example, "c84". Default names place, transition, arc, are variable "e#", "p#", "c#", and "m#", where # is a number. The inhibitory regulations from miRNAs are emphasized with thick arcs

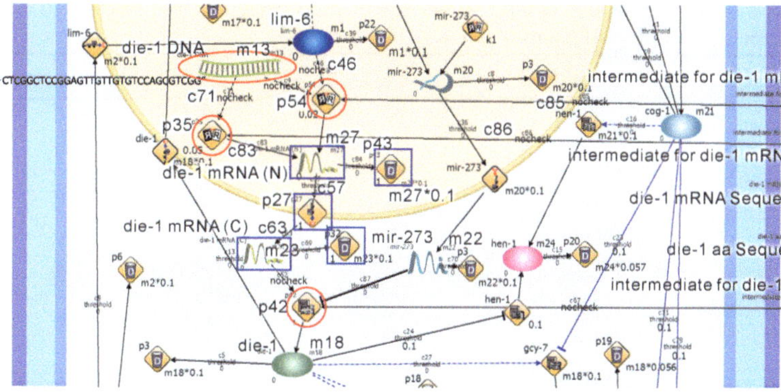

Fig. 6.7 A part of the HFPNe model of the cell fate determination using discrete, continuous, and generic places and transitions. Generic places and transitions are rounded by *red circles*. Discrete places and transitions are rounded by *blue rectangles*. Detailed kinetics and types of transitions in p27, p35, p42, p43, and p54 are summarized in Table 6.2

are not only the density of some molecule at some location in the cell but also the sequence information of the molecule, e.g. the copy number of some specific RNA in the cytoplasm, it is inevitable to include both information to understand the mechanisms. It is known that the gene *die-1* has a SNP on one of the exons which changes Methionine to Leucine [418]. Variations in miRNAs may also affect the translational regulation quantitatively. Thus, we consider that dynamic pathway modeling should deal with not only abstract continuous events but also discrete and sequence involved events.

Figure 6.7 shows a part around the gene *die-1* of the HFPNe model build with Cell Illustrator that elaborates continuous, discrete and generic places and transitions. In this model, the detailed sequence translation steps are extended to be modeled. Updated kinetics in Fig. 6.7 are summarized in Table 6.2. The model in Fig. 6.6 is modified as follows: First, a generic place of type `string` is added with name "die-1 DNA" that holds a single DNA sequence of *die-1* (m13). The place "die-1 mRNA(N)" (m27) with icon  is changed to a discrete place so that we can count the number of copies of *die-1* mRNA in the nucleus. The transition "p35" with icon  is a generic transition input arc c71 from "die-1 DNA" and output arc c83 to "die-1 mRNA(N)" that replaces the continuous transitions *die-1* in Fig. 6.6. This transition represents the basal transcription from "die-1 DNA". The transition "p54" with icon  transcripts "die-1 DNA" with the activation by continuous place "lim-6" through arc c46. In order to observe and use the *die-1* mRNA being produced, generic places named "intermediate die-1 mRNA(1)" (m4) and "intermediate die-1 mRNA(2)" (m25) are newly added and connected to transitions "p54" and "p35" by arcs c85 and c86, respectively. The complete *die-1* mRNA sequence is stored in "die-1 mRNA sequence" (m26). Transition "p43" with icon  is a discrete transition which degrades "die-1 mRNA(N)" with the speed of m27*0.1. Then "die-1 mRNA(N)" is translocated to the cytoplasm. This process is performed

Table 6.2 The kinetics presentation of discrete and generic transitions. The row order in this table corresponds directly to the flow of pathway in Fig. 6.7

Name/Icon	Type	Speed (Kinetics)	
p54	generic	arc: c4	m13
		arc: c46	m1

		arc: c85	import("gon.Transcription");
			totalnum = m13.length();
			num = m4.length();
			if (totalnum > num) {
			nextcode = m13.substring((totalnum - num) - 1, totalnum - num);
			newsequence = (m4 + Transcription::Trans(nextcode));
			} else {
			newsequence = "";
			m26 = m4;
			};
			newsequence;

		arc: c50	if(m4.length() == m13.length()){
			m27++;
			};
			m27;

p35	generic	arc: c71	m13
		arc: c86	m1

		arc: c83	import("gon.Transcription");
			totalnum = m13.length();
			num = m25.length();
			if (totalnum > num) {
			nextcode = m13.substring((totalnum - num) - 1, totalnum - num);
			newsequence = (m25 + Transcription::Trans(nextcode));
			} else {
			newsequence = "" ;
			m26 = m25;
			};
			newsequence;

		arc: c50	if(m25.length() == m13.length()) {
			m27 = m27 + 1;
			};
			m27;

p43	discrete		m27*0.1
p27	discrete		1
p42	generic	arc: c52	m23
		arc: c87	m22

		arc: c61	if(m28.length() == m29.length() {
			m18 = m18 + 1;
			};
			m18;

by a discrete transition "p27" with icon with input arc c57 and output arc c63. The number of copies of *die-1* mRNA in the cytoplasm is represented by a discrete place "die-1 mRNA(C)" (m23). On the other hand, we use a continuous place "die-

1" (m18) for the *die-1* protein. A generic transition "p42" with icon  translates "die-1 mRNA(C)" to the protein with an inhibition of "mir-273" (m22). The amino acid sequence of *die-1* protein is stored in "die-1 aa sequence" (m29). The amino acid sequence being produced is represented by generic place "intermediate die-1 protein" (m28). The readers can refer to the web site [61] for the HFPNe model and the details of the functions and parameters defined for transitions and arcs in the HFPNe model.

6.5 Concluding Remarks

In order to carry out simulation, continuous or discrete models are easier to be built than hybrid or higher models. Toy models are useful to understand the principles in the dynamic systems as was shown in physics. However, for deeper understanding of complex biological systems, we consider that more detailed descriptions of the systems are at least needed but with intuitive biological understanding. Design principles in biological systems are not yet well understood. Therefore, abstraction of systems is premature at our current stage. Our approach with HFPNe is one direction towards this aim. In order to realize this for practice, we also developed Cell Illustrator [59] that implemented HFPNe as a biological pathway modeling architecture.

The available analysis to elementary Petri nets, for example, reachability analysis and invariant analysis, can not be directly applied to our high level Petri net. Instead, other approaches are developing by using model checking [231] and data assimilation based approaches [278, 281, 386].

Moreover, for simulation oriented ontology, we developed both Cell System Ontology (CSO) [171] and Cell System Markup Language (CSML) [61] as description languages for dynamic pathway models. A database containing dynamic pathways written in CSML was also developed [279] to the contents rich biological pathway database named TRANSPATH [215]. An introductory textbook on pathway modeling is also published [280]. We do not think that any single concept is enough for dynamic pathway modeling but the concept of Petri net has a very high affinity for biological pathways.

6.6 Problems

6.1

1. Create the same plots in Fig. 6.4, when the delay of Fig. 6.4 in the transition $t1$ is changed to 3.
2. When will the token number of $p2$ becomes 4, if the speeds of $p1$ to $t1$ and $t1$ to $p2$ are changed to 1 and 2, respectively?

6.2 List three suitable biological events that can utilize discrete, continuous or generic elements in HFPNe.

6.3 Find the DNFL components (1)–(5) shown in Figs. 6.5 and 6.6. The same model can be accessed via http://www.csml.org/models/csml-models/ase-cell-fate-simulation/.

6.4 Compare the simulation results of model having continuous elements only and model having continuous, discrete and generic elements. These models can be accessed via http://www.csml.org/models/csml-models/ase-cell-fate-simulation/. Pay attention to the difference in simulation results of highlighted elements in Fig. 6.7.

6.5 Signal transduction pathways, gene regulatory networks and metabolic pathways modeled in HFPNe are registered in http://www.csml.org/models/csml-models/. Select a few models and launch them using CIOPlayer (or CIO) to further your knowledge of HFPNe.

Acknowledgement We would like to thank André Fujita and Koh Chuan Hock for their helpful comments and critiques.

Chapter 7
Stochastic Modeling

Ivan Mura

Abstract This chapter provides an introduction to the concepts underlying the stochastic modeling of biological systems with Petri Nets. It introduces a timed interpretation of the occurrence of transitions in a net that suites the randomness observed in biochemical reactions occurring in living matter. Thanks to the foundational work of Gillespie in the 70s, this randomness can be easily accounted for by the representative power of Stochastic Petri Nets. The chapter illustrates the Stochastic Petri Net model specification process, the possibilities of analytical and numerical evaluation of model dynamics as well as the basic concepts underlying the simulative approaches, through the application to simple instances of biological systems to help the reader familiarizing with this discrete stochastic modeling formalism. Additional examples of larger scale models are presented, and exercises suggested to consolidate the understanding of the main concepts.

7.1 Introduction

The intuitive graphical formalisms of Petri net coupled with the considerable amount of theoretical results supporting their analysis has greatly favored the spread of this modeling formalism in various scientific and technical fields, among them being Systems Biology.

Because Petri net easily lend themselves to represent state/transition models, there is quite an immediate mapping between reaction oriented descriptions of biological systems and places, arcs and transitions. However, describing a biological system through a set of reactions is far from being a complete specification of a system in quantitative terms, as the exact dynamics of how the transformations occur is still to be defined. As a result, it is possible to build a Petri net that encodes the

I. Mura (✉)
The Microsoft Research–University of Trento Centre for Computational and Systems Biology, Piazza Manci 17, 38123 Trento, Italy
e-mail: mura@cosbi.eu

I. Koch et al. (eds.), *Modeling in Systems Biology*, Computational Biology 16, DOI 10.1007/978-1-84996-474-6_7, information described in a set of reactions, but then the issue still remains of how the transitions modeling the reactions fire over time, whether there is a specific order to be imposed on possible simultaneous firings, and more generally how to deal with concurrency.

Providing an answer to the questions above requires a further step of interpretation of a Petri net model, and different interpretations are in fact possible. The choice among them depends on the level of abstraction at which we intend to study a biological systems, but also on the amount of information that is available. As already presented in previous chapters of this book, a discrete interpretation is possible, as well as a continuous one. This chapter deals with a specific subcase of the discrete interpretation, namely that one in which the transitions firings are timed, and the firing times are random variables.

This interpretation leads us to consider the class of Petri nets that we call here *Stochastic Petri net* (SPN, hereafter). An important remark here is about terminology: the SPN acronym was initially proposed for a specific stochastic extension of Petri net [267], and other variants have taken different names: Generalized Stochastic Petri net [243], Stochastic Activity Networks [260], Stochastic Reward Nets [269], Stochastic Well-Formed Colored Nets [73], just to mention some of them. In this chapter, we shall use the generic acronym SPN to refer to Petri net for which the occurrence times of transition firings are random variables following a negative exponential distribution (cumulative density function $F(t) = 1 - e^{-\lambda t}$, $t \geq 0$, 0 elsewhere, $\lambda > 0$). As long as the allowed distributions are restricted to this type, the time-dependent evolution of an SPN model can be analyzed through the solution of its underlying Continuous-Time Markov Chain.

Various tools were developed for the SPN variants mentioned above, to provide graphical user interfaces and analytical and simulation support to model definition and solution, see the Related Work section at the end of this chapter for a list of some of them. Several of these proposals originated from attempts to limit the sometime cumbersome graphical notation of basic Petri nets, with the addition of more expressive modeling constructs. Quite interestingly, apart few exceptions as in the case of [140], there has been no application of such advanced Petri net modeling tools in the research communities of biologists.

The SPN formalism is still based on a discrete state-space modeling approach, hence it has the expressive power to capture the discrete molecular dynamics of the system. Various applications to biology exist in the literature, see the Related Work section of this chapter, and they all date to the last decade. This recent deployment of SPN models in system biology is due to the growing interest being paid to the important results produced by Gillespie in mid 70s [130, 131] about stochastic molecular dynamics. Before then, dealing with the evolution of a biological system at the molecular level was considered to be infeasible apart for toy models of very simple systems, because tracking the state of each molecule in terms of its position and speed is indeed a complex task, both in mathematical and computational terms. Hence, for a long time the preferred modeling approach has been the one that abstracts discrete state variables representing number of molecules into continuous variables that account for concentration of molecules. Gillespie proved that there exist a lower-level abstraction that is able to account for the behavior of individual molecules, but that do not require dealing explicitly with their positions and speeds. Moreover, this abstraction is mathematically and computationally approachable as it allows exploiting the vast repertoire of methods and tools developed for Markov processes. The formulation of stochastic molecular dynamics proposed by Gillespie nicely fits into the expressive capabilities of the SPN modeling approach.

7.2 Basic Concepts

This section introduces the fundamental concepts upon which the stochastic modeling of biological systems is based on. First, it presents the results obtained by Gillespie in mid 70s about the stochastic kinetics of chemical systems, which provide the theoretical ground for most of the stochastic modeling research conducted today in the systems biology domain. Second, it introduces the SPN modeling formalism through some simple examples of biological systems, taking care of discussing how and when the specific interpretation of transition firing in SPN models matches with the conditions prescribed by Gillespie for a sound modeling of randomness in reaction occurrence times.

7.2.1 The Theoretical Basis of Stochastic Molecular Dynamics

The results provided in Gillespie's first two papers of 1976 [130] and 1977 [131] stem from a precise characterization of the stochastic behavior of chemical systems in terms of the probability of the occurrence of a reaction.

In the following, we shall consider a biochemical system composed of N species and M reactions. We model the state of such system with a vector $\overline{x}$ whose component $x_i \in \mathbb{N}$ represents the number of molecules of chemical species i, $i = 1, 2, \ldots, N$. Then, the evolution of the system can be modeled by a time-indexed sequence of states, which we denote as $\{\overline{x}_t\}_{t \geq 0}$. In the interpretation we consider in this chapter $\overline{x}$ is a random vector and $\{\overline{x}_t\}_{t \geq 0}$ is a stochastic process, probabilistically moving from one state to the other in continuous time.

The interpretation given above does not restrict $\{\overline{x}_t\}_{t \geq 0}$ to be a stochastic process of a specific type. The most crucial point of Gillespie's work was to show that, for many instances of biological systems, $\{\overline{x}_t\}_{t \geq 0}$ can be a very simple process, as pointed out by the following result:

> *For every reaction j, $j = 1, 2, \ldots, M$ in the system, if $\overline{x}$ is the state of the system at time t, $t \geq 0$, the probability that a reaction of type j occurs in the next infinitesimal time interval $t + \delta t$ can be expressed as $a_j(\overline{x}) \cdot \delta t$, where the only dependence of $a_j(\overline{x})$ on t is through the state vector $\overline{x}$.*

Functions $a_j(\overline{x})$ are called *propensity* functions. The property stated above implies that, in every state of the system and for every reaction j, $j = 1, 2, \ldots, M$ the time to the next occurrence of reaction j is a random variable following a negative exponential distribution. This allows describing the evolution of the system over time from an initial state, through a simple system of first order linear differential equations, known as the Chemical Master Equation (CME, hereafter). In probability theory, the CME describes the evolution of a Continuous-Time Markov Chain (CTMC, hereafter), whose state space corresponds to the set of possible states of the chemical system, and whose transitions correspond to the occurrence of reactions. A rich set of exact analysis techniques for both transient (time-dependent) and steady-state analysis of CTMC exists [98], which can be exploited to investigate the dynamic evolution of a biochemical system.

Gillespie took care of demonstrating that this fundamental property holds indeed of non-trivial biochemical systems. He did not define directly the borders of its applicability, but rather identified a set of sufficient conditions that ensure the validity of his results. Specifically, Gillespie introduced the following two hypotheses:

Hypothesis 1 *The chemical system is under thermal equilibrium conditions.*

Hypothesis 2 *The chemical system is such that, at any time t, the concentration of each species is homogeneous in the reaction vessel (i.e., does not depend on space).*

It is important to notice that the homogeneity described in Hypothesis 2 is in fact achieved if nonreactive collisions are much more frequent than reactive ones, which ensures diffusion processes proceed at much higher rate than any reaction in the system. Another major hypothesis was introduced for bimolecular reactions:

Hypothesis 3 *In a bimolecular reaction, the time to the occurrence of the reaction is largely determined by the time to the reactive collision whereas the time necessary for the chemical transformation of the colliding species into the reaction products is negligible.*

If a system satisfies Hypotheses 1, 2 and 3, then we can rely on Gillespie's results, the same as to say that the evolution of the state of the system over time is described by a CTMC whose transition rates from any state $\overline{x}$ are given by the propensity functions $a_j(\overline{x})$, for any j.

Gillespie analytically computed the propensity functions $a_j(\overline{x})$ for some template reactions, assuming that the molecules are approximately spherical. For instance, when this is true, function $a_j(\overline{x})$ for a bimolecular reaction $A + B \rightarrow C$ turns out to be as follows:

$$a_j(\overline{x}) = V^{-1}\pi r_{AB}\sqrt{\frac{8kT}{\pi m_{AB}}} \cdot x_A \cdot x_B \tag{7.1}$$

where x_A and x_B are the number of molecules of A and B in current state of the system $\overline{x}$, V is the volume of the reaction vessel, r_{AB} is the distance between the geometrical centers of two reactant molecules A and B at which the reaction happens, k is the Boltzmann's constant, T is the absolute temperature and m_{AB} the reduced mass defined as $m_{AB} = m_A m_B/(m_A + m_B)$, m_A and m_B being the mass of molecule A and B, respectively. Notice that the propensity function (7.1) is in the form $a_j(\overline{x}) = k_j \cdot f_j(\overline{x})$, that is the product between a constant and a function of $\overline{x}$. The constant k_j is called the *rate constant* of the reaction and it accounts for the physical and chemical properties of the reactant species and of the reaction environment, whereas the form of function $f_j(\overline{x})$ only depends on the reaction type (bimolecular, in this case) and on the stoichiometry. As heavy biochemicals species usually feature complex shapes, the spherical assumption is rarely justified. However, the rate constants appearing in propensity functions may still be determined from the outcomes of wet-lab experiments that measure the kinetic characteristics of reactions. The propensity function $a_j(\overline{x})$ of a dimerization reaction $A + A \to A_2$ takes the following form:

$$a_j(\overline{x}) = k_j \cdot \binom{x_A}{2} \tag{7.2}$$

whereas that one of a monomolecular reaction $A \to B$ will be in the form $a_j(\overline{x}) = k_j \cdot x_A$.

The fundamental property may also hold of systems that do not satisfy the set of Hypotheses 1, 2 and 3, as it seems to be proved by the successful validation of many stochastic models of biochemical systems against experimental data. However, Gillespie's results only allow us taking it for granted for systems that are well-mixed, under thermal equilibrium, and only for specific types of chemical reactions, often termed *elementary*. In this context, an elementary reaction is one that does not abstract any intermediate species. There are only two types of elementary reactions, namely:

1. *Monomolecular* reactions, where a molecule of species A transforms itself into a set of reaction product molecules.
2. *Bimolecular* reactions, where a molecule of species A and a molecule of species B, where species B may be the same as species A, bind to generate a new molecule.

Examples of reactions of type (1) reactions include the isomerization $A \to B$, and the split of a molecule into sub-molecules $A \to B + C$. Examples of type (2) reactions include complexation $A + B \to C$, and, as a special case, the dimerization $A + A \to A_2$. Any reaction that involves more than two reactant molecules is not elementary. A simple statistical argument shows that the likelihood of more than two molecules simultaneously colliding in a reaction vessel is infinitesimal, and any reaction involving more than two reactant molecules, such as $A + B + C \to D$, must occur indeed as a sequence of consecutive bimolecular reactions, for instance $A + B \to E$, $E + C \to D$.

7.2.2 SPN Modeling of Stochastic Molecular Dynamics

Being an abstract modeling formalism, Petri nets do not refer per se to any specific aspect of the biological domain, but rather a meaning has to be associated by the modeler to places, tokens and transitions, and the same applies to SPNs. However, a natural interpretation of Stochastic Petri net elements can be considered when modeling biological phenomena at the discrete molecular level, as we are going to describe in this section.

Let us assume that we are interested in modeling a biochemical system composed by N biological entities, the abundances or the state of which are modified by the occurrence of M reactions. Here, biological entity means a chemical or biochemical species, as well as more complex entities, such as ribosomes, receptors, genes. Similarly, the word reactions is used to indicate any event that can change the multiplicity or the state of entities. Thus, the transfer of a methyl group onto a residue, the degradation of a messenger RNA, the binding of a ligand to a receptor and the translocation of a transcription factor molecule from the cytoplasm into the nucleus, are all examples of what we summarily call here reactions. As SPN models allow us to freely select the abstraction level, all of these different scenarios can be easily encoded. We also assume that the system we want to model satisfies Gillespie's hypotheses for stochastic molecular kinetics, so that the occurrence times of reaction j are random variables following a negative exponential distribution whose state-dependent parameter is specified by a propensity function $a_j(\overline{x})$, $j = 1, 2, \ldots, M$.

An SPN model of this generic system is built as follows:

1. A place is added to the model to represent each of the various states of biochemical entities that we are interested in distinguishing. If the activity of a protein A is different from that of its phosphorylated form A^{-P}, then we may want to use two distinct places in the Petri net to be able to keep track of the respective amounts and activity of A and A^{-P} as the model evolves.
2. A transition is added to the model to represent each of the various reactions. The transition firing time is chosen to be a random variable distributed as the negative exponential distribution of parameter $a_j(\overline{x})$, where $\overline{x}$ is the vector marking of the SPN model.
3. Each transition is connected to input (resp. output) places by arcs according to the reactant (resp. products) it consumes (resp. produces).
4. Stoichiometry of reactions is straightforwardly represented by associating positive integer weights to arcs.
5. Tokens are assigned to places—to provide the initial marking of the SPN—according to the initial number of molecules of the species or of the entity represented by the place. Notice that, apart from the cases in which there are a few of them, tokens are not graphically depicted.

The graphical notation of SPN does not differ from the basic Petri Net one. The distinguishing characteristics of SPN are indeed observed when the dynamics of marking evolution are considered. The enabling of transitions is subject to the standard rule described in Part I of this book, that prescribes that a transition is enabled if each input place holds a number of tokens greater or equal to the weight of the connecting arc. Let us now describe the firing of SPN transitions through some models of biological examples. Suppose we are building an SPN model of the simple reaction $A \xrightarrow{k} B$, where k is the rate constant of the reaction, and consequently

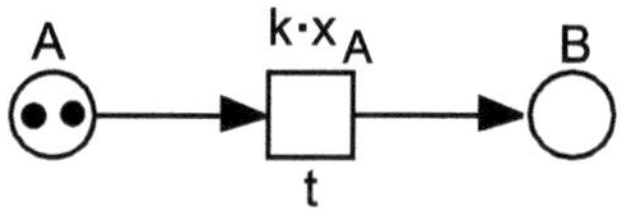

Fig. 7.1 SPN model of the monomolecular reaction $A \xrightarrow{k} B$. Tokens in place A and B represent the number of molecules of species A and B, respectively, whereas transition t models the reaction that transforms one molecule of A into one molecules of B. The firing rate of t is, according to the mass-action law, proportional to x_A, the amount of reactant

$a(\overline{x}) = k \cdot x_A$ is the propensity function computed as per Gillespie. Let us also assume that the initial state of the system is one in which $x_A = 2$ molecules of species A and $x_B = 0$ molecules of species B are present at time $t = 0$. The modeling steps 1–5 listed above would result in the two places, single transition SPN shown in Fig. 7.1.

The exponential rate associated to a transition expresses the speed at which the modeled reaction occurs. In the initial marking of the net in Fig. 7.1, transition t is enabled because its input place A is marked. A firing time τ_1 is thus chosen for t, drawn from the negative exponential distribution of parameter $k \cdot x_A = 2k$, and a clock starts to countdown from τ_1 to 0. When the clock reaches 0, transition t fires and the marking of the net is changed from $x_A = 2$, $x_B = 0$ to $x_A = 1, x_B = 1$. After the firing, transition t is still enabled, but its rate has now become $k \cdot x_A = k$. Consequently, its new firing time τ_2 will be selected from an exponential random variable different from the one out of which τ_1 was sampled. Again, a clock is set to countdown until the new firing time, that is, $\tau_1 + \tau_2$ is reached. At that time, the marking of the net is changed to $x_A = 0, x_B = 2$, where no transitions are enabled anymore and the evolution stops.

It is interesting to notice that the marking-dependent firing rate of SPN transitions can be interpreted as being a result of the maximal parallelism of molecular transformations. Indeed, in the example above, in the initial marking of the SPN each of the two molecules of A is being transformed into one molecule of B at the same time. Because these two reactions are independent from each one and both reactions happen in a time that is exponentially distributed with parameter k, the time to the occurrence of the first of them is a random variable distributed as the minimum between the times of the reactions, which is again[1] a negatively distributed one with parameter $2 \cdot k$. More generally, the rate constant is multiplied by the number of simultaneous copies of the reactions that take place in parallel, which is indeed a simple way of modeling chemical reactions obeying the mass-action law. A similar

[1]The very useful closure property of statistically independent negative exponential random variables under the minimum operator is easily demonstrated. Suppose X and Y are negative exponential random variables of rate λ_X and λ_Y, respectively, and let the random variable Z to be defined as $Z = \text{MIN}\{X, Y\}$. Then, the cumulative density function $F_Z(t)$ of Z can be computed as $F_Z(t) = \mathbb{P}[Z \leq t] = 1 - \mathbb{P}[Z > t] = 1 - \mathbb{P}[\text{MIN}\{X, Y\} > t] = 1 - \mathbb{P}[X > t, Y > t] = 1 - \mathbb{P}[X > t]\mathbb{P}[Y > t] = 1 - e^{-\lambda_X} e^{-\lambda_Y} = 1 - e^{-(\lambda_X + \lambda_Y)}$, which is the same as to say that Z is distributed as a negative exponential random variable of parameter $\lambda_X + \lambda_Y$.

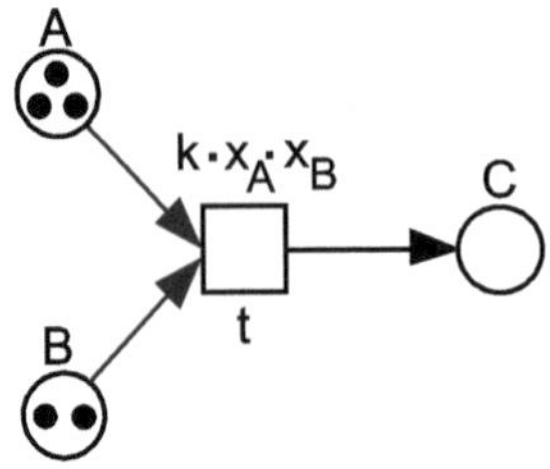

Fig. 7.2 SPN model of the bimolecular reaction $A + B \xrightarrow{k} C$. Each firing of transition t models the occurrence of the bimolecular reaction. According to the incoming and outgoing arcs, at each firing one token is removed from both places A and B and one is added to place C. Notice that, for t to be enabled to fire, both places A and B have to hold a non-zero number of tokens

argument explains the marking-dependent form of the transition rate of bimolecular reactions. Suppose we are building an SPN model of the reaction $A + B \xrightarrow{k} C$, where the initial state of the system is $x_A = 3$, $x_B = 2$ and $x_C = 0$. By applying the modeling steps 1–5 we obtain the SPN model shown in Fig. 7.2.

In the initial marking of the model, there are six several independent ways in which the bimolecular reaction can occur, each one associated to one specific selection of the pair of molecules A and B that react. Thus, the rate associated to transition t in the initial marking is $k \cdot x_A \cdot x_B = 6k$. After the firing, the marking of the SPN model in Fig. 7.2 is changed to $x_A = 2, x_B = 1, x_C = 1$, and the subsequent firing of transition t will occur at a rate that is $k \cdot x_A \cdot x_B = 2k$. Again, the parallelism intrinsically present in our interpretation of the system results in a mathematical form of the transition rates that nicely matches the deterministic mass-action law.

Let us now consider a more complex system, where molecules of a substrate species S are transformed into molecules of a product species P by the action of an enzyme E following a Michaelis–Menten reaction scheme, as described in more detail in Chap. 8.

Assuming we do not want to abstract the intermediate species formed by the enzyme-substrate complex, the set of biochemical reactions included in our systems are the following ones: $S + E \xrightarrow{k_{\text{on}}} S:E$, $S:E \xrightarrow{k_{\text{off}}} S + E$, $S:E \xrightarrow{k_{\text{cat}}} P + E$. Our SPN modeling procedure described in steps 1–5 produces the model shown in Fig. 7.3. In the SPN model on Fig. 7.3, tokens contained in place $S:E$ represent molecules of the complex enzyme-substrate. One token is added to place $S:E$ at each firing of transition t_{on}, which is modeling the bimolecular binding reaction between molecules of species S and molecules of species E. Each molecule of the complex may undergo catalysis producing one molecule of the product species P and releasing back the enzyme molecule. This transformation is modeled by transition t_{cat}. Moreover, each molecule of the complex may also spontaneously unbind back into its constituent molecules, as modeled by transition t_{off}. We use this example of competition between reactions (unbinding and catalysis) to explain the way competition among conflicting transitions is handled in SPN models.

Let us consider the fate of a molecule of enzyme-substrate in the system. Whether it will unbind or rather undergo catalysis is determined according to the rates of the

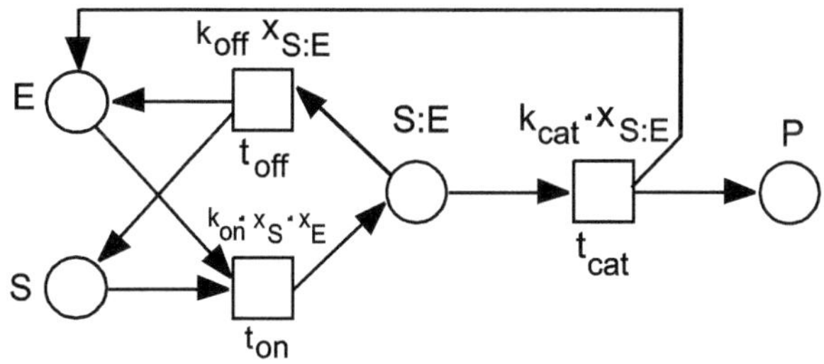

Fig. 7.3 SPN model of an enzymatic reaction. The two transitions t_{on} and t_{off} model the fast binding-unbinding of the enzyme E to the substrate S, whereas transition t_{cat} represent the catalysis steps. Given a nonzero initial amount of tokens in places E and S, the network dynamics moves the tokens from place S to place P, modeling the transformation of substrate molecules into product ones. Notice that the number of tokens in place E plus the number of tokens in place $S:E$ stays constant as the net marking evolves, as the number of enzyme molecules is not changed by the enzymatic reaction

two possible reactions. If the catalysis occurs to happen at a slower rate than the unbinding (this is usually the case), the enzyme-substrate molecule will have more chances to break into its constituent molecules. This rate-dependent dynamics gets faithfully encoded into the marking evolution of the SPN. As soon as one token reaches place $S:E$, the two transitions t_{off} and t_{cat} get both enabled. This means that both of them may happen, but only either of them will happen. The selection of which one will occur is determined by the result of the random selection of firing times. More, precisely, both transition draw a random time for their firing according to the firing rates they have in current marking of the net, and the one that has drawn the minimum firing time is selected for firing. This way of handling concurrent firings in SPN models is called *race policy*, and extends itself naturally to more general cases where multiple transitions are involved.

It is interesting to make some quantitative considerations about the result of the race policy application. In the example we are considering, the firing rates of transitions t_{off} and t_{cat} are respectively $k_{\text{off}} \cdot x_{S:E}$ and $k_{\text{cat}} \cdot x_{S:E}$. The likelihood that the value of a random variable distributed as the firing time of t_{off} exceeds the value of a random variable distributed as the firing time of t_{cat} is easily computed with some basic maths, and is given by $p_{\text{cat}} = k_{\text{cat}}/(k_{\text{off}} + k_{\text{cat}})$. This expression is the probability that, every time a competition among a catalysis and an unbinding reaction occurs, the catalysis reaction occurs first, whereas $p_{\text{off}} = 1 - p_{\text{cat}}$ is the probability that the unbinding occurs first. Because $k_{\text{cat}} < k_{\text{off}}$ in Michaelis–Menten enzymatic transformations, we have that $p_{\text{cat}} < p_{\text{off}}$, and thus, each time a competition between the two reactions occurs, the unbinding has a higher chance to occur with respect to the catalysis. However, in the long run there will be instances of this race that will be won by the catalysis reaction, which ensures the enzymatic reaction proceeds toward the accumulation of the molecules of the product species P.

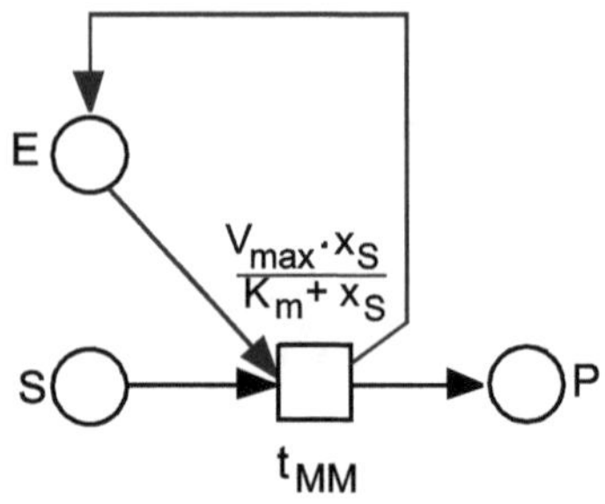

Fig. 7.4 SPN model of an abstract enzymatic reaction, in which an equivalent product production rate has been mathematically worked out as a function of the substrate and enzyme concentration. Notice that the Michaelis–Menten firing rate of transition t_{MM} will vary at each firing

7.2.3 Beyond Pure Mass-Action

The Michaelis–Menten system provides a useful example to discuss how modeling abstractions that go beyond the pure mass action can be accounted for with the SPN formalism. It is indeed obvious that a flexible selection of the level of abstraction is welcome when modeling biological systems. In some cases, the details of molecular interactions may not be known and may even not be observable, whereas in many other cases we may not want to include all the available knowledge to limit the size of the models being built.

When dealing with an enzymatic transformation, both the considerations above may apply. Specifically, on one side the rates k_{on}/k_{off} of binding/unbinding between enzyme and substrate are typically not known, and only the k_{cat} rate can be experimentally measured. On the other hand, an elegant, abstract model exists whose dynamical properties have been mathematically investigated and are known to provide a very good approximation of system behavior. This model is built by abstracting the intermediate species formed by the complexation of the substrate S and the enzyme E. An SPN model that incorporates such an abstraction is shown in Fig. 7.4.

Transition t_{MM} is now modeling a direct transformation of the substrate molecules S into molecules of the product species P. The two parameters V_{max} and K_m characterize the effectiveness of the enzyme in transforming the substrate, and can be determined experimentally. Notice that the number of tokens in place E is not affected by the firing of t_{MM}, as this model is indeed abstracting the direct participation of the enzyme molecules into the reaction. The presence of the enzyme is still necessary, but whether one or more tokens are present in place E does not appear to make any difference for the evolution of the net. Obviously, the reaction is indeed dependent on the enzyme concentration, but this dependency is captured in the rate, through parameter K_m. This is a good approximation when the enzyme E is much less abundant than the substrate S, as it usually happens. In this case, most of molecules of E are bound to molecules of S and the enzymatic reaction proceeds at a rate that is almost completely determined by the abundance of the substrate. This situation is exactly the one captured by the model in Fig. 7.4.

It is worthwhile commenting on whether this abstract reaction meets the necessary conditions that Gillespie defined for stochastic kinetics. In other words, how accurate is the assumption that each occurrence time of the reaction $S + E \to P + E$

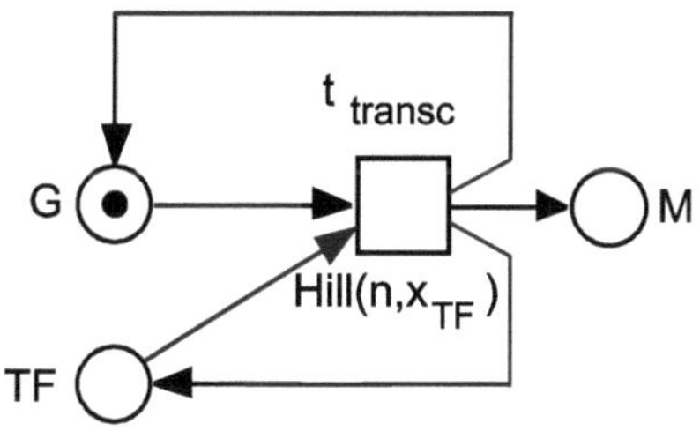

Fig. 7.5 Abstract SPN model of a cooperative gene activation phenomenon, modeled through a Hill rate function of transition t_{transc}. The number of tokens in place *TF* affects the rate of firing of transition t_{transc} in a nonlinear way

in the model in Fig. 7.4 is a random variable following a negative exponential distribution? This point is quite delicate and requires careful treatment. From a quantitative point of view, it is obvious that an approximation is entailed when we make the abstraction that brings us from the SPN in Fig. 7.3 to the one in Fig. 7.4. Indeed, if we assume that all reactions in the model in Fig. 7.3 are elementary, that is we can justifiably assume a negative exponential distribution of the reactions times, then it follows that using exponentially distributed firing times in the SPN in Fig. 7.4 would introduce an approximation. Replacing a sub-model of exponential transitions with a single aggregate one does not preserve in general the stochastic characteristics of the event occurrence times. Thus, this abstraction is only valid when certain assumptions are met. For instance, it can be shown that when the catalysis is the rate limiting step of the enzymatic transformation, then the abstract model shown in Fig. 7.4 is a very accurate model that can replace the more detailed one in Fig. 7.3. Indeed, in this case the time to the production of each molecule of product species *P* is given by the sum of the short times spent in the binding/unbinding plus the time necessary for the catalysis reaction. Because the latter one is much slower, its stochastic characteristics largely determine those of the overall sum of reaction times.

SPN models that include general rate functions on transitions can account for other common features of biological networks. For instance, the Hill dynamics is usually employed to abstract mechanisms of cooperativity. Consider a gene *G* that to be transcribed needs the binding of a dimer or a tetramer of a transcription factor *TF* to the promoter region. Consider the simple SPN model in Fig. 7.5, where tokens in place *TF* are not shown for the sake of clarity.

The rate of transition t_{transc} in Fig. 7.5 is a shorthand notation for the *Hill function*, which is defined as follows:

$$Hill(n,x) = \frac{Q \cdot x^n}{K_m^n + x^n} \tag{7.3}$$

This function expresses the regulation effect of the transcription factor *T* on the transcription rate. More generally, it can be used to express the dependency of a reaction rate from an activator species whose effect gets amplified through cooperativity. The parameter n is called *Hill coefficient* and is related to the cooperativity factor, that is the number of molecules involved in the cooperative effect. The two parameters Q and K_m are usually experimentally determined and they are analogous to those for Michaelis–Menten enzymatic reactions. Indeed, in the case when $n = 1$, no cooperativity, the Hill function reduces to the Michaelis–Menten kinetic rate. Another form of the Hill function exists, which accounts for negative regulation, that is, repression, and is given by $HillNeg(n,x) = Q/(K_m^n + x^n)$. Hill functions provide a sigmoidal form of the transition rate as a function of the driving species and allow modeling complex non-linear behaviors in a very compact way.

7.3 Methods to Determine SPN Evolution

In a well-defined SPN model, the firing of enabled transitions according to the race policy determines the subsequent markings of the net. Thus, an SPN model can be studied to determine the evolution of its marking from the initial one. Because SPN models encode the concept of firing time, we are obviously referring to the marking of the net at a certain time, and due to the randomness of the firing times, we are more precisely talking about the probability distribution of markings over time.

Since the focus of this chapter is on quantitative evaluation of SPN models of biological systems, we will not be dealing with structural checks of models. It is however important to notice that some of the analysis techniques defined in Part I of this book are still applicable to SPN models, provided that the information about time is disregarded. Indeed, the fact that firing time distributions span over the whole $(0, \infty)$ subset of real numbers ensures that any sequence of transition firings that is possible to happen in the untimed version of an SPN is also possible in the SPN timed version, although with low probability. Therefore, the reachability set of the untimed and timed version of the SPN are exactly the same. The analysis techniques based on T-invariants and P-invariants are thus applicable for checking properties of the SPN evolution, and also provide very useful tools for model debugging.

The quantitative evaluation of SPN models is based on the analysis of the discrete-state continuous-time stochastic process governing the evolution of the marking, which we denoted with $\{\overline{x}_t\}_{t\geq 0}$. The reachability set $RS(\mathbb{M})$ of an SPN model $\mathbb{M}$ defines the set, not necessarily finite, of possible model markings. For any time $t \geq 0$ and $\overline{x} \in RS(\mathbb{M})$, we denote by $P[\overline{x}, t]$ the probability that at time t the marking of the SPN model is $\overline{x}$. The marking probability distribution at time $t \geq 0$ is the vector that collects probabilities $P[\overline{x}, t]$ for all $\overline{x} \in RS(\mathbb{M})$. At time $t = 0$ the marking of SPN model $\mathbb{M}$ is $\overline{x}_0 \in RS(\mathbb{M})$ (the initial marking) and the marking probability distribution vector entries are all 0 except for $P[\overline{x}_0, 0]$ that is equal to 1.

Given that the evolution of the model starts in a defined marking, the problem to be solved to quantitatively determine the evolution of the model is the one of calculating the conditional probability $P[\overline{x}, t \mid \overline{x}_0, 0]$, that is the probability that the SPN marking is $\overline{x}$ at time $t > 0$ given that it was $\overline{x}_0$ at time $t = 0$. Measures of interest for the system, such as the likelihood of reaching a certain states within a certain time, the average number of molecules of a species in a given time window, the rate of occurrence of certain reaction event, can all be estimated from the marking probability distribution of an SPN model of the system.

Obtaining a probability distribution over the possible states of a stochastic process is in general a complex task. However, because of the specific choice of transition firing time distribution in SPN models, the stochastic process of marking evolution is a Markov chain, and a vast repertoire of analysis techniques exist for this class of stochastic processes. The two following types of analysis are commonly used when evaluating the probability distribution of markings of an SPN model:

- Time-dependent, also called *transient* analysis, which aims at evaluating the measures of interest at given time points or time windows, and is mostly based on the marginalization of the marking probability distribution.
- Time independent, also called *steady-state* analysis, which looks at equilibrium measures and is based on the limiting probability distribution of the marking, that is, $\lim_{t\to+\infty} P[\overline{x}, t \mid \overline{x}_0, 0]$.

Transient analysis of an SPN model $\mathbb{M}$ can be used to determine time courses of modeled biochemical species. For instance, suppose we are interested in the average number of molecules of a biochemical species A at a discrete set of time points $t_1, t_2, \ldots, t_n$, which we denote by $E[A]_{t_i}$, $i = 1, 2, \ldots, n$. These measures can be computed from the marking distribution probability as follows:

$$E[A]_{t_i} = \sum_{\overline{x} \in RS(\mathbb{M})} P\big[\overline{x}, t \mid \overline{x}_0, 0\big] \cdot x_A \tag{7.4}$$

Given that we have available the whole marking probability distribution, more insights can be obtained into the modeled system by not restricting the evaluation to the average values. For instance, we can compute the variance of the number of molecules of species A at time t, which we denote by $\mathrm{Var}[A]_t$ as follows:

$$\mathrm{Var}[A]_{t_i} = E\big[A^2\big]_t - E^2[A]_t = \sum_{\overline{x} \in RS(\mathbb{M})} \big(P\big[\overline{x}, t \mid \overline{x}_0, 0\big] \cdot x_A^2\big) - E^2[A]_t \tag{7.5}$$

Another measure that is often of interest is the rate of occurrence of a given reaction j at a given time instant t, which we denote by $r_{j,t}$, computed as follows:

$$r_{j,t} = \sum_{\overline{x} \in RS(\mathbb{M})} P\big[\overline{x}, t \mid \overline{x}_0, 0\big] \cdot a_j\big(\overline{x}\big) \tag{7.6}$$

Weighting propensities by the marking probability distribution as in (7.6) provides the instantaneous rate of reaction j, that is the effective rate of the reaction at time t, which takes into account the probabilities of the SPN model to be in the various markings in which the transition modeling the reaction is enabled. From $r_{j,t}$, with some additional calculation, it is possible to obtain the average number of times reaction j occurs in a given time interval $[t_1, t_2]$, which we denote as N_{j,t_1,t_2} as follows:

$$N_{j,t_1,t_2} = \int_{t_1}^{t_2} r_{j,t}\, dt \tag{7.7}$$

The steady-state analysis of an SPN model can only be applied to models whose marking probability distribution possess an equilibrium distribution. Whether or not an SPN model $\mathbb{M}$ has a steady-state marking probability distribution depends on some properties of the stochastic process $\{\overline{x}_t\}_{t\geq 0}$, which has to be *ergodic* . The precise definition of ergodicity is beyond the scope of this chapter, the interested reader can see for instance [98]. Informally, if for any two markings $\overline{x}_1, \overline{x}_2 \in RS(\mathbb{M})$

there exist a sequence of transition firings such that $\overline{x}_1$ is reachable from $\overline{x}_2$ in an average finite time and vice versa, then the stochastic process underlying SPN model $\mathbb{M}$ is ergodic. Notice that such condition excludes from having a steady-state marking distribution all SPN models that have *absorbing* markings, that is, markings that once reached can not be left. Those places with absorbing markings correspond to traps.

For SPN models that have a steady-state marking probability distribution, the $\lim_{t\to+\infty} P[\overline{x}, t \mid \overline{x}_0, 0]$ provides a time-independent distribution, denoted as $\Pi(\overline{x})$, which apart from being independent from time, is also independent from the initial state. Indeed, in the long run the bias induced by the choice of the initial marking is removed. Distribution $\Pi(\overline{x})$ provides the probability of finding, at a random observation time under equilibrium conditions, the SPN model in marking $\overline{x}$. Therefore, the definitions provided in (7.4), (7.5) and (7.6) are still valid at steady-state provided that the transient marking probability distribution is replaced by the steady-state one. Computing the average number of occurrences of a reaction j over a given interval $[t_1, t_2]$ in steady state only requires multiplying the steady-state rate of reaction j by the interval width.

Obviously, calculating the steady-state probability through its limit definition would be totally impractical. As we show in the following, there are much more efficient ways of doing it. In the next sections, we present the methods available to calculate or approximate the marking probability distribution of an SPN model $\mathbb{M}$.

7.3.1 Analytical and Numerical Approaches

Analytical approaches to the computation of the marking probability distribution of SPN models of biological systems are based on the explicit solution of the CME, a set of linear differential equations that describe the evolution of the marking occupation probabilities over time. The CME includes one equation for each possible marking of the model, which already suggests that this approach has limited applicability. Very few SPN models that have large or infinite sizes of their reachability set are indeed amenable to this type of analysis. However, there are notable exceptions, and closed-form for the CME solution can be obtained. We will see some examples in the following. For models that resist to the exact analytical solution (the majority of models of interest), but for which the reachability set is of finite size, the solution to the CME can be computed through numerical algorithms (either exact or approximate). Again, the size of the reachability set determines the limits of applicability of this approach. Presently, this limit is set to about 10^5–10^6 distinct markings (with some larger values for particular models), a number that is easily reached by models of biological systems including many copies of the same biochemical species.

7.3.1.1 Transient Analysis

Let us consider a very simple biological system, where molecules of a species A cross the membrane between two compartments, say cytoplasm and nucleus. We can

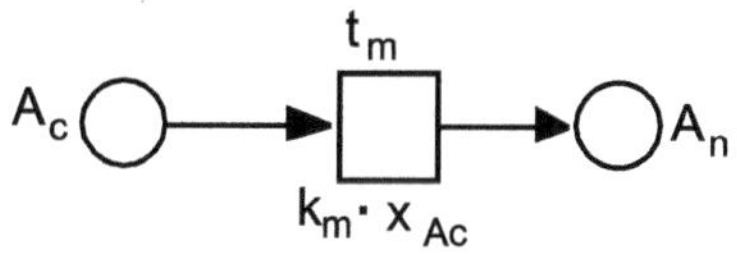

Fig. 7.6 SPN modeling molecules of species A crossing a membrane. The membrane itself is not explicitly represented in the model, which only captures the time necessary for crossing it with the firing of transition t_{in}. To distinguish molecules of species A that are outside the membrane from those that are inside the membrane and additional place, A_n is included in the model. Indeed, because tokens in SPN models do not carry any qualifying attribute, this additional place is necessary to keep track of their state, in this case the compartment they are enclosed in

build a very simple SPN model $\mathbb{M}$ of this phenomenon, as the one shown in Fig. 7.6, where tokens contained in place A_c and A_n represent molecules of species A in the cytoplasm and in the nucleus, respectively. Transition t_m models the membrane crossing, at a rate $k_m \cdot x_{A_c}$ that is determined by a pure mass action kinetics.

Let as assume that there are initially $x_{A_c} = N$ molecules of species A in the cytoplasm and no molecules of A in the nucleus at time $t = 0$. We want to determine the average number of molecules in the nucleus at time $t > 0$. To this, we can compute the conditional marking probability distribution at time t given the initial state at time $t = 0$, and then use (7.4) to compute the average value of the number of molecules in the nucleus. There are $N + 1$ possible markings for the model in Fig. 7.6, which define the reachability set $RS(\mathbb{M}) = \{(N - i, i) \mid i = 0, 1, \dots, N\}$.

We start our analytical calculation of the marking probability distribution by observing that each molecule of A is crossing the membrane independently from the other ones, that is the overall phenomenon is the result of the superimposition of the single contributions given by each molecule. This implies that the single SPN model shown in Fig. 7.6 is indeed equivalent to a set of N identical simpler SPN models structurally equivalent to the one in Fig. 7.6 where place A_c contains exactly one token. Let us consider one instance $\mathbb{M}'$ of these simpler SPN models. Only the two markings $(1, 0)$ and $(0, 1)$ are possible for the model, the reachability set $RS(\mathbb{M}')$ has thus cardinality 2. The only possible transition among markings is the one that brings the model from $(1, 0)$ to $(0, 1)$, occurring at a rate k_m. The time to the transition follows the negative exponential distribution of rate k_m. The conditional probability of $\mathbb{M}'$ to be in marking $(1, 0)$ at time $t > 0$, given that the model is in marking $(1, 0)$ at time $t = 0$ is equal to the probability that the transition t_m has not fired yet at time t, that is

$$P\big[(1,0), t \mid (1,0), 0\big] = e^{-k_m t} \tag{7.8}$$

whereas the conditional probability $P[(0, 1), t \mid (1, 0), 0]$ is equal to $1 - P[(1, 0), t \mid (1, 0), 0] = 1 - e^{-k_m t}$. Therefore, at any time $t > 0$, the conditional marking probability distribution over $RS(\mathbb{M}')$ is given by the vector $(e^{-k_m t}, 1 - e^{-k_m t})$. Because of the independence among subnets, we can easily compute the conditional marking probability distribution for the original model $\mathbb{M}$, observing that the probability of being in marking $(N - i, i)$ at time $t > 0$ is exactly the probability that i molecules out of the N initially present in place A_c have moved to place A_n. Which molecules have moved and which have not is actually not important, as we are not interested in distinguishing them. The fact that exactly i molecules have moved is accounted for by a probabilistic choice expressed through a binomial distribution, as follows:

$$P\big[(N-i,i),t \mid (N,0),0\big] = \binom{N}{i}\left(1-e^{-k_m t}\right)^i \cdot \left(e^{-k_m t}\right)^{N-i},$$
$$i = 0,1,\ldots,N \tag{7.9}$$

The vector $(P[(N,0),t \mid (N,0),0], P[(N-1,1),t \mid (N,0),0], \ldots, P[(0,N),t \mid (N,0),0])$ provides the transient marking distribution probability of the model $\mathbb{M}$. According to (7.4), the average number of molecules in the nucleus at time t, which we denote by $E[A_n]_t$ is therefore analytically calculated as follows:

$$\begin{aligned} E[A_n]_t &= \sum_{i=0}^{N} P\big[(N-i,i),t \mid (N,0),0\big] \cdot i \\ &= \sum_{i=0}^{N} \binom{N}{i}\left(1-e^{-k_m t}\right)^i \cdot \left(e^{-k_m t}\right)^{N-i} \cdot i \end{aligned} \tag{7.10}$$

which can be simplified to $N \cdot (1-e^{-k_m t})$. Notice that the value computed in (7.10) is the average value of a binomial distribution of parameters $N, 1-e^{-k_m t}$. Its variance $\mathrm{Var}[A_n]_t$ is given by $N \cdot (1-e^{-k_m t}) \cdot e^{-k_m t}$.

Thus, for model $\mathbb{M}$ we can relatively easily compute the transient marking distribution probability. However, this is possible only because of the simplicity of the model and the independence among the events that change the state of the different molecules. In general, calculating the transient marking distribution probability through the analytical approach requires manipulating the CME of the stochastic model, through a process that we now describe using again model $\mathbb{M}$ as an example.

One formal way of representing the transitions among markings in model $\mathbb{M}$ is through the *infinitesimal generator* matrix $Q^{\mathbb{M}}$, a squared matrix of size $(N+1)\times(N+1)$, where $N+1$ is the cardinality of the reachability set $RS(\mathbb{M})$. The infinitesimal generator matrix $Q^{\mathbb{M}} = \|q^{\mathbb{M}}_{\overline{x}',\overline{x}''}\|$ is as follows:

$$\|q^{\mathbb{M}}_{\overline{x}',\overline{x}''}\| = \begin{cases} x'_{A_c} \cdot k_m & \text{if } \overline{x}'-\overline{x}'' = (1,0) \\ -x'_{A_c} \cdot k_m & \text{if } \overline{x}'-\overline{x}'' = (0,0) \\ 0 & \text{otherwise} \end{cases} \tag{7.11}$$

Matrix $Q^{\mathbb{M}}$ allows writing in a compact form the CME that governs the transient conditional marking probability distribution $P[(N-i,i),t \mid (N,0),0]$. Let us denote by $\overline{P_{t|0}}$ the vector of conditional marking probabilities at time $t>0$ for the model, and by $\overline{P_0}$ the initial marking probability distribution at time $t=0$, that is, the vector that assigns probability 1 to the initial marking and 0 to each other marking of the model, so that we can write the CME as follows:

$$\frac{d}{dt}\overline{P_{t|0}} = \overline{P_{t|0}} \cdot Q^{\mathbb{M}},\ \overline{P_{0|0}} = \overline{P_0} \tag{7.12}$$

Equation (7.12) consists of a set of linear differential equations, one for each marking of the model. These equations and are also called the Chapman–Kolmogorov

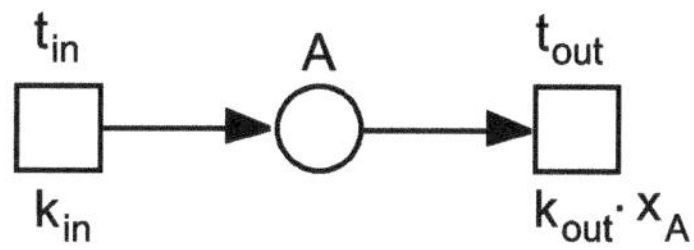

Fig. 7.7 SPN model for cytoplasm-nucleus shuttling. The two transitions t_{in} and t_{out} model, respectively, the addition of one molecule of A and the removal of one molecule of A from the number of molecules of A that are inside the nucleus, represented by the tokens in place A. Because the model is not explicitly considering the number of molecules in the cytoplasms, the incoming rate of molecules of A stays constant, as modeled by the zero-order firing rate of transition t_{in}. This corresponds to assume that the concentration of A in the cytoplasm stays constant, unaffected by the shuttling process

forward equations. Because (7.12) is formed by linear differential equations, a general solution formula is available, and the vector of functions $\overline{P_{t|0}}$ can be calculated as follows:

$$\overline{P_{t|0}} = \overline{P_0} \cdot e^{Q^{\mathbb{M}} \cdot t} \tag{7.13}$$

The expression $e^{Q^{\mathbb{M}} \cdot t}$ in (7.13) is the *matrix exponential* of $Q^{\mathbb{M}} \cdot t$. The exponential of a squared matrix M is defined as the following sum:

$$e^M = \sum_{i=0}^{\infty} \frac{M^i}{i!} \tag{7.14}$$

Formula (7.13) provides a general solution for the transient marking distribution probability, however at the expense of computing the matrix exponential of the infinitesimal generator matrix. Various algorithms exist for this purpose, which are implemented, for instance, by the tools R, Maple and Mathematica. Moreover, most tools for SPN modeling and evaluation implement a very efficient algorithm known as *randomization* [145] to compute the transient marking distribution probability, which again is based on a numerical approximation scheme for solving (7.13). It is important to remark that the approximation is entailed by the necessary truncation of the infinite series in (7.14), and that very accurate solutions can be however obtained for models that have large (but finite) cardinalities of the reachability set, up to 10^5 and in some cases, depending on the sparsity of the infinitesimal generator matrix, up to 10^6 distinct markings.

7.3.1.2 Steady-State Analysis

Let us now consider another simple biological system, in which protein A crosses the nuclear membrane at a constant rate k_{in}, and molecules of A that are in nucleus cross-back the membrane at a rate k_{out}, proportional to the amount of proteins in the nucleus as per the mass-action law. We are interested in determining analytically the steady-state average and variance of the number of molecules of A that are in the nucleus. The SPN model $\mathbb{M}$ shown in Fig. 7.7 captures the two processes that affect the number of molecules of species A in the nucleus (represented by the number of tokens in place A), namely the translocation from the outside of the nucleus to the inside (transition t_{in}) and the opposite movement (transition t_{out}). Let us assume that at time $t=0$ no molecules of species A are in the nucleus. Because we are interested in the equilibrium value of the number of molecules, we must first ask ourselves whether model $\mathbb{M}$ has a steady-state marking probability distribution. The set of possible markings is infinite, and consists of all natural numbers, each one corresponding to a possible value of the amount of molecules of protein A. The model can move away from the initial marking through a firing of t_{in}, and come back to it through one or multiple firings of transition t_{out}, and the same applies for any other possible marking. Moreover, the fact that the rate of transition t_{out} grows linearly with the number of tokens in place A ensures that the model will return to the initial marking in an average finite time. These considerations informally justify our believe that model $\mathbb{M}$ has a steady-state distribution of the number of tokens in place A.

To calculate the steady-state probability distribution vector of the number of molecules in place A, which we denote by $\Pi=(\Pi(0),\Pi(1),\Pi(1),\ldots)$, we start from the CME equation given by (7.12), and we impose the equilibrium condition, which entails setting to zero the left-hand side, that is the derivative of the marking probability distribution, so that we obtain the following simpler system of linear equations:

$$\overline{0}=\Pi\cdot Q^{\mathbb{M}} \tag{7.15}$$

The steady-state marking distribution probability for model $\mathbb{M}$ is a solution to (7.15). However, it is immediate to notice that (7.15) has multiple solutions (for instance, the null vector $\overline{0}$ is a trivial one), hence we add to the linear system the following normalization condition for a probability measure $\sum_{i=0}^{\infty}\Pi(i)=1$.

By manipulating algebraically the infinite set of equations described in matrix form by (7.15) we can explicitly calculate the steady-state probability distribution, as we explain the following. Let us first describe the entries of the infinitesimal generator matrix $Q^{\mathbb{M}}=\|q_{i,j}^{\mathbb{M}}\|$, $i,j\geq 0$, which are as follows:

$$\left\|q_{i,j}^{\mathbb{M}}\right\|=\begin{cases} k_{\text{in}} & \text{if } i=j-1\\ j\cdot k_{\text{out}} & \text{if } i=j+1\\ -k_{\text{in}}-i\cdot k_{\text{out}} & \text{if } i=j\\ 0 & \text{otherwise}\end{cases} \tag{7.16}$$

Let us start with the equation for marking 0, which is as follows:

$$0=-\Pi(0)k_{\text{in}}+\Pi(1)k_{\text{out}} \tag{7.17}$$

From (7.17), we obtain that $\Pi(1)=\Pi(0)k_{\text{in}}/k_{\text{out}}$. The linear equation for the steady-state probability $\Pi(1)$ of marking 1 is as follows:

$$0=\Pi(0)k_{\text{in}}-\Pi(1)(k_{\text{in}}+k_{\text{out}})+\Pi(2)2k_{\text{out}} \tag{7.18}$$

From (7.18) and (7.17), we obtain $\Pi(2)=\Pi(0)(k_{\text{in}}/k_{\text{out}})^2/2$. The equation for any other marking i, $i\geq 2$ is as follows:

$$0=\Pi(i-1)k_{\text{in}}-\Pi(i)(k_{\text{in}}+i\cdot k_{\text{out}})+\Pi(i+1)(i+1)k_{\text{out}} \tag{7.19}$$

and it easy to check that the substitution process allows obtaining each $\Pi(i)$, $i > 2$ as a function of $\Pi(0)$, as follows:

$$\Pi(i) = \Pi(0)\frac{1}{i!}\left(\frac{k_{\text{in}}}{k_{\text{out}}}\right)^i \tag{7.20}$$

From the normalization condition we can obtain the value of $\Pi(0)$, as follows:

$$\Pi(0) + \sum_{i\geq 1}\Pi(0)\frac{1}{i!}\left(\frac{k_{\text{in}}}{k_{\text{out}}}\right)^i = 1 \implies \Pi(0) = \left[1 + \sum_{i\geq 1}\frac{1}{i!}\left(\frac{k_{\text{in}}}{k_{\text{out}}}\right)^i\right]^{-1}$$
$$\implies \Pi(0) = e^{-k_{\text{in}}/k_{\text{out}}} \tag{7.21}$$

Hence, we can calculate $E[A]$, the average steady-state number of proteins in the nucleus as follows:

$$E[A] = \sum_{i=0}^{\infty} e^{-k_{\text{in}}/k_{\text{out}}}\frac{1}{i!}\left(\frac{k_{\text{in}}}{k_{\text{out}}}\right)^i \cdot i = \frac{k_{\text{in}}}{k_{\text{out}}} \tag{7.22}$$

The result in (7.22) is well-known one. It turns out from the Poisson steady-state distribution of the number of molecules in the biochemical system modeled by the SPN in Fig. 7.7, and can be obtained from a deterministic continuous model based on ODEs of the same system. The steady-state variance Var$[A]$ of the number of molecules of A, which can not be determined through a deterministic model, is again given by Var$[A] = k_{\text{in}}/k_{\text{out}}$.

The purpose of this detailed modeling is to show that an analytical solution approach can be viable even for SPN models whose cardinality of the reachability set is infinite. Of course, in this particular example, the analytical treatment was greatly simplified by the simple structure of the CME, which allows solving the linear system in (7.15) by a straightforward substitution procedure. In the general case, provided that the size of the reachability set is finite, the solution to (7.15) can be carried out numerically. Many stable implementations of algorithms from linear systems exist, which can take advantage of sparseness or regularity properties of the infinitesimal generator.

It is worthwhile noticing that the possibilities of analytical and numerical solution of SPN models go well beyond the current capabilities of automated tools to support them. For instance, no general purpose tool for SPN is supporting the analytical derivation of the steady-state marking probability distribution for models that have an infinite cardinality of the reachability set. This is because checking the applicability of the analytical or numerical solution requires the identification of specific structural properties, a task that can hardly be automated.

The importance of analytical and numerical solution approaches lies in the quality of the results they can produce. The explicit availability of the exact (or of a reliable estimation) of the marking probability distribution allows calculating many measures of interest with the finest possible accuracy. However, these approaches in general have to cope with the state-explosion problem, namely the exponential growth of the reachability set cardinality. As we saw in the example above, even extremely small SPN models may have infinite reachability sets, and more complex models easily turn out to be intractable with these approaches. When analytical and numerical solutions are out of reach, simulation becomes the tool of choice, as we are going to explain in next section.

7.3.2 Simulation Approaches

As we saw in Sect. 7.3.1, directly attacking the analytical solution of the CME is a nontrivial task, which can be done only when the set of reactions is very small or it results in some structural property of the CTMC that can be exploited to simplify the mathematical treatment. Alternatively, the CME solution can be numerically approximated through the randomization method, but for this approach to be applicable the state space of the underlying CTMC must be finite and of limited size. A totally different way of computing the transient state probability distribution function described by the CME is through the computer-assisted generation of realizations of the stochastic process $\{\overline{x}_t\}_{t>0}$ ruling marking evolution, which can be done through a simulative approach.

Simulation is a powerful technique that allows estimating measures of interest of an SPN model without calculating explicitly the marking probability distribution. The basic idea of SPN simulation is to generate possible evolutions of the marking of the model. Starting from the initial marking $\overline{x}_0$ of the net, it is possible to determine which transition firing event will happen next in the model, and at what time. Thus, the next future marking of the SPN is chosen by updating the initial marking with the respective changes of tokens in the input and output places of the transition that was selected for firing, and the process repeats. In this way, a simulation of an SPN model determines a sequence σ of pairs $(\overline{x}_{t_i}, t_i)$, $i = 0, 1, \ldots,$ where $\overline{x}_{t_i}$ is the marking of the system at time t_i, and t_i is the time at which the SPN model reaches that marking. The sequence σ is called a *history* or a *trace* of model evolution.

An efficient and exact simulation algorithm for studying biological systems described by a CME was proposed by Gillespie in the already mentioned paper [130], and is known as the Stochastic Simulation Algorithm (SSA, hereafter). The success of SSA for the study of biochemical systems dynamics is well demonstrated by the huge number of studies, papers and computational tools based on it that have appeared since the original publication. The main reasons for this widespread acceptance stem from the simplicity of the SSA, which easily lends itself to straightforward (though not necessarily optimized) implementations, and from the clear link that is maintained with the intuitive descriptive language of chemical reactions.

The SSA algorithm provides a computational scheme for generating realizations of the Markovian stochastic process described by a CME. The SSA is exact, in the sense that it generates only possible realizations of a CTMCs with correct probability. The exactness we are talking about here has in fact nothing to do with whether Hypotheses 1, 2 and 3 are valid or not for a biochemical system. Assuming that these hypotheses hold of a biochemical system described as a set of reactions, the SSA provides a way to explore the dynamic evolution of the system, without requiring any additional assumption and without introducing any further approximation.

We then assume that the firing times of each transition tr_j, $j = 1, 2, \ldots, M$ in the SPN model follow a negative exponential distribution whose parameter is determined by the propensity function $a_j(\overline{x})$, and that ν_j is the marking change vector, that is, the column of the SPN incidence matrix that specifies how many tokens are removed from the input places and how many tokens are added to the output places of transition tr_j, $j = 1, 2, \ldots, M$. Then, the following steps describe the core of the *direct method* algorithm proposed by Gillespie in 1976 to generate traces of the form $\sigma = (\overline{x}_{t_i}, t_i)$, $i = 1, 2, \ldots$ for the evolution of the SPN over time.

(0) $t = 0$;
(1) $\overline{x} =$ initial marking of the SPN model;
(2) Out($\overline{x}$,0);
(3) while (TRUE) do
(4) $\quad a_0(\overline{x}) = \sum_j a_j(\overline{x})$;
(5) $\quad u =$ Uniform[0, 1];
(6) $\quad \tau = -\ln(u)/a_0(\overline{x})$
(7) $\quad v =$ Uniform[0, 1];
(8) $\quad j = \mathrm{Min}\{h \mid \sum_{i=1}^{h} a_i(\overline{x}) \geq v \cdot a_0(\overline{x})\}$
(9) $\quad t = t + \tau$;
(10) $\quad \overline{x} = \overline{x} + \nu_j$
(11) $\quad$ Out($\overline{x}, t$);
(12) endwhile

As specified in step (6), the time τ to the next reaction is an exponential random variable of mean $1/a_0(\overline{x})$, where $a_0(\overline{x})$ is the total propensity, calculated as per step (4). The sample of the negative exponential random variable is computed through the cumulative function inversion method using a [0, 1] uniform random number generated at step (5). The probability that next transition to fire is tr_j, $j = 1, 2, \ldots, M$ is $a_j(\overline{x})/a_0(\overline{x})$, and step (8) selects the index of the transition to fire according to their weighted propensities. Step (2) outputs the initial marking of the SPN model at time 0, and step (11) outputs each subsequent pair $(\overline{x}, t)$ generated by the simulation algorithm. The while loop would continue forever until a termination condition is added, which normally is given either in terms of a bound on the simulation time, or a bound on the number of firings.

The SSA scheme has given rise to a family of computationally efficient algorithms for simulating CTMCs that represent the evolution of a set of coupled biochemical reactions. These algorithms are much more efficient than the general event-driven simulation algorithms, which also apply to CTMCs. A first variant of the direct method algorithm was already proposed by Gillespie in 1976, called the *first reaction* method. According to the first reaction simulation algorithm, at each simulation cycle one uniform random number is selected for each of the enabled M transitions, and which reaction is to fire is determined by the minimum of the putative firing time (straightforward race-policy). Obviously, this approach is costlier than the direct method in computational terms, because at some steps may require generating several random numbers, however it saves the cost of computing the sum in step (8) of the direct method algorithm. A more efficient version was proposed in

2000 by Gibson and Brucke [127], called the *next reaction* method. In the next reaction algorithm, the putative times of the enabled transitions are saved in an indexed binary tree so that the minimum is always at the top. Moreover, a dependency graph is used to keep track of the coupling (in other terms, of dependencies) among transition firing times to determine when putative times in the tree have to be re-sampled. Other variants were produced in 2004 (*modified direct* method, [55]) and in 2006 (*sorting direct* method, [256]).

A direct inspection of the simulation trace provides the input data to estimate the measures of interest. For instance, if we were interested in determining the number of molecules of a species A at time $t > 0$, we would need to focus on the element of the trace whose time τ is the larger one such that $\tau \leq t$, and extract the value x_A from the marking $\overline{x}_\tau$. Hence, a simulation approach circumvents the state-explosion problem that plagues the analytical and numerical approaches by looking at a single realization of the marking stochastic process $\{\overline{x}_t\}_{t\geq 0}$.

As the firing of transitions is determined by the pseudo-random number sequence generated during simulation, different simulation runs starting with different seeds of the random number generator lead to different simulated traces of the model. Therefore, when using simulation to estimate measures of interest of an SPN model, multiple runs are necessary. Suppose for instance that we are interested in determining the average value of the number of molecules of species A at a certain time point $t > 0$. Each simulation trace only provides a single sample of the value x_A at time t, and multiple simulation traces are to be obtained to determine the distribution over the possible values of the number of molecules of A. Therefore, a number N of simulation is to be performed, each one providing a sample $x_{A,i}$ of the number of molecules at time t, $i = 1, 2, \ldots, N$, and the average value $E[A]_t$ is to be determined by averaging the samples as follows:

$$E[A]_t = \sum_{i=1}^{N} \frac{x_{A,i}}{N} \tag{7.23}$$

A different approach can be used when we are interested in steady-state measures. In this case, a single simulation run can be used to collect multiple samples. Care must be put in ensuring that simulation has indeed reached an equilibrium behavior and that the biasing effect of the initial marking has been eliminated.

A crucial point here is the determination of how many samples are necessary to compute a reliable estimation of the measure of interest. This matter is very important and has been studied with great attention. The interested reader may see [8] for a more complete treatment of the topic, here we just provide the basic guidelines on how to proceed in obtaining reliable estimates. Again, statistics supports us in this task. From the samples, besides the estimation of the average value, it is possible to compute the sample variance, which provides an indication of the spread of the distribution of the number of molecules of species A at time t. Intuitively, the larger the variance, the wider is the set of possible values of the molecules of A at time t. Consequently, the number of samples required to provide statistically meaningful estimations of the measure of interest grows with the variance. Operatively, together with the estimated averages it is common practice in simulation to calculate *confidence intervals* for the estimations. Confidence intervals are computed for selected confidence levels, usually 95%, 98% or 99%. A confidence interval $[\tilde{E}[A]_t - \delta, \tilde{E}[A]_t + \delta]$ estimated for the estimated average $\tilde{E}[A]_t$ at the confidence level 95% indicates that the exact value $E[A]_t$ (which is indeed unknown) is within an interval centered at $\tilde{E}[A]_t$ and of amplitude δ with probability 0.95. The amplitude of the confidence level is proportional to the squared root of the sample variance (the standard deviation), and inversely proportional to the squared root of N, the number of samples. Therefore, increasing the number of samples reduces the statistical uncertainty expressed by the confidence interval width. A standard approach commonly used in stochastic simulation is to generate enough samples to reduce the relative width of the confidence interval to 10% of the estimated value. The selection of the confidence level also affects the width δ of the confidence interval. For a given number of samples, the higher the confidence level, the larger δ.

Finally, we want to attract the attention of the reader to the basic concept of statistical independence of the samples. When computing statistics of measure of interest, the various samples used must be collected in a way to guarantee their statistical independence. Let us observe that any two pairs $(\overline{x}_{t_i}, t_i)$ and $(\overline{x}_{t_j}, t_j)$, $t_j > t_i$, from the same simulation trace σ are in fact not independent, because there is clearly a correlation between the fact that at time t_i the marking of the SPN model was $\overline{x}_{t_i}$ and the fact that later at time t_j the marking is $\overline{x}_{t_i}$. Intuitively, this correlation decreases as the distance in time between the two samples increases. Therefore, spacing enough observations of the marking allows collecting multiple samples from inside the same simulation run. This is important when we are interested in evaluating steady-state values of the measure of interest. Instead, when we want to collect samples from the transient behavior of a model, different simulation runs must be used, each one fed by a distinct sequence of pseudo-random numbers.

7.4 Examples of Modeling and Evaluation

This section provides a few examples of SPN models of biological systems, to help the reader fixing the most important concepts about model specification and evaluation and to provide further hints on the modeling capabilities of the SPN formalism.

7.4.1 A Biochemical Oscillator

A very simple and well known chemical system is the auto-catalytic one known as Lotka–Volterra, whose formulation was first proposed by Lotka and then independently modeled by Volterra [236]. This system is interesting because it shows how even simple SPN models of biological systems can give raise to unexpectedly complex behaviors.

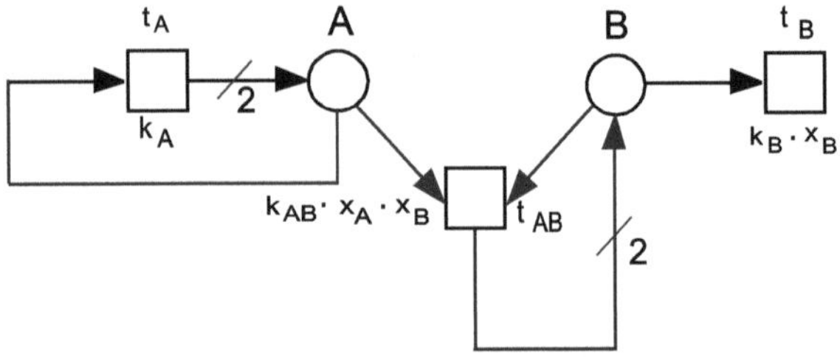

Fig. 7.8 SPN model for the Lotka–Volterra auto-catalytic system. According to the standard graphical notation of SPN models, small dashes and an integer are depicted on those arcs that have a weight different from 1. Molecules of A are produced through a self-catalytic process represented by the firing of transition t_A. Each firing of transition t_A consumes one token from place A and deposits two tokens in place A, with a net production of one new molecule. Firing of t_A is disabled if the number of tokens in place A reaches 0. Similarly, transition t_{AB} models the self-catalytic production of molecules of B, which consumes one token from place A and from place B to produce a net increase of one token in place B. Again, if the number of tokens of either places A or B becomes 0, the transition gets disabled

The system consists of only two species A and B. Species A is produced through an auto-catalytic reaction of the form $A \to 2A$, whose rate constant is $k_A = 20.0$, whereas species B is produced by an auto-catalytic reaction in which one molecule of B converts one molecule of A into another molecule of B, in a bimolecular reaction of the form $A + B \to 2B$, whose rate constant is $k_{AB} = 0.01$. Each molecule of species B is degraded at a rate $k_B = 10.0$. We assume that Gillespie's hypotheses about thermal equilibrium, fast diffusion and speed of bimolecular reactions are satisfied by the Lotka–Volterra system, so that an SPN model of the system will provide an accurate representation of system dynamics.

The SPN shown in Fig. 7.8 models the Lotka–Volterra system described above. Notice the usage of the weight equal to 2 on the output arcs of transitions t_A and t_{AB}, which models the stoichiometry of reactions $A \to 2A$ and $A + B \to 2B$.

We assume that the initial marking of the net at time $t = 0$ is $(x_A = 100, x_B = 100)$. The set of possible markings of the SPN model is infinite, as there is not a limit on the number of molecules of species A and B in the system. Also, there is not an obvious structure of the infinitesimal generator of the stochastic process underlying the model that makes it possible to compute analytically the transient marking probability distribution. Therefore, we resort to simulation to estimate the measures of interest. To grab an idea about the SPN marking evolution, we first inspect the evolution of the number of molecules of B, as produced by a single run of simulation. The measure x_B, that is, the number of tokens in place B, in the time window $[0, 32]$ is plotted in the chart shown in Fig. 7.9.

The chart demonstrates a very interesting oscillating behavior of the number of molecules over time. Around 60 oscillations cycles are completed in the considered time window. Moreover, at $t \approx 31.1$ the oscillations stop abruptly, because the number of molecules of species B becomes 0, and transition t_{AB} can not be enabled anymore. A chart of the number of molecules of species A would also show a similar behavior. In this case as well, should the number of molecules of A reach 0,

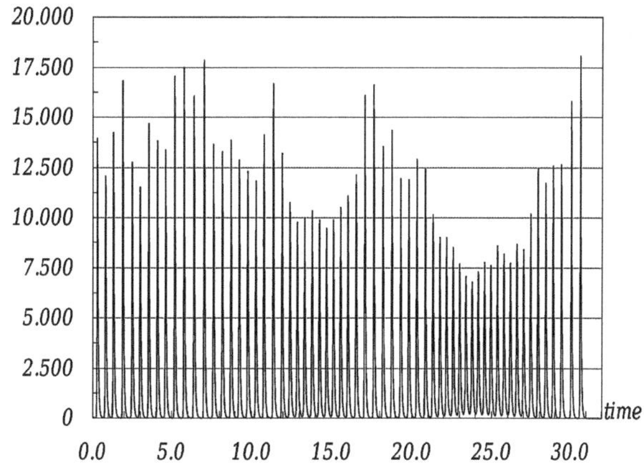

Fig. 7.9 Simulated number of molecules of species B. Time is over the horizontal axis, simulated number of molecules on the vertical one. The oscillations stop at around 31.1 time units. Notice that the frequency of oscillation appears to show much less variability than its amplitude

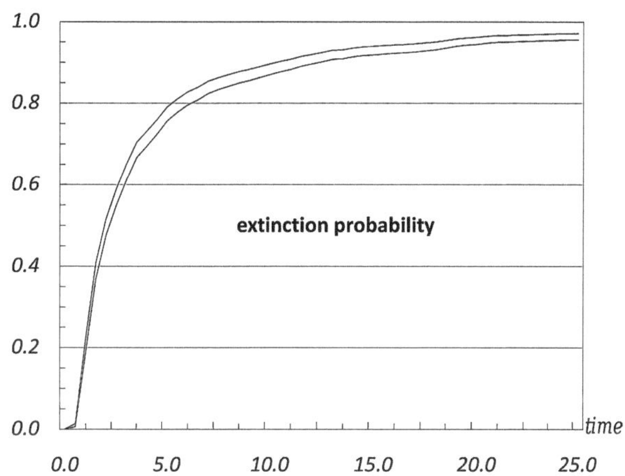

Fig. 7.10 Extinction probability of species B over time, as obtained with 2,000 simulation runs at a confidence level of 95%. The average simulated value is not shown, only the upper and lower limits of the confidence intervals for the simulated measure. Extinction is very likely ($\geq$0.95) after 25 time units

transition t_A would not fire anymore. However, if B is the first species to reach 0, t_A would continue to be enabled, and x_A would grow continuously during simulation.

This behavior is constantly shown across different simulation runs, with the notable difference that the time at which oscillations stop is random. From model simulation, we can also estimate the probability that the oscillation stops within a certain time, that is we can compute the distribution of species B *extinction* time. We show in Fig. 7.10 the upper and lower extremes of the confidence interval (95% confidence level) for the distribution of the extinction time of species B, computed through 2,000 runs of simulation.

As it can be observed from the plot in Fig. 7.10, the probability of species b getting extinct before $t = 25.0$ is fairly high, around 0.95, which tells us that the simulated trace in Fig. 7.9 is indeed an unlikely one, as the oscillation stops at around $t = 31.0$. It is interesting to observe that the finite duration of the oscillations is a property that can not be studied through a continuous deterministic interpretation of the set of reactions defining this instance of the Lotka–Volterra system [161].

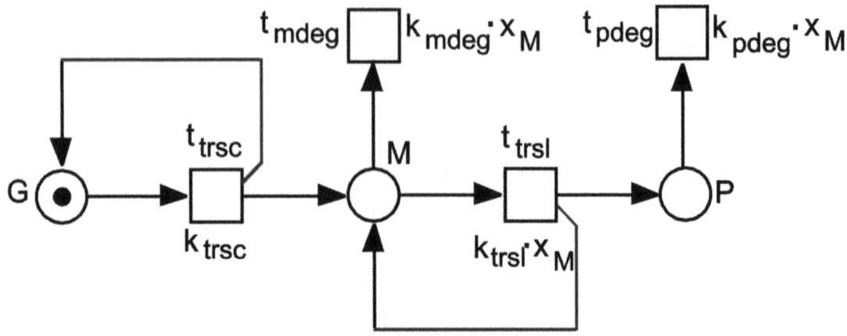

Fig. 7.11 SPN model of the gene→mRNA→protein synthesis chain. Transition t_{trsc} models the transcription process. It adds tokens to the place M, whose marking represents the number of mRNA molecules. Molecules of mRNA are removed from place M by the firing of transition t_{mdeg}, modeling mRNA degradation. Transition t_{trsl} models the translation of mRNA molecules into protein molecules, whose number is given by the number of tokens in place P. Transition t_{pdeg} models protein degradation. The number of tokens in place M at steady-state follows a Poisson distribution of parameter $k_{\mathrm{trsc}}/k_{\mathrm{mdeg}}$. Under equilibrium conditions, the average translation rate is $k_{\mathrm{trsl}}k_{\mathrm{trsc}}/k_{\mathrm{mdeg}}$, and thus the average number of tokens in place P is given by $(k_{\mathrm{trsl}}k_{\mathrm{trsc}})/(k_{\mathrm{mdeg}}k_{\mathrm{pdeg}})$

7.4.2 A Protein Synthesis Network

Let us now consider a different example system, which consists of a gene expression network. We consider a gene G coding for protein P. The synthesis chain gene→mRNA→protein can be modeled by the SPN model shown in Fig. 7.11, where place G contains exactly 1 token, modeling the gene, and tokens contained in places M and P (initial marking set to 0) represent molecules of mRNA and protein P, respectively. Transition t_{trsc} and t_{trsl} model the gene transcription process and the mRNA translation process, respectively, whereas transition t_{mdeg} and t_{pdeg} represent the degradation of mRNA and of protein molecules, respectively.

We assume that at time $t = 0$ no molecules of mRNA and protein are present and we are interested in computing the number of protein molecules at equilibrium. This can be done fairly easily through analytical considerations. Indeed, notice that the sub-model consisting of places G and M and of transitions t_{trsc} and t_{mdeg} is structurally equivalent to the one shown in Fig. 7.7. Therefore, the number of mRNA molecules will follow in equilibrium a Poisson distribution of parameter $k_{\mathrm{trsc}}/k_{\mathrm{mdeg}}$. This means that, in steady state, the average number of mRNA molecules is given by the Poisson distribution average value, which is again $k_{\mathrm{trsc}}/k_{\mathrm{mdeg}}$. In turn, this implies that the average rate of firing of transition t_{trsl} is $k_{\mathrm{trsl}}k_{\mathrm{trsc}}/k_{\mathrm{mdeg}}$, and thus, the steady-state average number of protein molecules in given by $k_{\mathrm{trsl}}k_{\mathrm{trsc}}/(k_{\mathrm{mdeg}}k_{\mathrm{pdeg}})$. The rate constant values for the SPN model in Fig. 7.11 are as follows: $k_{\mathrm{trsc}} = 0.09$, $k_{\mathrm{mdeg}} = 0.01$, $k_{\mathrm{trsl}}0.05$, $k_{\mathrm{pdeg}} = 0.009$, which result in a number of steady-state protein molecules $k_{\mathrm{trsl}}k_{\mathrm{trsc}}/(k_{\mathrm{mdeg}}k_{\mathrm{pdeg}}) = 50$.

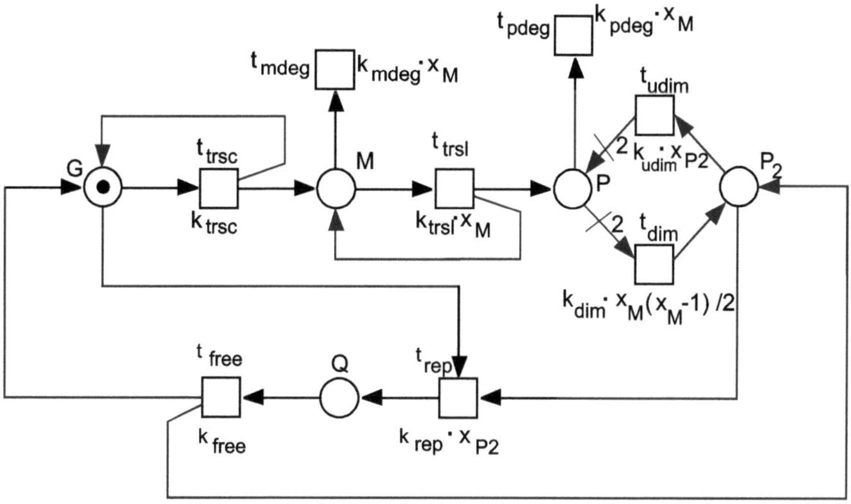

Fig. 7.12 SPN model of cooperative repression. A reversible dimerization process of molecules of protein P is modeled by transitions t_{dim} and t_{udim}. Notice the usage of non-unitary weights on the incoming and outgoing arcs of place P. Two tokens are removed from place P at each firing of transition t_{dim}, and two tokens are added to P at each firing of transition t_{udim}. Tokens in place P_2 represent molecules of the dimer, which can bind the gene and suppress transcription. This reversible binding is modeled by the pair of transitions t_{rep} and t_{free}

7.4.3 A Gene Regulatory Network

Let us now consider a gene regulation mechanism for the example system modeled above, where the synthesis of protein P is negatively regulated by a dimer P_2 of the same protein. The dimer works as a repressor, binding to a DNA region close to the gene and contrasting its transcription. This example is a slight adaptation of the one found in [15] for the regulation of protein cI in phage λ.

In the SPN model in Fig. 7.12, molecules of protein P dimerize through a reversible reaction (transitions t_{dim} and t_{udim}) to form molecules of a stable dimer species represented in the model by tokens contained in place P_2. Notice the integer weights assigned to the input arc to transition t_{dim} and the output arc of transition t_{udim} to correctly account for the molecule number change in dimerization and undimerization events. The dimer molecules can reversibly bind DNA (transitions t_{rep} and t_{free}) repressing the gene. The state of the repressed gene is represented by the presence of a token in place Q. Apart from place G, all places of the SPN model are initially empty.

The dimerization process modeled by transition t_{dim} results in a reduction of the effective rate of firing of transition t_{trsc} and the model in Fig. 7.12 reaches an equilibrium between protein production and protein degradation different from the one we estimated for the model in Fig. 7.11. Figure 7.13 shows the simulated transient average levels of protein P over the time window $[0, 2000]$ for three different sets of rate constant values, all sharing the same values $k_{\text{trsc}} = 0.09$, $k_{\text{mdeg}} = 0.01$,

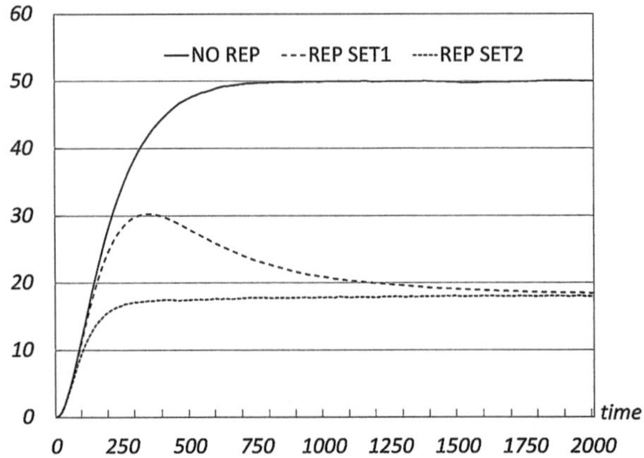

Fig. 7.13 Simulated time courses for the time-dependent number of protein molecules in the gene regulation network of Fig. 7.12, for three different set of model parameters. The number of simulation runs is 10,000; confidence intervals, computed at 95% confidence level, are extremely small and therefore not shown in the plots for the sake of readability. In set NO-REP, the dimerization process is suppressed, and thus the amount of protein molecules reaches the predicted steady-state level $k_{trsl}k_{trsc}/(k_{mdeg}k_{pdeg}) = 50$. In set REP-SET1, the gene repression effect is able to bring down the equilibrium level of protein P to around 18 molecules. This same equilibrium value is achieved by the model with set REP-SET2, in which the dimerization process proceeds ten times faster. As a result the gene expression is controlled in a more efficient way, and the transient peak of the number of protein molecules visible in the plot for set REP-SET1 disappears

$k_{trsl} = 0.05$, $k_{pdeg} = 0.009$ (no change with respect to the model in Fig. 7.11), and $k_{rep} = 0.05$, $k_{free} = 0.9$. The simulation results are obtained with 10,000 runs of simulation, and the confidence intervals, computed at a confidence level of 95%, are within 0.5% of relative width (not shown for the sake of clarity). A first set (plot NO-REP), with no repression, reaches the expected steady-state level of protein equal to 50 molecules. In a different rate constant assignment (set REP-SET1), we use the values $k_{dim} = 0.0001$ and $k_{udim} = 0.0005$, whereas in the third one (set REP-SET2) we use $k_{dim} = 0.001$ and $k_{udim} = 0.005$. As it can be observed from Fig. 7.13, both REP-SET1 and REP-SET2 result in a reduction of the steady-state level of protein P. Moreover, because the ratio k_{dim}/k_{udim} does not change between the two sets, the equilibrium level of the dimer is the same in the two simulated time courses, and consequently the equilibrium level of protein P is also the same. The difference is in the speed with which the steady-state level is reached, as well as in the profile of the time course, which achieves a peak in the plot for REP-SET1 (slow repression) not visible in the plot for REP-SET2.

7.5 Related Work

Application of the Petri nets modeling formalism to the study of biological systems only started in recent years, but it already generated a variety of different areas of active research. Modeling studies based on the quantitative evaluation of various timed variants of Petri nets have also been appearing. The first works exploiting the representative power of Stochastic Petri nets date to the end of the 90s. The work in [140] is more a kind of advertisement for the existence of an intuitive graphical modeling tool that can represent both structure and stochastic dynamics of biological networks. In [141], the same authors present a more detailed application case dealing with a genetic regulation network.

The appearance of these works basically stem from the fruitful combination of Gillespie's formulation of stochastic kinetics to a well-developed core of Petri net research results already available at that time. Indeed, the computer science and engineering research communities had been working on Petri net since the mid 70s, building up a consistent body of knowledge about analytical and simulative approaches to the evaluation of SPN models. For instance, the two research works mentioned above [140, 141], exploited the modeling and simulation capabilities of the stochastic activity networks tool Möbius [77], whose first releases were already available in mid 80s. Other popular SPN tools developed outside the systems biology community are the Stochastic Petri Net Package [76] and GreatSPN [74]. Another tool supporting the Petri net modeling formalism, which has a good record of applications to the modeling and evaluation of biological systems is the Snoopy tool [154].

The work in [369] proposes another application of SPNs to the modeling of the *Escherichia coli* stress circuit, and [399] presents the modeling and evaluation of the estrogen production pathway, again using the Möbius tool. In [21], the authors use colored stochastic petri nets to model dynamics of epidemic diseases spread.

The publications in [149, 295] are the first surveys about possible applications of Petri net to systems biology, covering both qualitative and quantitative analysis opportunities. This type or papers demonstrate the growing maturity of this new application field, which leads to the development of original approaches, seeking for extensions of the representative power, definition of model refinement approaches and attempting to combine both structural analysis and dynamic evaluation, clearly a must in a domain were the complexity of systems can easily overcome the capabilities of modeling tools.

The work in [244] uses Petri net models at various levels of abstraction for reconstructing the plausible regulatory networks of the plasmodium *Physarum polycephalum*, and employs simulation to incrementally validate (or disprove) candidate models against experimental data coming from various mutants. In [154], an attempt to reconciliate the qualitative and quantitative (discrete and continuous) modeling paradigms in Petri net is presented, using the classical example provided by the ERK/MAPK pathway. This paper shows well how the debate about the respective merits/disadvantages of diverse modeling approaches is received in the Petri net community. In [270], the authors deal with the protein interaction network that regulated cell-cycle in yeast *Saccharomyces cerevisiae*, and explain how it is possible to augment the predictive capabilities of continuous deterministic models expressed as systems of ordinary differential equations by translating them into SPNs, with an informal, semi-automatic mapping process. That work makes clear the need for transition rates that go beyond pure mass-action, as a way to encode into SPNs arbitrary abstractions. A further extension, which will push the modeling formalism outside the class of SPNs, is the one advocate by [156], which calls for both nonmass action transition rates as well as additional firing distributions types. This last extension greatly extends the expressivity of models, but of course precludes any possibility of analytical treatment, leaving simulation as the only technique of choice.

7.6 Summary

We presented in this chapter a practical and concise introduction to the concepts underlying the stochastic modeling of biological systems with Petri net. The stochastic modeling approach has been gaining momentum after the widespread acceptance of the important contribution provided by Gillespie, thanks to which the Stochastic Petri net modeling formalism has found a novel application field.

Through some simple examples of biochemical systems, we described in this chapter the SPN model specification process, in terms of a direct mapping between entities and their states into places, reactions and kinetic rates into arcs and transitions, stoichiometry into arc weights. We also took care of discussing how and when the stochastic interpretation of the firing in Stochastic Petri net can be reconciled with the stochastic kinetics model proposed by Gillespie, a topic often disregarded.

The most relevant approaches for Stochastic Petri net models evaluation have been presented. Analytical and numerical evaluation of models in terms of their transient and steady-state marking probability distribution were described, to provide examples of the existing exact analysis opportunities. Then, we introduced the Stochastic Simulation Algorithm, which has become a de-facto standard in simulation approaches for biological systems. The logic of simulation and the principles underlying statistical aggregation of simulation samples have been presented.

Additional examples of larger scale models have been defined and evaluated at the end of the chapter, to consolidate the understanding of the main concepts and provide further hints on the representative power and analysis tools made available by the Stochastic Petri net modeling formalism.

7.7 Problems

We propose in this section some SPN modeling and evaluation exercises to help the reader putting into practice the concepts introduced in the chapter.

7.1 Consider a biochemical species A that can be in two different states, active and inactive. Inactive molecules of A become active at a rate k_{on}, whereas active molecules of A revert to the inactive state at a rate k_{off}. The total amount of molecules of A stays constant, and is A_0. Define an SPN model for this system.

7.2 Consider the following additional information for the system in Exercise 1. Activation of A is caused by the addition of a phosphate group by a kinase KA, and deactivation of A is caused by the removal of a phosphate group by a phosphatase PA. Define an SPN model that includes this information.

7.3 For the SPN model built in Exercise 1, determine analytically the average and variance of the number of active molecules of species A at equilibrium. Hint: follow the same steps used to calculate (7.22).

7.4 For the SPN model in Fig. 7.11, determine new values of the rate constants for transitions t_{mdeg} and t_{trsl} so that the average number of proteins at steady state is still 50, but the average number of mRNA molecules is 5.

7.5 Evaluate, by simulation, the variance of the steady-state number of proteins in the example model worked out in Fig. 7.11 and the modified one defined in Exercise 4.

7.6 Modify the example provided in Fig. 7.12 to account for another additional negative feedback loop exerted by tetramers $P4$ of protein P. Add a reversible tetramerization reaction $P2 \rightarrow P4$, and a reversible binding between $P4$ and G.

7.7 Evaluate by simulation the equilibrium level of protein P in the modified model worked out in Exercise 6. Compare it with the result of the original example model in Fig. 7.12. Assume that the rate constant for the tetramerization reaction $P2 \rightarrow P4$ is $2k_{\mathrm{dim}}$, whereas the rates of the reverse reaction $P4 \rightarrow P4$ is the same as k_{udim}. Finally, consider a different rate constant value assignment where the rate of DNA binding of the tetramer is $5k_{\mathrm{rep}}$, whereas the unbinding rate value is the same as k_{free}.

Chapter 8
Quantitative Analysis

Jörg Ackermann and Ina Koch

Abstract The final aim of modeling biochemical processes is to gain a theoretical model which explains and predicts the dynamic behavior of the system in terms of quantities. The limitation of this type of modeling lies rather in the lacking of necessary kinetic data than in the mathematical concepts which are mostly based on coupled ordinary differential equations (ODEs). Whereas kinetic data can be found for some reactions, for the vast majority of pathways kinetic data have not been identified. For many biochemical processes, it still is a task to produce significant experimental data. Continuing efforts in well-designed experiments and data analysis have made kinetic data available for some pathways and some organisms, and with these data at hand quantitative methods become more and more useful. All quantitative methods, applied in modeling of biochemical processes, can easily be adapted to the Petri net formalism. The Petri net formalism offers the advantage of a combination of methods of classical systems biology with discrete Petri net modeling techniques, including an intuitive description of biochemical networks.

The aim of this chapter is to provide an introduction to basic methods for quantitative modeling of biochemical networks and a description in terms of the Petri net formalism. This includes for example, the classical principles of chemical reaction kinetics, the mass action, steady-states, stability and bifurcation analysis, Michaelis Menten kinetics, and Hill kinetics. Moreover, we provide extensive references for further reading and give references to standard tools in this field.

J. Ackermann (✉)
fluIT Biosystems GmbH, Mikroforum-Ring 1, 55234 Wendelsheim, Germany
e-mail: joerg.ackermann@fluit-biosystems.com

I. Koch
Max Planck Institute for Molecular Genetics, Ihnestrasse 73, 14195 Berlin, Germany
e-mail: ina.koch@molgen.mpg.de

I. Koch
Institute for Computer Science, Johann Wolfgang Goethe University Frankfurt am Main, Robert-Mayer-Strasse 11-15, 60325 Frankfurt am Main, Germany
e-mail: ina.koch@bioinformatik.uni-frankfurt.de

I. Koch et al. (eds.), *Modeling in Systems Biology*, Computational Biology 16,
DOI 10.1007/978-1-84996-474-6_8, © Springer-Verlag London Limited 2011

8.1 Introduction

Quantitative modeling plays a crucial role for the interpretation of experimental data in biology. Petri net formalism is classically applied to qualitative analysis and considers discrete entities called tokens. Quantitative analysis in system biology, on the other hand, is based on reaction rates and continuous concentrations. To convert Petri nets to quantitative models, tokens have to carry real number values and represent concentrations. Transitions have to describe the conversion of concentrations of substances at the pre-places to concentrations of the products at the post-places. The firing rules are no longer simple discrete events, but contain functions to describe chemical reactions. Depending on the kinetic model various functions for reaction rates have to be applied. Petri net formalism does not provide new kinetic concepts, but is useful through the possible combination of discrete with continuous modeling at different levels of abstraction as applied in hybrid Petri nets, see Chaps. 3 and 6.

In the modeling of biological systems, Michaelis–Menten kinetics and Hill kinetics are of great importance to describe enzymatic reactions and allosteric effects, respectively. Both of these kinetic concepts are based on the law of mass action. The derivation of the corresponding coupled differentiation equations is known for many decades, and their concepts are described in detail in many textbooks, for further reading see textbooks by Atkins [18], Voet and Voet [405], and Stryer et al. [378]. For numerical treatment of large reaction systems and parameter identification, see [91, 203, 376]. Since mathematicians and computer scientists are usually not very familiar with chemical reaction kinetics, we start with the description of the concepts of kinetics for biochemical systems providing the translation to the Petri net formalism. Let us first consider the *Law of Mass action* and molecular reaction types.

8.2 Mass Action

Chemical reactions describe the conversion of chemical substances which are called *reactants*, *educts*, or *substrates* in case of catalyzed reactions. The resulting chemical substances are the *products*. Reactions can take place spontaneously, in which case they are called to be *exothermic*. Otherwise, reactions are *endothermic*. According to the number of participating reactants two fundamental types of chemical reactions can be distinguished, unimolecular reactions and bimolecular reactions.

The *Law of Mass action* states that the rate of conversion of masses in generic chemical reaction is proportional to the product of the masses of the reacting substances. For a generic reaction

$$mA + nB \xrightarrow{k} C, \tag{8.1}$$

with reactants, A and B, and stoichiometry coefficients, m and n, the reaction rate is given by

$$r = k[A]^m[B]^n, \tag{8.2}$$

where r denotes the reaction rate, $[A]$ the concentration of Substance A, and k the reaction constant.

Mass action describes the behavior of reactants and products in an generic chemical reaction as an equation where the velocity or rate of a chemical reaction is directly proportional to the concentration of the reactants. It offers the advantage of well-developed rigorous methods and powerful numerical tools at hand. When coming to biochemical reaction, either in vivo or in vitro, the modeler must be aware that essential effects may result from the low number of copies of proteins, and that such effects are ignored in an approach based on mass action kinetics.This law is useful to obtain the correct equilibrium equation for a broad range of types of reactions. In the case of dynamic studies, however, the expressions for the rate is applicable to elementary reactions only. The following elementary reaction types can be distinguished.

8.2.1 Zero-Order Reactions

The simplest type of reactions are zero-order reactions in which the reaction rate does not depend on the concentrations of reactants. Such type of reactions are used to model the synthesis from a null species or to add a source species to the system

$$NULL \longrightarrow B. \tag{8.3}$$

The reaction rate is constant and the reaction rate coefficient k is given in units of $mole\ litre^{-1}\ second^{-1}$. In the case that a null species with constant concentration is explicitly modeled, the reaction rate coefficient k must be defined in units of $second^{-1}$.

8.2.2 First-Order Reactions or Unimolecular Reactions

The first-order or unimolecular reaction describes the elementary conversion of an *unstable* molecule of type A to a molecule of type B

$$A \longrightarrow B. \tag{8.4}$$

The conversion may stand for various complex processes, as for example, chemical modifications or structure changes, whose details are intentionally ignored or unknown. The molecule may have become unstable because temperature or pH-value of the environment have changed, or the reaction may be part of a larger reaction system which produces unstable molecules of type A. The classical example of a first-order reaction is radioactive decay. The reaction rate is proportional to the concentration $[A]$ of the reactant and the reaction rate coefficient k is given in units of $second^{-1}$.

The conversions of the individual molecules of type A are not synchronized. Instead, the transition of a molecule of type A to a molecule of type B is independent of other transitions and takes place with a certain fixed probability. We do not want to distinguish individual molecules of type A and type B and, instead, count only the number of molecules of type A and number of molecules of type B, denoted in the following by integers N_A and N_B, respectively. On the description level of the counts N_A and N_B, the conversion of each individual molecule of type A and B according to (8.4) represents a Markov process

$$N_A \longrightarrow N_A - 1, \tag{8.5}$$

$$N_B \longrightarrow N_B + 1. \tag{8.6}$$

Since the conversion of each individual molecule of type A is not influenced by the conversion of other molecules in the system this process occurs with high probability if N_A is high and with low probability if N_A is low. Quantitatively, the transition probability for the Markov chain above can be written as

$$P_{(A,B)\rightarrow(A-1,B+1)} = kN_A, \tag{8.7}$$

where k is a positive and real constant. Consequently, during a small time period δt the average change of the numbers N_A and N_A, respectively, is described by

$$\delta N_A = -kN_A\delta t, \tag{8.8}$$

$$\delta N_B = +kN_A\delta t. \tag{8.9}$$

In chemical reactions, a huge amount of molecules is participating, for example, one gram of matter contains up to 10^{23} particles. The unit for substance amount is *mole* which is defined as the number of atoms contained in 12 gram of the isotope ^{12}C. The Avogadro number $N_A = 6.02214179(30) \times 10^{23}$ gives the exact number of elementary entities in one mole [232].

The absolute number of individual protein inside a living cell may be below thousand copies. This number is far less than the Avogadro number mentioned above. Global cytoplasmic concentrations are found to range from 0.04 (formin Cdc12p) to 63 micromolar (actin). The proteins are not homogeneously distributed in a cell but concentrate up to 100 times in contractile rings and 7500 times in spindle pole bodies at certain times in the cell cycle [164, 419].

To derive equations valid for arbitrary dimension of the reaction chamber, we divide the equations by the volume V of the system. The limes of infinitesimal small time period δt yields a system of ordinary differential equations for the concentrations $[A] =_{\text{def}} N_A/V$ and $[B] =_{\text{def}} N_B/V$

$$\frac{d[A]}{dt} = -k[A], \tag{8.10}$$

$$\frac{d[B]}{dt} = +k[A]. \tag{8.11}$$

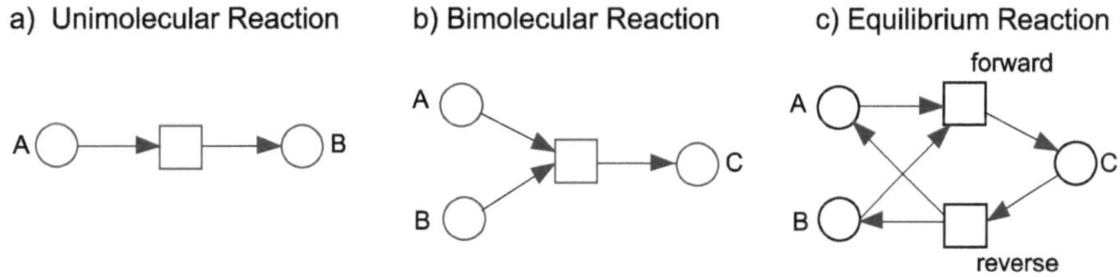

Fig. 8.1 The Petri net representations for (**a**) a unimolecular reaction corresponding to (8.4), (**b**) a bimolecular reaction corresponding to (8.13) and (8.14), and (**c**) an equilibrium reaction corresponding to (8.18). Equilibrium reactions are represented by two transitions, one for the forward and one for the reverse reaction

Concentrations are given in units mole/liter (also molar or M), denoting the number of molecules (in units of N_A) in one liter of solution. Values for concentrations are typically given in mM (millimolar, one thousandth of a molar), μM (micromolar, one millionth of a molar), and nM (nanomolar, one billionth of a molar). Values for the reaction rate constant k are in the order of fractions of a second up to several minutes depending on whether, for example, a combustion process or a complex biochemical reaction is modeled. Commonly, the notation

$$A \xrightarrow{k} B \tag{8.12}$$

is understood as a synonym for the ODE system (8.10)–(8.11), where k defines the rate constant coefficient for the speed of the chemical reaction.

Typical rate constants coefficients k of biochemical processes vary over a broad range. For example, in RNA folding processes, individual stem loops form on the microsecond time scale at physiological temperatures and folding of tRNA proceed with a time constant of about 0.1–1 s, whereas a folding of long RNA strands via multiple kinetic intermediates or via traps of misfolded three-dimensional structures can be surprisingly slow (several minutes) [388]. For the Petri net representation of a unimolecular reaction, see Fig. 8.1.

8.2.3 Second-Order Reactions or Bimolecular Reactions

Second-order or bimolecular binding reactions

$$A + B \xrightarrow{k} C \tag{8.13}$$

or

$$2A \xrightarrow{k} C \tag{8.14}$$

describe interactions between two entities of different type or of the same type in a reaction system. The reaction rates are then proportional to the concentrations of the two reactants, $r \sim [A][B]$, or to the square of the concentration of the one reactant, $r \sim [A]^2$, respectively.

The reaction (8.13) is equivalent to the system of ODEs

$$\frac{d[A]}{dt} = -k[A][B], \tag{8.15}$$

$$\frac{d[B]}{dt} = -k[A][B], \tag{8.16}$$

$$\frac{d[C]}{dt} = +k[A][B], \tag{8.17}$$

where the kinetic rate constant coefficient k now is given in units of litre mole^{-1} second^{-1}. A typical value for the rate constant is for example, $k = 10^8$ litre mole^{-1} second^{-1} for RNA-polymerase or reverse transcriptase reaction [36]. Note that, bimolecular reactions result in a nonlinear system of ordinary differential equations and, hence, are responsible for complex behavior as for example, explosion, oscillation, bifurcation, waves or spatiotemporal pattern formation. Unimolecular and bimolecular reaction terms are the elementary reaction types to model any mass action reaction system. For the Petri net representation of a unimolecular reaction, see Fig. 8.1.

8.2.4 Reversible Mass Action or Equilibrium Reactions

In principle, each chemical reaction is reversible. Whether the reaction prefers a conversion of mass to one direction or the reverse direction depends on its thermodynamic parameters. A reversible reactions can be equivalently described by two separate reactions or by a single "for and back" reaction.

In the notation of a single "for and back" reaction a bimolecular reaction going in forward and reverse direction is given by

$$A + B \underset{k_r}{\overset{k_f}{\rightleftharpoons}} C, \tag{8.18}$$

and yields the equation system

$$\frac{d[A]}{dt} = -k_f[A][B] + k_r[C], \tag{8.19}$$

$$\frac{d[B]}{dt} = -k_f[A][B] + k_r[C], \tag{8.20}$$

$$\frac{d[C]}{dt} = +k_f[A][B] - k_r[C]. \tag{8.21}$$

Starting with initial concentrations of species A, B, and C the reaction will go on until a kind of equilibrium, which is called *steady-state*, is reached. At steady-state the derivation of the concentration with respect to time becomes zero, that is,

$$\frac{d[A]}{dt} = \frac{d[B]}{dt} = \frac{d[C]}{dt} = 0. \quad (8.22)$$

Inserting this condition in the ODE above leads to the definition of the equilibrium constant

$$K_d =_{\text{def}} \frac{k_r}{k_f} = \lim_{t\to\infty} \frac{[A]_t[B]_t}{[C]_t}. \quad (8.23)$$

The equilibrium constant, K_d, has the unit of a concentration and determines whether the reaction tends more toward the reactants or the products of the reaction. The range of typical values of K_d is 10^{-8} M to 10^{-11} M. By mass conservation, the relations follow

$$[A] + [C] = [A]_0, \quad (8.24)$$

$$[B] + [C] = [B]_0, \quad (8.25)$$

where $[A]_0$ and $[B]_0$ are the initial concentrations of the two binding partners. We apply the mass conservation condition to eliminate $[A]$ and $[B]$ from the equilibrium equation (8.23):

$$K_d = \frac{[A][B]}{[C]} = \frac{([A]_0 - [C])([B]_0 - [C])}{[C]}. \quad (8.26)$$

This leads to the quadratic equation

$$[C]^2 - [C]\big[K_d + [A]_0 + [B]_0\big] + [A]_0[B]_0 = 0 \quad (8.27)$$

with the solution

$$[C]_s = \frac{K_d + [A]_0 + [B]_0}{2} - \sqrt{\left(\frac{K_d + [A]_0 + [B]_0}{2}\right)^2 - [A]_0[B]_0}. \quad (8.28)$$

The fraction of bound species A, given by $[C]_s/[A]_0$, is plotted versus $[B]_0$ in Fig. 8.2 for $K_d = 10^{-9}$ M and $[A]_0 = 10^{-10}$ M. For growing concentration of the binding partner, $[B]_0$, the fraction of bound species increases and becomes 50% when $[B]_0$ is around the value of K_d.

8.2.5 Example: A Simple Predator Prey Model

Motivated by the oscillatory fish catches in the Adriatic, in 1926 Volterra proposed a model system for the predation of species by another. Motivated originally by a problem in population dynamics this system—well-known as Lotka–Volterra model—nowadays serves as general model system for oscillatory behavior. The classical example for oscillatory behavior in chemical systems, however,

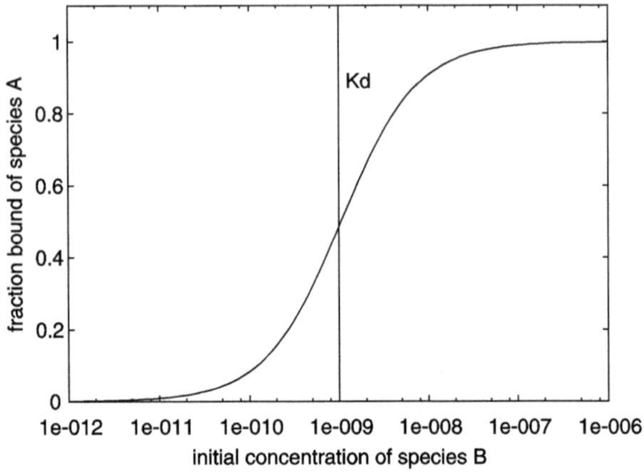

Fig. 8.2 Fraction of bound species A versus initial concentration $[B]_0$. The initial concentration of species A is set to the constant value $[A]_0 = 10^{-10}$ M. The equilibrium constant ($K_d = 10^{-9}$ M) is shown as a *vertical line*. The fraction of bound species A becomes 50% when the initial concentration of the binding partner $[B]_0$ becomes equal to K_d

is the Belousov–Zhabotinsky reaction which has been demonstrated in a vast number of classroom lectures. Predator-prey interaction in simple biomolecular reaction systems have been studied as well [1, 411].

The Lotka–Volterra model consists of a prey species A which, in absence of predation, grows uncontrolled by an autocatalytic self-replication

$$A \xrightarrow{k_A} A + A. \tag{8.29}$$

Such a reaction leads to an exponential growth and is, of course, a strong approximation for the replication of fishes, because in real life space and resources of food would present upper limits for the concentration of the prey. As long as the dynamic of interest is not influenced by such upper limit, the approximation of the replication by such a simple reaction is well satisfied. More complicated Lotka–Volterra models with resource limitations are well studied, see [272] and literature therein. For a formal definition of a self-reproducing system using Petri nets, we refer to [359].

The predator species B is able to feed on the prey for its own reproduction

$$B + A \xrightarrow{k_{AB}} B + B, \tag{8.30}$$

but will die if no prey is available

$$B \xrightarrow{k_B} . \tag{8.31}$$

The kinetic parameters k_A, k_{AB}, and k_B are positive constants. For known kinetic parameters, the dynamics of the concentration is readily determined by a system of ODEs

$$\begin{aligned} \frac{d[A]}{dt} &= k_A[A] - k_{AB}[A][B], \\ \frac{d[B]}{dt} &= +k_{AB}[A][B] - k_B[B]. \end{aligned} \tag{8.32}$$

A conventional abbreviation is to write such an ODE in the form

$$\frac{d[A]}{dt} = f_A([A],[B]) \quad \text{and}$$
$$\frac{d[B]}{dt} = f_B([A],[B]) \tag{8.33}$$

with

$$f_A([A],[B]) =_{\text{def}} k_A[A] - k_{AB}[A][B]$$

and

$$f_B([A],[B]) =_{\text{def}} k_{AB}[A][B] - k_B[B].$$

Introducing further a concentration vector $\vec{x}$ with component $x_1 = [A]$ and $x_2 = [B]$ as well as a rate vector $\vec{f}(\vec{x})$ with components $f_1(\vec{x}) = f_A([A],[B])$ and $f_2(\vec{x}) = f_B([A],[B])$ we yield the general form

$$\frac{d\vec{x}}{dt} = \vec{f}(\vec{x}). \tag{8.34}$$

8.2.6 Steady States

In general, it is substantial for any reaction system to learn as much as feasible about the variety of dynamics the system is able to show. In this context, the steady-state analysis represents an essential method. Steady states (or fixed points) are the solutions of the algebraic equation

$$\vec{f}(\vec{x}) = 0 \tag{8.35}$$

and define values of the concentration at which their time derivation becomes zero in (8.34). For the predator-prey model above, the steady-states are easily computed, see Problem 8.6.2. The trivial fixed point is $[A] = [B] = 0$, and a second fixed point is reached for $[A] = k_B/k_{AB}$, and $[A] = k_A/k_{AB}$. In the following, such singular points are denoted by a lower index s, as for example, $\vec{x}_s$.

8.2.7 Stability Analysis

Dynamically the systems may remain stable at the steady-state concentration $\vec{x}_s$ or otherwise may be driven away from it by an arbitrary small perturbation $\delta\vec{x}$, see Fig. 8.3. Linearization of (8.34) about the singular points $\vec{x} = \vec{x}_s$ reads

$$\frac{d(\vec{x}_s + \delta\vec{x})}{dt} = \vec{f}(\vec{x}_s + \delta\vec{x}) \doteq \vec{f}(\vec{x}_s) + \mathbf{J}(\vec{x}_s)\delta\vec{x} \tag{8.36}$$

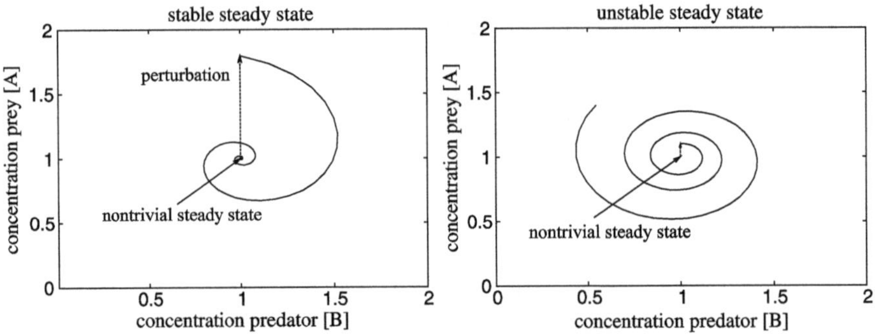

Fig. 8.3 Response to a perturbation from steady-states sketched in the phase plane of concentrations. *Left part*: A perturbation from a stable steady located at $[A] = [B] = 1$ decreases with time and the concentrations relax to the steady-state again. The steady-state is an *attractor*. *Right part*: A perturbation from an unstable steady at $[A] = [B] = 1$ increase with time. The steady-state is *repulsive*. The concentrations will not go back to this state. Instead, the concentrations will either end up in another (stable) steady-state or otherwise will increase to infinity

with the Jacobian matrix $\mathbf{J}(\vec{x}_s) =_{\text{def}} [\nabla_{\vec{x}} \vec{f}(\vec{x})]_{\vec{x}=\vec{x}_s}$. Note that, by definition (8.35) the rate vector $\vec{f}(\vec{x})$ is equal to zero at $\vec{x} = \vec{x}_s$ yielding

$$\frac{d\delta\vec{x}}{dt} = \mathbf{J}(\vec{x}_s)\delta\vec{x}. \tag{8.37}$$

This ODE describes the response of the dynamical system to a small perturbation $\delta\vec{x}$, and the formal solution is given by

$$\delta\vec{x}(t) = e^{\mathbf{J}(\vec{x}_s)t}\delta\vec{x}_0. \tag{8.38}$$

Obviously, we would be able to find an initial perturbation, $\delta\vec{x}_0$, which grows exponential with time if and only if the Jacobian matrix has at least one positive real eigenvalue. The trivial steady-state of the Lotka–Volterra model (8.32) is unstable, whereas the nontrivial steady-state is neutrally stable, see Problem 8.6.2.

The continuous rate equations for the Lotka–Volterra model have two steady-states: one trivially fixed, unstable, and a second, neutrally stable steady-state. The neutrally stable steady-state gives rise for a limit cycle. Limit cycles are closely connected to a *Hopf bifurcation*. The Hopf bifurcation plays an exceptional role for dynamical systems and describes a parameter value at which the Jacobian of the linear equation (8.36) has purely imaginary eigenvalues, see Problem 8.6.2. The connection of such a bifurcation to limit cycle solutions was formulated by E. Hopf in 1942. Let

$$\frac{dx}{dt} = \vec{f}(\vec{x}, \nu) \tag{8.39}$$

be a second-order autonomous system of differential equations which has a singular point $x_s(\nu)$ for each value of the real parameter ν. Suppose that the linearized (Jacobian) matrix $\mathbf{J}(\nu)$ of (8.39) about the singular point, $x_s(\nu)$, has purely imagi-

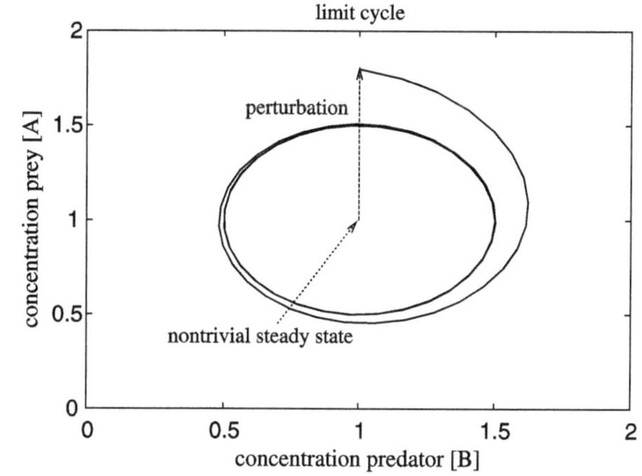

Fig. 8.4 Sketch of a limit cycle in the phase plane of concentrations. The limit cycle emerges by a perturbation from the neutrally stable steady-state at $[A] = [B] = 1$ and converges to a stable periodic orbit

nary eigenvalues $\pm i\omega$ for $\nu = \nu_0$. If the Jacobian matrix (family) $\mathbf{J}(\nu)$ decomposed in the form

$$\mathbf{J}(\nu) = \mathbf{J}(\nu_0) + (\nu - \nu_0)\mathbf{K}(\nu) \tag{8.40}$$

fulfills the condition

$$\mathrm{Tr}\,\mathbf{K}(\nu_0) \neq 0, \tag{8.41}$$

then there exists a periodic solution of (8.39) for ν in some neighborhood of $\nu = \nu_0$ and $\vec{x}$ in some neighborhood of $\vec{x} = \vec{x}_s$ with an approximate frequency $\omega/(2\pi)$.

For the majority of reaction systems of practical interest, the number of steady-states as well as their stability depends on the values of the kinetic parameters. Identification and characterization of critical parameter values at which the characteristics of the system (as the number of steady-states or their stability) changes instantaneously is the task of bifurcation analysis [72]. There exists a whole bunch of different bifurcation types. An example is a saddle point bifurcation point at which two steady-states collide and annihilate each other. At a pitchfork bifurcation two steady-states emerge from one steady-state, and at a transcritical bifurcation a steady-state loses its stability. Obviously, the knowledge of the bifurcation points of a system is very helpful. A detailed discussion of the rich variety of bifurcation types is out of the scope of this contribution.

Interacting entities in real-life systems, as population of animals, organs, cells, proteins, DNA, RNA, etc., are always countable and discrete. Usually, the number of individuals is huge and an description in term of concentrations and mass action kinetics is well satisfied. Nevertheless, there exist fundamental differences between continuous and discrete models. Concerning the stability of the Lotka–Volterra model this can easily be demonstrated.

The dynamical behavior of the discrete model is characterized by the time evolution of the probability distribution $P(A, B, t)$ to find A individuals of prey and B individuals of predator at time t in the system. Instead of a system of ordinary differential equations (8.32) for continuous concentrations we have to solve a *Master equation* for the probability $P(A, B, t)$, see Problem 8.6.3.

The neutral stable steady-state observed for the continuous system vanishes in the discrete model, see Problem 8.6.3. Nevertheless, the discrete system shows oscillations in analogy to the continuous model. The "limit cycles" of the stochastic model are, however, no stable trajectories because stochastic fluctuations cause transitions from one trajectory to neighboring ones. Because of the stochastic fluctuations there is no hope for the predator to survive. Once, by change, the prey has died out the system will end in the only possible steady-state, $A = B = 0$. Note that, stability analysis of continuous systems predicts a limit cycle behavior, which is also very helpful for the discussion of a more realistic discrete model.

In nature, the conception of space plays a crucial role and a prey can often survive by wandering to regions where the predator can not follow so quickly. This is an effect observed only for discrete systems, because no regions with zero predator exist in continuous models. Similar effects as reaction waves, complex pattern formation, and stochastic resonance play a crucial role for spatially extended systems, but will not discussed here; for further reading, see [150].

8.2.8 *Spatial Instability*

Reactions systems are applied to model a broad diversity of interacting systems. Such interactions of entities are above assumed to take place in well-mixed and homogeneous environments. On the microscopic level the process of mixing is driven by the random walk of the particles. The appropriate method to describe the random walk of particles on the abstraction level of mass action kinetics is the classical partial differential equation for diffusion:

$$\frac{\partial \vec{x}}{\partial t} = \vec{f}(\vec{x}) + \nabla_{\vec{r}}(\vec{D}\nabla_{\vec{r}}\vec{x}), \tag{8.42}$$

where the concentration vector, $\vec{x}$, is a function of time, t, and space, $\vec{r}$. The diffusion matrix, $\vec{D}$, may be a function of $\vec{x}$ and $\vec{r}$. A sophisticated way to derive this equation is the Fokker–Plank equation [122] using a probability density function with a Markov process that describes the random walk movements of the particles. If the diffusion is very fast on the time scale of reactions, the concentration vector, $\vec{x}$, becomes independent from the spatial dimension, $\vec{x}(t,\vec{r}) = \vec{x}(t)$, and the reaction-diffusion equations above will be reduced to the ODE (8.34).

Diffusion is usually considered as a stabilizing process. Under certain conditions, however, a reaction system which in the absence of diffusion would tend to a linear stable steady-state may show spatial inhomogeneous patterns evolving by diffusion driven instability. The idea is that, the linearized system is stable against any homogeneous perturbation, $\delta\vec{x}(t)$, but is unstable to perturbation, $\delta\vec{x}(t,\vec{r})$, with given spatial (periodic) modulation. Such a *Turing mechanism* can be shown for various reaction systems including many activator-inhibitor systems and the Belousov–Zhabotinsky-reaction. There exists a rich variety of complex and beautiful spatial-temporal patterns evolving for reaction systems. A prominent example is the simple system studied by Pearson [293] in which a Hopf bifurcation interacts with a Turing instability.

There exist numerous methods to treat reactions in compartmented systems. One approach is the direct numerical solution of the reaction–diffusion equation with continuous space, time, and concentration by applying finite element or finite volume methods. The coupled map lattice simulates continuous concentration on a discrete grid of space and time points [415], whereas delay differential equations are often used to model the transport of substances or pharmaceutical agents from one body part to another. Cellular automata with discrete time, space, and concentration represent a classical method connected to complexity theory, Turing machine, dynamic systems, and self-replicating systems.

8.3 Michaelis–Menten Kinetics

Michaelis–Menten kinetics developed by Leonor Michaelis and Maud Menten [261] describes the kinetics of enzyme-catalyzed reactions which do not exhibit allosteric effects, such as cooperativity. It is based on mass action kinetics under two assumptions. First, the Michaelis–Menten model applies when a meta-stable enzyme–substrate complex is formed, and the concentration of this enzyme–substrate complex changes much slower than those of the substrate and product. The enzyme-catalyzed reaction is described by

$$E + S \underset{k_{-1}}{\overset{k_1}{\rightleftharpoons}} ES \overset{k_2}{\rightarrow} E + P, \tag{8.43}$$

where E stands for enzyme, S for substrate, ES for the enzyme–substrate complex, P for product, and k_1, k_{-1}, and k_2 for the rate constants. For the corresponding Petri net, see Fig. 8.5.

According to mass action, see Sect. 8.2, we can formulate the following differential equations for the concentration changes of each component over time t

$$\frac{d[S]}{dt} = -k_1[E][S] + k_{-1}[ES], \tag{8.44}$$

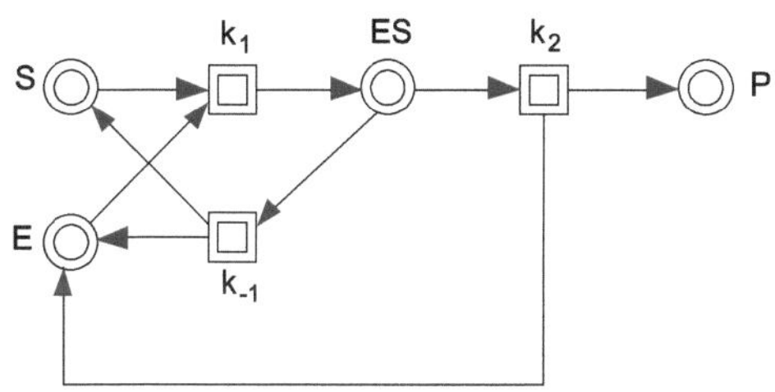

Fig. 8.5 The continuous Petri net representing (8.43). Places depicted by concentric cycles describe concentrations, and the firing rules of the transitions depicted by two concentric rectangles correspond to the underlying differential equations

$$\frac{d[E]}{dt} = -k_1[E][S] + k_{-1}[ES] + k_2[ES], \tag{8.45}$$

$$\frac{d[ES]}{dt} = +k_1[E][S] - k_{-1}[ES] - k_2[ES], \tag{8.46}$$

$$\frac{d[P]}{dt} = +k_2[ES] = v_0, \tag{8.47}$$

where $[X]$ denotes the concentration of a component X. The catalytic rate, v_0, depends on the concentration of the enzyme–substrate complex.

The second assumption is that a *quasi steady-state approximation* (*QSSA*) is valid. The QSSA states that the concentration of the enzyme–substrate complex ES remains constant even if the concentrations of substrates and products vary. Consequently, the rates for formation and breakdown of [ES] have to be identical

$$k_1[E][S] = (k_{-1} + k_2)[ES]. \tag{8.48}$$

A Michaelis–Menten constant, K_M, is defined by

$$\frac{[E][S]}{[ES]} = \frac{(k_{-1} + k_2)}{k_1} = K_M. \tag{8.49}$$

This constant characterizes enzyme–substrate interactions and the activity of the catalytic function. When the concentration of substrate reaches the Michaelis–Menten constant half of the active sites of the enzyme are filled, that is, K_M is a measure for substrate concentration required for a substantial catalytic activity. By mass conservation, it follows that the total enzyme concentration $[E]_T$ is identical to the sum of concentrations of free enzyme $[E]$ and the concentration of enzyme substrate complex $[ES]$

$$[E]_T = [E] + [ES]. \tag{8.50}$$

Inserting (8.50) and (8.49) into (8.48) and some algebra yield

$$[ES] = \frac{[S][E]_T}{K_M + [S]}. \tag{8.51}$$

For the catalytic rate, v_0, in (8.47), we get

$$v_0 = k_2[ES] = k_2[E]_T \frac{[S]}{K_M + [S]}. \tag{8.52}$$

The catalytic rate, v_0, reaches its maximal value, $v_{\max} = k_2[E]_T$, when the concentration of substrate $[S]$ is given in excess ($[S] \gg K_M$). In terms of $v_{\max}$, the Michaelis–Menten-equation appears in the well-known form

$$v_0 = v_{\max} \frac{[S]}{K_M + [S]}. \tag{8.53}$$

8.3.1 Enzyme Kinetics with Inhibitors

Many medical and biotechnological applications are interested in inhibiting specific enzymes and knocking out selected reactions. Examples are *penicillin*, which covalently modifies *transpeptidase*, and *methotrexate*, which inhibits *dihydrofolate reductase*. Since the rate of dissociation of the enzyme–inhibitor complex is significantly greater than zero the inhibition of *dihydrofolate reductase* by *methotrexate* is called a *reversible* inhibition. In contrast, the covalent binding of *penicillin* to *transpeptidase* is denoted as *irreversible* inhibition. According to mechanism of inhibition, reversible inhibitors can be classified into three types: *competitive*, *uncompetitive*, and *noncompetitive* (*or mixed*). In the following, we briefly explain the three types of inhibition and present the corresponding Michaelis–Menten equations, for a derivation of the formulas and more details see [405].

In the case of a competitive inhibition, the substrate and inhibitor are structurally very similar and compete for the same binding site. Only enzyme–substrate complexes, ES, and enzyme–inhibitor complex, EI, can emerge but higher order complexes like an enzyme–substrate–inhibitor complex, ESI, can not be formed, see Fig. 8.6. As for the Michaelis–Menten kinetics above, we apply the QSSA

$$k_I = \frac{[E][I]}{[EI]}, \tag{8.54}$$

where k_I denotes the equilibrium concentration for the reversible binding of the inhibitor, I, to the enzyme. The enzyme–inhibitor complex EI is, of course, catalytically inactive. Accounting for the enzyme–inhibitor complex the mass conservation (8.50) now reads

$$[E]_T = [E] + [EI] + [ES]. \tag{8.55}$$

Inserting the QSSA (8.49) for the enzyme–substrate complex as well as Equation 8.54 for the enzyme–inhibitor complex yields (Problem 8.6.4)

$$v_0 = v_{\max} \frac{[S]}{\alpha K_M + [S]}, \tag{8.56}$$

with

$$\alpha = 1 + \frac{[I]}{k_I}. \tag{8.57}$$

In case of an uncompetitive inhibition, the inhibitor binds only to the enzyme-substrate complex. A catalytically inactive enzyme-substrate-inhibitor complex arises, but no enzyme–inhibitor complex will be formed, see Fig. 8.7. The QSSA for binding of the inhibitor gets the form

$$k_I' = \frac{[ES][I]}{[ESI]} \tag{8.58}$$

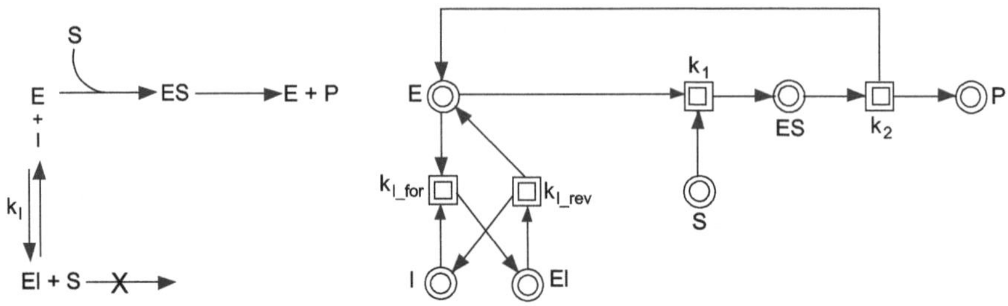

Fig. 8.6 Competitive inhibition: Either an enzyme-substrate complex or an enzyme–inhibitor complex can be formed, because substrate and inhibitor bind to the same binding site of the enzyme

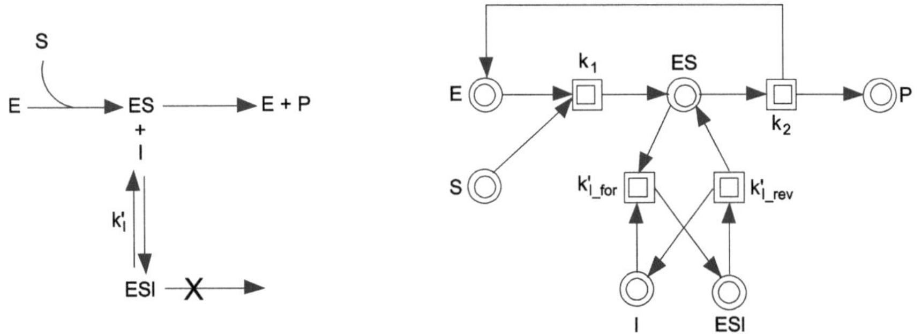

Fig. 8.7 Uncompetitive inhibition: The inhibitor binds directly to the enzyme-substrate complex, but not to the free enzyme

leading to (see Problem 8.6.4)

$$v_0 = v_{\max} \frac{[S]}{K_M + \alpha'[S]}, \tag{8.59}$$

where

$$\alpha' = 1 + \frac{[I]}{k'_I}. \tag{8.60}$$

In the noncompetitive or mixed inhibition both, the enzyme–inhibitor complex and enzyme-substrate-inhibitor complex, are formed. Both binding steps are assumed to be in a quasi steady-state, each with its individual dissociation constant, see Fig. 8.8.

Combining the corresponding QSSAs with mass conservation gives (Problem 8.6.4)

$$v_0 = v_{\max} \frac{[S]}{\alpha K_M + \alpha'[S]}. \tag{8.61}$$

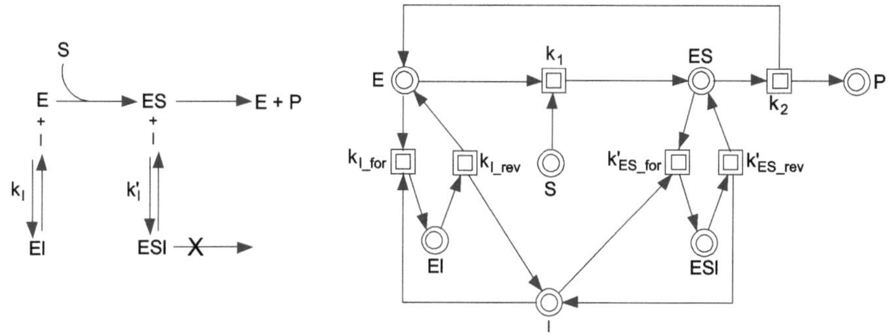

Fig. 8.8 Noncompetitive inhibition: The inhibitor can combine with either the enzyme or the enzyme-substrate complex

8.4 Hill Kinetics

In a bimolecular reaction, one species may have several binding sites. In some cases, the binding of a ligand to a binding site induces a change of the tertiary structure. The new tertiary structure may present the binding sites more accessible for the binding of ligands and hence, may be advantageous for the binding of further ligands to free binding sites. Whereas the binding of the first ligand may be slow the binding of a second, third etc. ligand can become much faster. Such an effect is called *cooperative binding*. The Hill equation is a classical method [160] to describe cooperative binding.

A prominent example of a structure with several binding sites is the streptavidin molecule which consists of four identical subunits and binds four biotin molecules. The binding of biotin to streptavidin is the strongest noncovalent reaction known on nature and plays an exceptional role in medical diagnostics. The cooperativity of the binding of biotin to streptavidin is a disputed question in literature, for example, see [138, 336]. The qualitative Petri net model is depicted in Fig. 8.12. In the following, the biotin binding by streptavidin serves as an illustrative example to demonstrate how cooperative binding can be modeled.

8.4.1 Neutral Binding

Let us assume that the binding of biotin to a binding site of streptavidin is independent of the number of biotin already bound to the streptavidin molecule. Consequently, we can ignore the aggregation of binding sites to the tetrameric structure of an streptavidin molecule and, hence, a reaction

$$A + B \underset{k_r}{\overset{k_f}{\rightleftharpoons}} C \tag{8.62}$$

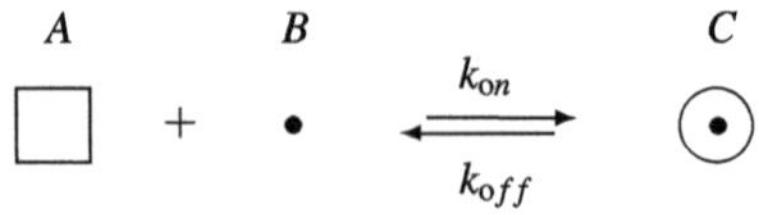

Fig. 8.9 Microscopic binding reaction of biotin B to a binding site A of streptavidin. The species C denotes the complex of the binding site and biotin. Upon binding of biotin, the tertiary structure of the binding site changes significantly. The two different structures of the binding site are indicated by a *square* and a *circle*, respectively. This notation has nothing to do with the graphical representation of places and transitions in Petri nets

determines the steady-state of free binding sites A, biotin B, and bound sites C, see Fig. 8.9. The steady-state concentration satisfies the condition

$$\frac{[A][B]}{[C]} = K_A, \tag{8.63}$$

where the *microscopic dissociation constant* K_A is defined by

$$K_A =_{\text{def}} \frac{k_r}{k_f}. \tag{8.64}$$

The mass conservation condition

$$[A] + [C] = [A]_0 \tag{8.65}$$

can be applied to transform the steady-state condition (8.63) into

$$\frac{[A]_0 - [A]}{[A]} = \frac{[B]}{K_A} \tag{8.66}$$

or

$$\frac{1 - Y}{Y} = \frac{[B]}{K_A}, \tag{8.67}$$

where Y is the fraction of free sites

$$Y =_{\text{def}} \frac{[A]}{[A]_0}. \tag{8.68}$$

For initial biotin given in excess, for example, $[B] \approx [B]_0 \gg [A]_0$, we get the *Hill function* for neutral binding

$$\ln\left(\frac{1 - Y}{Y}\right) = \ln[B]_0 - \ln K_A. \tag{8.69}$$

Consequently, for neutral binding the Hill curve approaches a straight line with slope 1 and crosses the y-axes at $y = -\log_{10}(K_A)$, see Fig. 8.10.

Alternatively, we may consider a mixture of streptavidin and biotin in which we may find streptavidin molecules with no (S), one (S_1), two (S_2), three (S_3), or

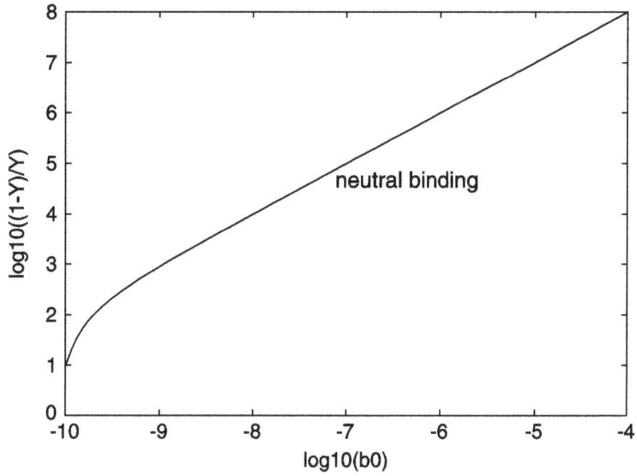

Fig. 8.10 Hill curve for neutral binding. The microscopic dissociation constant is set to $K_A = 10^{-12}$ M, and the initial concentration of streptavidin is set to the value $[A]_0 = 10^{-10}$ M. The fraction of free binding sites, Y, is computed via solution (8.28). For initial concentration of biotin in excess ($[B]_0 \gg [A]_0$), the Hill curve approach a straight line with slope 1 and crosses the y-axes at $y = -\log_{10}(K_A) = 12$

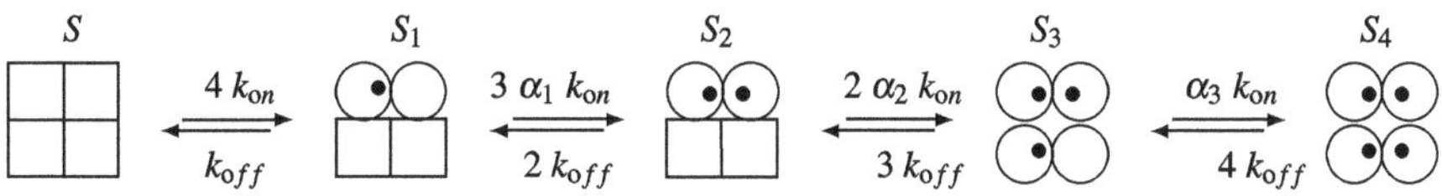

Fig. 8.11 The binding of biotin to streptavidin. The tetrameric streptavidin consists of four subunits each of which can bind one biotin. The binding of a biotin may induce a change of the tertiary structure of streptavidin. The different structures of the binding site are distinguished by *squares* and *circles*, respectively. The new structure (*circle*) may favor the binding of biotin. In this case, the process is called to be cooperative, and the parameters α_i are greater than one

four (S_4) biotin molecules bound to it, see Fig. 8.11. Since S has four free active sites the molecular on-rate for the binding of one biotin molecule is four times the microscopic on-rate constant k_{on}. Analogously, the pre-factors for the on-rate of S_1, S_2, and S_3 are three, two, and one, respectively. The steady-state condition reads

$$\frac{4[S]b}{[S_1]} = \frac{3\alpha_1[S_1][B]}{2[S_2]} = \frac{2\alpha_2[S_2][B]}{3[S_3]} = \frac{\alpha_3[S_3][B]}{4[S_4]} = \frac{k_{\mathrm{off}}}{k_{\mathrm{on}}} = K_A. \tag{8.70}$$

The factors, α_1, α_2, and α_3, are introduced to account for variations of the microscopic binding rates due to possible differences in the three-dimensional structures of S, S_1, S_2, S_3, and S_4, respectively. No cooperativity corresponds to the choice $\alpha_1 = \alpha_2 = \alpha_3 = 1$, whereas the case $\alpha_i < 1,\ i = 1,2,3$ describes negative regulation. Cooperative binding is given for all factors $\alpha_i,\ i = 1,2,3$, greater than one. The steady-state condition leads to

$$\begin{aligned} [S_1] &= \frac{4[S][B]}{K_A}, & [S_2] &= \frac{6\alpha_1[S][B]^2}{K_A^2}, \\ [S_3] &= \frac{4\alpha_1\alpha_2[S][B]^3}{K_A^3}, & [S_4] &= \frac{\alpha_1\alpha_2\alpha_3[S][B]^4}{K_A^4}. \end{aligned} \tag{8.71}$$

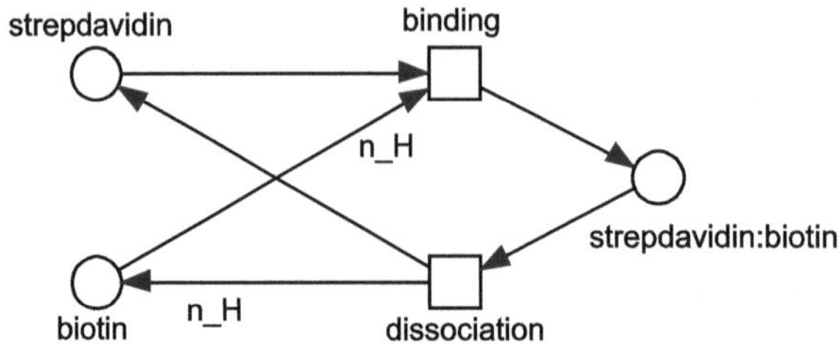

Fig. 8.12 The qualitative Petri net model for binding and dissociation of biotin to the tetrameric streptavidin. The Hill constant, n_H, corresponds to the weight of the edges connected to the place *biotin*. A Hill constant greater than one corresponds to cooperative binding

The fraction of free sites is given by

$$Y = \frac{4[S]+3[S_1]+2[S_2]+[S_3]}{4[S]_0}$$
$$= \frac{[S]}{[S]_0}\left(1+\frac{3[B]}{K_A}+\frac{3\alpha_1[B]^2}{K_A^2}+\frac{\alpha_1\alpha_2[B]^3}{K_A^3}\right)$$

and the fraction of saturated sites by

$$1-Y = \frac{[S_1]+2[S_2]+3[S_3]+4[S_4]}{4[S]_0}$$
$$= \frac{[S]}{[S]_0}\frac{[B]}{K_A}\left(1+\frac{3\alpha_1[B]}{K_A}+\frac{3\alpha_1\alpha_2[B]^2}{K_A^2}+\frac{\alpha_1\alpha_2\alpha_3[B]^3}{K_A^3}\right).$$

The choice, $\alpha_1 = \alpha_2 = \alpha_3 = 1$, leads, of course, to the Hill function for neutral binding

$$\frac{1-Y}{Y} = \frac{[B]}{K_A}.$$

8.4.2 Cooperative Binding

Co-operative binding is given for all factors α_i with $i = 1, 2, 3$ greater than one. In this case, we may find a concentration range

$$\frac{K_A}{\alpha} \ll [B] \ll \frac{K_A}{\alpha^{2/3}}$$

in which the fraction of free sites is well approximated by

$$Y \approx \frac{[S]}{[S]_0},$$

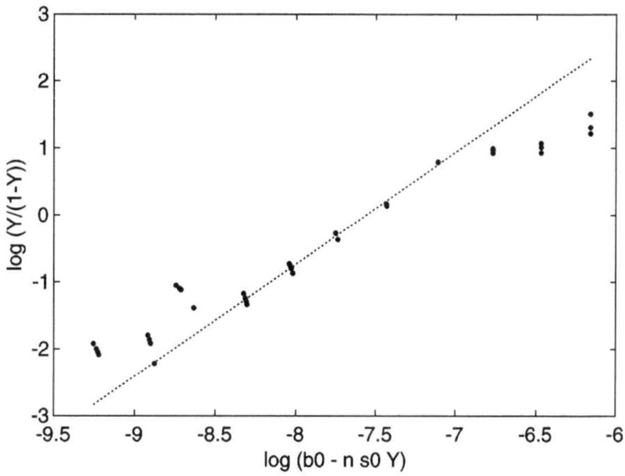

Fig. 8.13 Hill plot for the binding of Streptavidin–ATTO532 to mouse-anti-human-IgG-biotin. The fit of the experimental data to a straight line gives a Hill coefficient of $n_H = 1.67 \pm 0.09$ which is significantly greater than one and, hence, indicates cooperativity of binding

whereas the fraction of bound sites is dominated by

$$1 - Y \approx \frac{[S_4]}{[S]_0}.$$

Inserting the steady-state concentration (8.71) gives

$$\frac{1-Y}{Y} = \frac{[S_4]}{[S]} = \frac{\alpha^3 [B]^4}{K_A^4}$$

or the Hill equation for (ideal) cooperativity

$$\ln\left(\frac{1-Y}{Y}\right) = 4\ln\big([B]\big) - 4\ln(K_A) + 3\ln(\alpha).$$

The general form of the Hill equation is given by

$$\ln\left(\frac{1-Y}{Y}\right) = n_H \ln\big([B]_0\big) - C,$$

where any Hill coefficient $n_H > 1$ indicates cooperative binding [54, 139]. Figure 8.13 shows a typical experimental Hill plot.

8.5 Tool Support

Coming to quantitative modeling life can become complicated for nonexperts intending to construct a self-made model. There is a rich variety of knowledge to learn about dynamical systems, numerical implementation of stiff system, kinetic data, bifurcation analysis, network construction, databases, and many more. The way to approach a model depends on the background of the scientist (mathematical,

Table 8.1 List of kinetic modeling software packages

Name	Web link
GEPASI	http://www.gepasi.org/
Virtual Cell	http://www.nrcam.uchc.edu/
Jarnac/JDesigner	http://sbw.kgi.edu/
COPASI	http://www.copasi.org/
E-Cell	http://www.e-cell.org/
ProMoT/DIVA/Diana	http://www.mpi-magdeburg.mpg.de/
SBW	http://www.sys-bio.org/
PyBioS	http://pybios.molgen.mpg.de/
XPP-Aut	http://www.math.pitt.edu/~bard/xpp/xpp.html
CellDesigner	http://www.celldesigner.org/
CellWare	http://www.bii.a-star.edu.sg/achievements/applications/cellware/
CellIllustrator	http://www.cellillustrator.com/
Dizzy	http://magnet.systemsbiology.net/software/Dizzy/
Dynetica	http://www.duke.edu/~you/Dynetica_page.htm
Pasadena Twain	http://sbw.sourceforge.net/sbw/software/
PLAS	http://www.dqb.fc.ul.pt/docentes/aferreira/plas.html
JSim	http://nsr.bioeng.washington.edu/jsim/
SB Toolbox	http://www.sbtoolbox.org/
PLmaddon	http://www.sbi.uni-rostock.de/plmaddon/
MathSBML	http://sbml.org/Software/MathSBML
Cellerator	http://www.cellerator.org/
SBRT	http://www.bioc.uzh.ch/wagner/software/SBRT/
BIOSPICE	http://biospice.sourceforge.net/
Berkely Madonna	http://www.berkeleymadonna.com/
Calcium3D	http://personales.unican.es/gila/bbva.html
Chemesis	http://krasnow.gmu.edu/CENlab/software.html
CKS	http://www.almaden.ibm.com/st/computational_science/ck/
Dynafit	http://www.biokin.com/dynafit/
Genesis/Kinetikit	http://www.cns.atr.jp/~doi/GENESIS/index.html
KINSIM/FITSIM	http://www.biochem.wustl.edu/cflab/message.html
WebCell	http://webcell.kaist.ac.kr/
For many more see	http://sbml.org/SBML_Software_Guide/SBML_Software_Summary

biologa, from computer science, chemical, pharmacological, physical, or medical). The best way is to start with a small network of reactions and some basic understanding of the fundamental principles. The next step should be the construction of a simple fundamental Petri net model using some Petri net tool, one can find, for example, at [301]. The automated construction of reaction networks from experi-

Table 8.2 List of visualization tools

Name	Web link
BioTapestry	http://www.biotapestry.org/
BioUML	http://www.biouml.org/
CADLIVE	http://www.cadlive.jp/
Cytoscape	http://www.cytoscape.org/
Edinburgh Pathway Editor	http://www.bioinformatics.ed.ac.uk/epe/
GenMAPP	http://www.GenMAPP.org
Kohn Interaction Maps	http://discover.nci.nih.gov/mim/
NetBuilder	http://strc.herts.ac.uk/bio/maria/NetBuilder/
PathVision	http://www.PathVisio.org
PNK 2e	http://page.mi.fu-berlin.de/trieglaf/PNK2e/index.html
VisANT	http://visant.bu.edu/

Table 8.3 List of XML based standards

Name	Web link
BioPAX—Data exchange format for pathway data	http://www.biopax.org/
CellML—Cell Markup Language	http://www.cellml.org/
PSI-MI—Proteomics Standards for Molecular Interactions	http://www.psidev.info/
MATHML—Mathematical Markup Language	http://www.w3.org/Math/
MIRIAM—Minimal Information Required In Annotation of Models	http://www.ebi.ac.uk/miriam/
SBML—Systems Biology Markup Language	http://sbml.org/

mental data is still a great challenge. Consequently, simulating a self-made Petri net model and observing its basic behavior is advisable for the alignment of the structure of the model with the intention and idea of the user. In most cases, this intuitive and easy avenue is recommendable to avoid drawbacks in later stages of the task.

When feeling comfortable with the hopefully still simple model, the straight way to proceed is to select a standard tool and to transfer the net structure of the Petri net model to a kinetic model. There is a whole bunch of software packages available, but important aspects as for example, systems requirements, functionality, portability, capabilities of the user interfaces, reliability and compatibility have to be considered. A comparative review of 12 software packages for kinetic modeling of biochemical network can be found in [14]. For nonexpert users, this review identifies GEPASI as the best choice, and environments such as Jarnac/JDesigner and Virtual Cell are preferable for large-scale modeling and multi-compartmentalization, respectively. Table 8.1 compiles some references to other software packages.

After having gained some experience with numerical simulation, the user may like to read more about the background. Here, numerous textbooks are available. For mathematicians, the textbooks by Murray [272] and Haken [148] provide good introductions, including many references for further reading. If the numerics behind the simulation is of interest the book by Deuflhard and Bornemann [91] will be helpful. Within this book, you will find references to the packages LARKIN, PARKIN, MACRON, and FEM for chemical reaction kinetics, parameter identification, polymerization reaction kinetics, and spatial extended reaction systems, respectively. For detailed kinetic models of metabolic processes, we refer to the introductory review by Steuer and Junker [376], and for signal transduction to the book of Klipp et al. [203]. Numerous curated models can be found at http://www.ebi.ac.uk/biomodels-main/. Experimentally determined kinetic data of bimolecular reaction described in the literature are available via databases. A review of these databases, however, is out of the scope of this contribution. Useful sources are KDBI (http://bidd.nus.edu.sg/group/kdbi/kdbi.asp) and IntAct (http://www.ebi.ac.uk/intact). We also refer to the yearly database issue of the journal Nucleic Acids Research for further reading.

Continuing development and evolution of the model may raise awareness of topics as presentation, visualization, and publication. Then the list of links in Table 8.2 are worth to be visited. Cooperations between modelers become much easier if the standards in system biology are followed [200], see the links in Table 8.3. A further list of links to systems biology tools and resources can be found at http://www.sbi.uni-rostock.de/sbresources.html.

8.6 Problems

8.6.1 Continuous Versus Discrete Modeling

8.1 Give the main differences between discrete and continuous modeling?

8.2 How do you decide which modeling type you want to apply?

8.3 Give the stoichiometric equation system and the continuous Petri net for the following set of ODEs:

$$\frac{dA}{dt} = +k_2[C] - k_3[A][B]^2 + k_4[E],$$
$$\frac{dB}{dt} = +k_1[C] - k_3[A][B]^2 + k_4[E],$$
$$\frac{dC}{dt} = -k_1[C] - k_2[C],$$
$$\frac{dD}{dt} = +k_2[C],$$
$$\frac{dE}{dt} = +k_3[A][B]^2 - k_4[E].$$

8.6.2 Stability Analysis of the Continuous Lotka–Volterra Model

8.4 For the bifurcation analysis of the Lotka–Volterra model, it is advantageous to non-dimensionalize the system. Introduce a new time scale, τ, and scaled concentrations, a and b, to transform the ODEs (8.32) into the form

$$\begin{aligned} \frac{da}{d\tau} &= a(1-b), \\ \frac{db}{d\tau} &= \nu b(a-1). \end{aligned} \tag{8.72}$$

How does the bifurcation parameter, ν, depends on the kinetic parameters of ODE (8.32)?

8.5 Find all steady-state concentrations, a and b, at which the right-hand side of ODE (8.72) becomes zero.

8.6 Compute the Jacobian matrix

$$\mathbf{J}(a,b) =_{\text{def}} \begin{pmatrix} \frac{\partial f_a(a,b)}{\partial a} & \frac{\partial f_b(a,b)}{\partial a} \\ \frac{\partial f_a(a,b)}{\partial b} & \frac{\partial f_b(a,b)}{\partial b} \end{pmatrix} \tag{8.73}$$

for ODE (8.72).

8.7 Compute all eigenvalues and eigenvectors of the Jacobian for the trivial steady-state concentrations, $a = b = 0$. Is the trivial steady-state locally stable or unstable?

8.8 Compute all eigenvalues of the Jacobian for the nontrivial steady-state concentrations, $a = b = 1$. Is this steady-state locally stable or unstable?

8.6.3 Stability Analysis of the Discrete Lotka–Volterra Model

8.9 In a stochastic Lotka–Volterra, the number of individuals of the two species, prey and predator, are integers, denoted in the following by A and B, respectively. The multiplication of prey is described by a Markov process

$$A \longrightarrow A + 1, \tag{8.74}$$

which increases the number of prey by just one. Since each individual of prey has its own sporting change of replication the probability for this transition is proportional to the number of prey A

$$P_{(A,B)\to(A+1,B)} = k_A A. \tag{8.75}$$

A death process decreases the number of predators by one

$$B \longrightarrow B - 1. \tag{8.76}$$

Sad news for the predator is that, each of them can die and the reduction of predator occurs with a probability proportional to the number of predator

$$P_{(A,B)\to(A,B-1)} = k_B B. \tag{8.77}$$

The predator eating prey gives rise to transition

$$B \longrightarrow B + 1, \tag{8.78}$$

$$A \longrightarrow A - 1. \tag{8.79}$$

The predator is always hungry and eats whenever he meets a prey. He does not hunt actively but the prey, on the other side, is not very bright and unable to avoid the predator. Individuals of both species meet just by change. The probability for a unilaterally advantageous rendezvous is proportional to the number of different pairs of individuals of predator and prey, that is, proportional to the product AB

$$P_{(A,B)\to(A-1,B+1)} = k_{AB} AB. \tag{8.80}$$

The parameters k_A, k_B, and k_{AB} are positive constants. This stochastic model is the analogous to the rate equations stated above [148].

8.10 Sketch the possible stochastic transitions of the Lotka–Volterra model in a discrete phase plane of number of predators versus number of preys.

8.11 Formulate the Master equation for the time evolution of the probability $P(A, B, t)$ to find A individuals of prey and B individuals of predator at time t in the system.

8.6.4 Michaelis–Menten Kinetics

8.12 Give the single steps of the derivation of (8.56) for competitive inhibition, of (8.59) for uncompetitive inhibition, and of (8.61) for noncompetitive inhibition.

Chapter 9
Fuzzy Modeling

Lukas Windhager, Florian Erhard,
and Ralf Zimmer

Abstract Petri nets are very well suited for the representation of biological systems. Biological entities like proteins, metabolites, genes etc. can be defined as places; biochemical reactions, regulatory effects, modifications etc. can be defined as transitions. This 1 to 1 correspondence of molecules/reactions and places/transitions allows a very intuitive setup of a computational model framework. In this chapter, we will show how, additionally, the current states of biological entities and the reaction effects can be defined in a very intuitive and natural way using elements taken from fuzzy logic theory. Often exact data or detailed knowledge about concentrations, reaction kinetics or regulatory effects is missing. Thus, computational modeling of a biological system requires dealing with uncertainty and rough information provided by qualitative knowledge and linguistic descriptions. The Petri net and fuzzy logic (PNFL) approach allows natural language based descriptions of entities as well as (if-then) rule based definitions of reaction effects, both of which can easily and directly be derived from qualitative (linguistic) knowledge. PNFL bridges the gap between qualitative knowledge and quantitative modeling.

9.1 Introduction

Knowledge about biological entities and of most kinds of biological data is typically imprecise and incomplete. This is caused by the size, time-scale and complexity of

L. Windhager (✉) · F. Erhard · R. Zimmer
Institut für Informatik, Ludwig-Maximilians-Universität München, Amalienstr. 17, 80333 Munich, Germany
e-mail: lukas.windhager@bio.ifi.lmu.de

F. Erhard
e-mail: florian.erhard@bio.ifi.lmu.de

R. Zimmer
e-mail: ralf.zimmer@bio.ifi.lmu.de

I. Koch et al. (eds.), *Modeling in Systems Biology*, Computational Biology 16,
DOI 10.1007/978-1-84996-474-6_9,

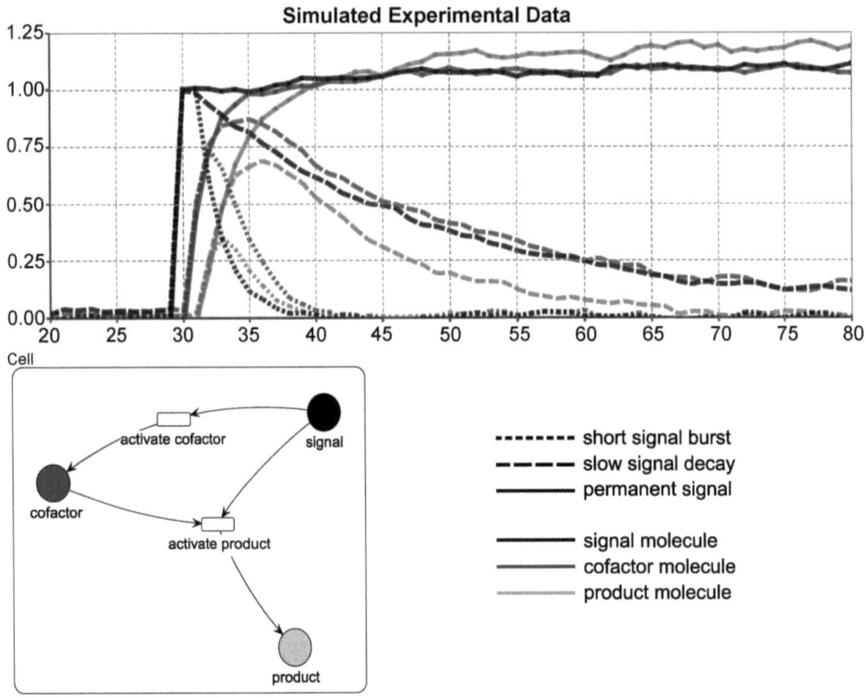

Fig. 9.1 A simple feed-forward motif. *Bottom*: A *signal* affects the concentration of a *cofactor*, while the concentration of a *product* is affected by both signal and cofactor. *Top*: The experimental data was simulated using a simple system of ODEs with mass-action kinetics, normal-distributed noise and three different decay rates for the signaling molecule. The reaction of the system after the addition of the signaling molecule is investigated. *Dotted*, *dashed* and *solid lines* indicate the measurements of the three different simulated experiments. Concentration (range axis) and time (domain axis) have arbitrary units

biological systems and processes (biological noise) and by the inexactness of measurements, postprocessing methods or other kinds of technical noise. So when considering biological data one typically works with average values characterized by smaller or larger variances. And most of the time, some kind of *best guess* (e.g., median) is considered as the truth, for example, as the true concentration of a protein. Often enough, one not even knows the correct scales of biological data or has only a rough idea about the concentrations or other properties of biological entities. In addition to the uncertain data, scales and exact quantities may not even be of great importance in a biological system.

Let us consider a small exemplary system consisting of three proteins (Fig. 9.1). The apparent information of the experiment consists of the measured concentrations during the observed time interval. However, the crucial point is to get an idea of the qualitative behavior of the system. The knowledge that can be deduced from this experiment is that the concentration of a cofactor started to increase after a signal was inserted into the system. And only when both signal and cofactor are present,

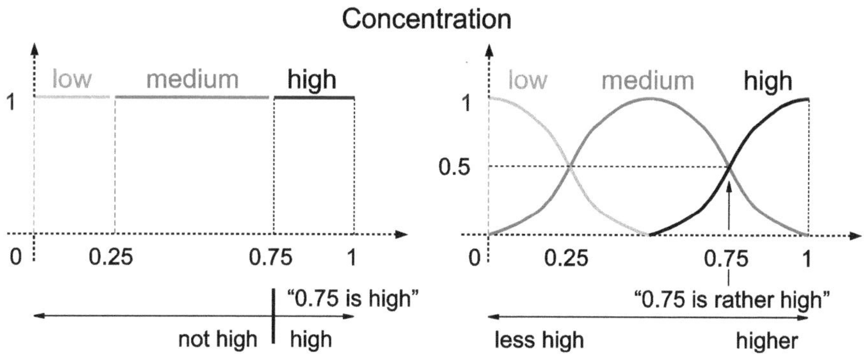

Fig. 9.2 Discretization of concentrations. *Left*: The (normalized) state-space of possible protein concentrations can be divided in sharp intervals, that is discretized by ordinary sets *low*, *medium* and *high*. A specific concentration, for example, 0.75, is assigned to a single set. *Right*: An inexact or fuzzy discretization can be done by defining fuzzy sets spanning the state space (universe of discourse). The degree of similarity of a specific concentration to a fuzzy set is indicated by the height of the according function curve

they induce the creation of the product. As soon as the signal is decayed, cofactor and product are decayed too until they reach a low concentration.

Such inexact quantitative knowledge and certainly the fact that humans do not reason in numbers lead to a more qualitative point of view of biologists and bioinformaticians. Typically, reasonable descriptions of knowledge explicitly or implicitly include or refer to discrete (and not continuous, real-valued) description of objects. And consequently, processes or dependencies are also described in natural language terms, which depend on the discrete description of the current states of biological objects. This is the very idea behind modeling systems with Petri nets and fuzzy logic [416].

It seems quite promising to develop a computational approach which is able to deal with uncertainty and rough information provided by qualitative knowledge and descriptions. The first step is to find a suitable mathematical or technical representation of discrete, inexact natural language terms. Let us assume the state space of possible concentrations of proteins is normalized to the interval $[0,1]$. It could be divided into three distinct intervals $[0,0.25)$, $[0.25,0.75)$ and $[0.75,1]$ such that concentrations within the borders of these intervals are then characterized by the terms *low*, *medium* and *high* (Fig. 9.2). But for example, defining the set of highly concentrated proteins as "the set of proteins present at a level of more than 0.75" is unsatisfactory as this strict border is artificial. It is difficult to argue, that a protein present at 0.750 is highly concentrated while it would *not* be highly concentrated at a level of 0.749. Therefore, it is not advisable to use classic (or discrete) sets as representatives of linguistic terms. It would be much more natural to define these sets with fuzzy borders.

The notion of fuzzy sets was introduced in the 1960s and the associated theory of fuzzy logic is well established and intensely used in many engineering applications [422]. Elements are not seen as being either part of a fuzzy set

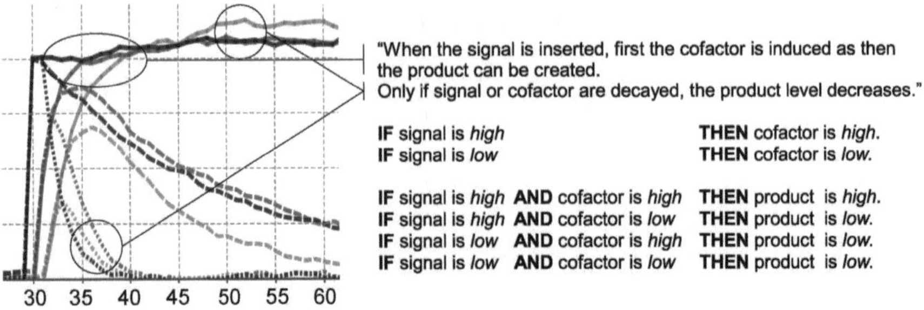

Fig. 9.3 If-then rules describe processes. Linguistic or "natural language" descriptions of (observed) processes can easily be reformulated to if-then sentences which serve as rule systems for a fuzzy logic based modeling. The behavior of the feed-forward motif observed in the three experiments can be described with two rule sets defining the effects of the two transitions indicated in Fig. 9.1. For the sake of simplicity, only two fuzzy sets are used for the discretization of the state space

or not (membership-value 1 *or* 0) but instead they are defined as being similar to elements described by a fuzzy set. The similarity is quantified by assigning a (membership-)value from 1 (equal) *to* 0 (dissimilar). For example, using fuzzy sets it is possible to define a concentration of 0.75 as "a rather high concentration" by defining the degree of similarity to the fuzzy sets *medium* and *high* as 0.5 each.

As the current states of entities are defined by linguistic terms implemented as fuzzy sets, the mathematical representation of processes in our framework has to be able to operate with fuzzy sets. Fuzzy logic systems are collections of simple if-then rules, which take a number of fuzzy sets as inputs (premises) and create fuzzy sets as outputs (conclusions). Such if-then rules can be derived from natural language descriptions of processes or dependencies in a straight-forward way; and vice versa, such rule sets can be easily interpreted by a human reader (Fig. 9.3).

In Sect. 9.2, we will give a theoretical introduction to fuzzy sets and fuzzy reasoning and show how fuzzy logic based inference is used in a Petri net context. First, an introduction to ordinary logic based reasoning is given in Sect. 9.2.1 and justified by the *modus ponens*. After defining the concept of a fuzzy set and of appropriate set theoretical operations in Sect. 9.2.2, we generalize the ordinary logic based reasoning to fuzzy logic based reasoning (Sect. 9.2.3). This subsection ends in the definition of *fuzzy logic systems*, which allow inference using natural language based rule systems and which are then embedded as functions in the Petri net context (Sect. 9.2.4). The definition of a *Petri net with fuzzy logic* (PNFL) discloses how fuzzy logic based inference is used to allow the simulation of (biological) systems.

9.2 Methods and Concepts

Before introducing fuzzy sets, we will first go through some aspects of ordinary logic reasoning which will then be generalized to fuzzy logic reasoning. This theoretical foundations justify the rule bases of fuzzy logic systems which are used in our framework [222, 257, 263].

9.2.1 Ordinary Logic Reasoning

We want to use rule bases to define the dynamics of a system, that is, to derive the new state of an entity from a given set of states of other entities. Those rule bases are given as linguistic if-then sentences. To implement such sentences in a computational framework, they have to be converted to (mathematical) functions first. These functions have to map several specified input sets to a single, specified output set. In the following section, we will show how such functions can be derived for ordinary sets using ordinary relations and logic implication and that the conclusions are backed by the modus ponens.

9.2.1.1 Modus Ponens

Definition 9.1 One of the most important inference rules in traditional propositional logic is the *modus ponens*

Premise 1 (Fact)	x is A
Premise 2 (Rule)	if x is A then y is B
Conclusion	y is B

which allows to derive a conclusion from given (combined) propositions.

If both premises are true, that is, the fact is true ("x *is* actually A") and the rule is valid, then it can be logically concluded that the consequent must also be true (the argument is *sound*). If one or both of the premises are false, the modus ponens does not apply. We want to emphasize that false premises *do not* imply that the conclusion has to be false as well. Modus ponens does not allow *any* deductions from false premises.

9.2.1.2 Ordinary Relations

Definition 9.2 A (two-valued) *relation* $R \subseteq X \times Y$ contains pairs (x, y) which are in some kind of relation described by R, that is, (x, y) fulfill conditions defining the relation.

E.g. let $G = \{g_1, g_2, g_3\}$ be a set of gene products and $C = \{c_1, c_2\}$ be a set of cell compartments. Then a relation $R \subseteq G \times C$ may describe that certain gene products can be found in certain cell compartments only: $R = \{(g_1, c_1), (g_2, c_2), (g_3, c_1), (g_3, c_2)\}$ where R contains an element (g_i, c_j) if and only if gene product g_i can be found in compartment c_j.

Definition 9.3 Let $R \subseteq X \times Y$ be a relation, then the *image* $R[M]$ of the relation with respect to the set $M \subseteq X$ is defined as

$$R[M] = \{y \in Y \mid \exists x \in X : (x, y) \in R \wedge x \in M\} \tag{9.1}$$

and contains all elements $y \in Y$ which are in relation to an element $x \in M$.

We want to point out that a set as well as a relation and its image can be represented by indicator functions

$$\begin{aligned} I_M : M \to \{0, 1\}, \quad & x \mapsto \begin{cases} 1 & \text{if } x \in M \\ 0 & \text{else} \end{cases} \\ I_R : X \times Y \to \{0, 1\}, \quad & (x, y) \mapsto \begin{cases} 1 & \text{if } (x, y) \in R \\ 0 & \text{else} \end{cases} \\ I_{R[M]} : Y \to \{0, 1\}, \quad & y \mapsto \begin{cases} 1 & \text{if } \exists x \in X : (x, y) \in R \wedge x \in M \\ 0 & \text{else} \end{cases} \end{aligned} \tag{9.2}$$

An equivalent definition of $R[M]$ can be given by using indicator functions instead of logical notations

$$I_{R[M]}(y) = \sup\{I_M(x) \cdot I_R(x, y) \mid x \in X\} \tag{9.3}$$

which yields $I_{R[M]}(y) = 1$ if and only if there is an $x \in M$ which is in relation R to the given $y \in Y$ and zero otherwise. The supremum (sup) of a set of real numbers is the smallest real number that is greater or equal to every number in this set. In contrast to the greatest element of a set, the supremum is not necessarily part of the set.

9.2.1.3 Logic Implication

An implication $x \in A \to y \in B$ for $A \subseteq X$ and $B \subseteq Y$ can be defined using the following relation:

$$R_{A \to B} = \{(x, y) \in X \times Y \mid x \in A \to y \in B\} = (A \times B) \cup \bar{A} \times Y \tag{9.4}$$

The image with respect to the set A is $R_{A \to B}[A]$ and its indicator function is defined as

$$I_{R_{A \to B}[A]} : Y \to \{0, 1\}, \quad y \mapsto \begin{cases} 1 & \text{if } \exists x \in A \text{ and } (x \in A \to y \in B) \\ 0 & \text{else} \end{cases} \tag{9.5}$$

or equivalently

$$I_{R_{A \to B}[A]} : Y \to \{0, 1\}, \quad y \mapsto \sup\{I_A(x) \cdot I_{R_{A \to B}}(x, y) \mid x \in X\} \tag{9.6}$$

The definition of the indicator function has the structure of a modus ponens with first premise (fact) $x \in A$, second premise (rule) *if* $x \in A$ *then* $y \in B$ and the conclusion $y \in B$. Obviously, $I_{R_{A \to B}[A]}$ indicates the set of all sound conclusions deduced by modus ponens. Therefore, it can be seen as a mapping of the input set A to the output set $R_{A \to B}[A] \subseteq B$ with respect to the given implication rule. This functions allows a general statement about conclusions of the rule. It is suited to answer the question, which conclusions are possible, given an unspecified element of X as premise (i.e., independent from a specific premise $x' \in X$).

We need to answer the question, which conclusion can be derived from a specific input $x' \in X$. This is easily done by restricting the image to the singleton set $\{x'\}$:

$$\begin{aligned} I_{R_{A \to B}[\{x'\}]} : Y \to \{0, 1\}, \quad & y \mapsto \sup\bigl\{I_{[\{x'\}]}(x) \cdot I_{R_{A \to B}}(x, y) \bigm| x \in X\bigr\} \\ \iff \quad & y \mapsto I_{R_{A \to B}}\bigl(x', y\bigr) \end{aligned} \tag{9.7}$$

The resulting set $R_{A \to B}[\{x'\}]$ then contains all valid conclusions given a specific x'. Obviously, this conclusion is still backed by the modus ponens if $x' \in A$, that is, the argument is sound. But unfortunately, a serious problem arises when applying this type of inference to any $x' \notin A$: the implication operation evaluates to truth whenever a false antecedent is given (independent of the consequent). Therefore, $I_{R_{A \to B}[\{x'\}]}$ will always yield 1 (true) if $x' \notin A$ is chosen. In this case the argument is not sound, as the modus ponens does not apply. Although the deduction is still (logically) correct, it violates common sense and is not desired in any practical application. It violates the cause and effect assumption, as here *noncause leads to anything*. We will point out a practical solution to this problem in Sect. 9.2.3 when considering fuzzy implication.

In the next sections, we will introduce a more formal definition of fuzzy sets, define operations on fuzzy sets and extend the relations and implications introduced above to finally define the rule bases working on fuzzy sets.

9.2.2 Fuzzy Sets

Definition 9.4 A *fuzzy set* μ_M defined over a universe of discourse X is a function $\mu_M : X \to [0, 1]$ which assigns to each element $x \in X$ a degree of membership $\mu_M(x)$ to the fuzzy set μ_M. The set of all fuzzy sets over X is denominated $F(X)$.

If $\mu_M(x) \in \{0, 1\}$ for all $x \in X$, then μ is the indicator function of a classical, or crisp, set $M \subseteq X$.

Definition 9.5 The process of mapping a crisp point x to $[0, 1]$ using a fuzzy set μ_M is called *fuzzification* and μ_M can be called a *fuzzifier*.

In most cases, the universe of discourse is the set of real numbers $X = \mathbb{R}$ or any interval on $\mathbb{R}$. Then μ is a real-valued function and can be displayed as in Fig. 9.2.

Typically, fuzzy sets describe linguistic terms like "nearly inactive", "at an average level" or "close to 100" and describe either an imprecise value or interval. Such fuzzy sets should be *convex*, that is, they should monotonically increase(decrease) when converging(diverging) to(from) a specific value or a specific interval, for example, when "getting closer to 100" or "leaving the average level". The most commonly used convex functions for fuzzy sets are triangular, trapezoidal or Gaussian functions. The choice of number, shape and parameters of fuzzy sets (e.g., median and standard deviation of a Gaussian) define the context and user dependent setup of the fuzzy model.

One of the greatest advantages of a discretization of a state space (universe of discourse) using fuzzy sets compared to the use of crisp sets is the fact that it is meaningful to overlap fuzzy sets. This allows us to express that for example, the expression of a gene is "somewhere between its normal level and its overexpressed state" instead of having to define it either as overexpressed or not (as both at the same time would be contradictory).

9.2.2.1 Operations on Fuzzy Sets

T-norms and T-conorms The classical logic assumes that propositions are either true or false, but not both true and false. In fuzzy logic, there is a gradual transition from truth to falsehood. Therefore, functions like the conjunction and disjunction have to be extended to operate on the unity interval $w_\wedge, w_\vee : [0,1]^2 \to [0,1]$ and the most basic requirement for such "fuzzy" conjunction and disjunction functions is to yield the same results as their classical counterparts when provided with either 0 or 1. Candidates for such conjunction and disjunction functions are T-norms (triangular norms, denoted $t(\ldots)$ or $\star$) and T-conorms (denoted $s(\ldots)$ or $\circ$), respectively. The most commonly used T-norms are

minimum	$t(\alpha,\beta) = \min\{\alpha,\beta\}$
Lukasiewicz-T-norm (bounded product)	$t(\alpha,\beta) = \max\{\alpha+\beta-1, 0\}$
(algebraic) product	$t(\alpha,\beta) = \alpha \cdot \beta$

Examples for commonly used T-conorms are

maximum	$s(\alpha,\beta) = \max(\alpha,\beta)$
Lukasiewicz-T-conorm (bounded sum)	$s(\alpha,\beta) = \min\{\alpha+\beta, 1\}$
algebraic sum	$s(\alpha,\beta) = \alpha+\beta-\alpha\cdot\beta$

All T-norms and T-conorms are commutative, associative and monotone, while additionally T-norms fulfill $t(\alpha, 1) = \alpha$ and T-conorms fulfill $s(\alpha, 0) = \alpha$ (see Fig. 9.4).

Union, Intersection and Complement of Fuzzy Sets The intersection of fuzzy sets can be derived from the respective definition for classical sets. An element x is in the intersection of two classical sets A and B if it is member of both sets:

$$x \in A \cap B \iff x \in A \wedge x \in B \tag{9.8}$$

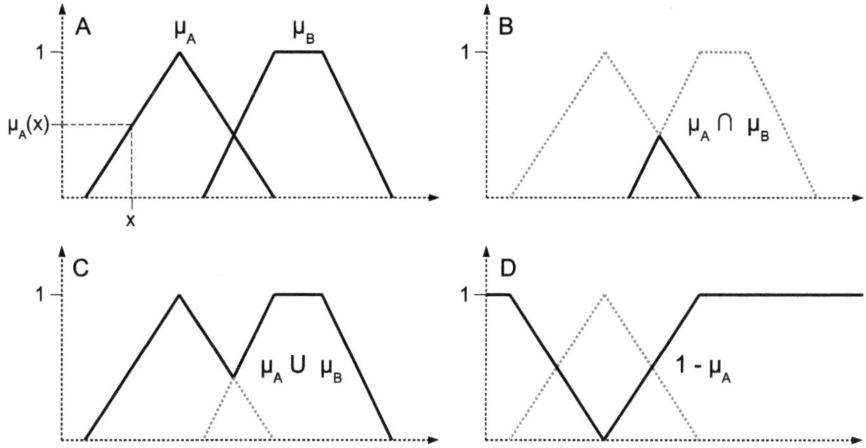

Fig. 9.4 Operations on fuzzy sets. (**a**) Two fuzzy sets μ_A and μ_B are defined with triangular and trapezoidal shape, respectively. A (crisp) point x is fuzzified by μ_A, that is, $\mu_A(x)$ is calculated using the triangular function. (**b**) The intersection $\mu_{A\cap_t B}(x)$ using the minimum T-norm. (**c**) The union $\mu_{A\cup_s B}(x)$ using the maximum T-conorm. (**d**) The complement of μ_A

The intersection of two fuzzy sets μ_A and μ_B with respect to a T-norm t can then be defined as fuzzy set $\mu_{A\cap_t B}$ with

$$\mu_{A\cap_t B}(x) = t\big(\mu_A(x), \mu_B(x)\big) = \mu_A(x) \star \mu_B(x) \tag{9.9}$$

Analogously, an element x can be in the union of two classical sets A and B:

$$x \in A \cup B \iff x \in A \vee x \in B \tag{9.10}$$

And the union of two fuzzy sets μ_A and μ_B with respect to a T-conorm s can be defined as fuzzy set $\mu_{A\cup_s B}$ with

$$\mu_{A\cup_s B}(x) = s\big(\mu_A(x), \mu_B(x)\big) = \mu_A(x) \circ \mu_B(x) \tag{9.11}$$

The complement of a fuzzy set is derived from the definition $x \in \bar{A} \Leftrightarrow \neg(x \in A)$ for classical sets and corresponds to the fuzzy set $\mu_{\bar{A}}(x) = 1 - \mu_A(x)$ (see Fig. 9.4).

9.2.3 *Fuzzy Logic Reasoning*

The use of fuzzy sets in logic reasoning leads to an extension of the classical modus ponens. This extension facilitates to reason with gradual truth values, that is, to infer conclusions from vague, imprecise knowledge.

9.2.3.1 Generalized Modus Ponens

Definition 9.6 In fuzzy logic, the modus ponens is extended to the *generalized modus ponens*

Premise 1 (Fact)	x is A^*
Premise 2 (Rule)	if x is A then y is B
Conclusion	y is B^*

The *generalized modus ponens* applies as long as there is a nonzero degree of similarity between the fuzzy set A^* of premise 1 (fact) and the fuzzy set A of the antecedent of premise 2 (rule), and as long as there is a nonzero similarity between the fuzzy set B of the consequent of premise 2 and the fuzzy set B^* of the conclusion. If additionally the degrees of truth of the premises are nonzero, a nonzero degree of truth for the conclusion can be deduced. In general, it holds that the higher the degree of truth of the premises, the higher the degree of truth of the conclusion is. When A, A^*, B, B^* are considered to be ordinary sets with $A = A^*$ and $B = B^*$ the generalized modus ponens reduces to the modus ponens.

9.2.3.2 Fuzzy Relations and Implication

A fuzzy set is a generalized ordinary set and analogously a fuzzy relation can be seen as a generalized ordinary relation. A two-valued fuzzy relation μ_R assigns a degree of membership to each pair $(x, y) \in X \times Y$ reflecting the strength of relation between x and y. Obviously, such a fuzzy relation can be defined as a fuzzy set with universe of discourse $X \times Y$.

The image of a fuzzy relation $\mu_R : X \times Y \to [0, 1]$ with respect to a fuzzy set $\mu_M : X \to [0, 1]$ is defined as a fuzzy set

$$\mu_R[\mu_M](y) = \sup\{\mu_M(x) \star \mu_R(x, y) \mid x \in X\} \tag{9.12}$$

with Y as universe of discourse and $\star$ as an intersection of fuzzy sets using a T-norm. The image of an ordinary relation (9.3) is a special case of this fuzzy set with the product as T-norm. For the sake of brevity, we will omit the notation of $[\mu_M]$ in the following.

Implication The fuzzy implication relation $\mu_{A\to B}(x, y)$ measures the degree of truth of the implication $x \in A \to y \in B$ and the image of this fuzzy implication relation with respect to the (premise) fuzzy set μ_{A^*} is

$$\mu_{B^*}(y) = \sup\{\mu_{A^*}(x) \star \mu_{A\to B}(x, y) \mid x \in X\} \tag{9.13}$$

which has the structure of a generalized modus ponens. The right-hand side of this equation defines the degree of truth $\mu_{B^*}(y)$ of a conclusion $y \in Y$ depending on an uncertain fact $\mu_{A^*}(x)$ and an uncertain rule $\mu_{A\to B}(x, y)$. As mentioned in Sect. 9.2.1.3, we want to derive a conclusion from a specific $x' \in X$. In ordinary logic, this is done by restricting the image to the set $\{x'\}$. In fuzzy logic, this restriction is performed by specifically defining μ_{A^*} as a fuzzifier for x', for example, $\mu_{A^*}(x) = 1$ for $x = x'$ and $\mu_{A^*}(x) = 0$ otherwise. Such a μ_{A^*} is called a *singleton fuzzifier*.

In general, the fact μ_{A^*} can be used to reflect our uncertainty about a crisp point $x' \in X$, which could for example be a measurement of a parameter of any type (e.g., a concentration). Therefore, one could choose μ_{A^*} to have its maximum at the observed value x' and could choose a wider support if the uncertainty about x' is large and a small support otherwise. The choice of μ_{A^*} influences the interval containing a possible supremum of (9.13). Notice that the supremum can be different from the actual parameter x'.

Singleton fuzzification is widely used as it leads to a tremendous reduction in computational cost due to the disappearance of the supremum operation:

$$\begin{aligned} \mu_{B^*}(y) &= \mu_{A^*}\big(x'\big) \star \mu_{A\to B}\big(x', y\big) \\ &= 1 \star \mu_{A\to B}\big(x', y\big) \\ &= \mu_{A\to B}\big(x', y\big) \end{aligned} \tag{9.14}$$

So in the case of singleton fuzzification the degree of truth of y is deduced by an uncertain rule from a (certain) measurement x'.

To evaluate (9.13), one has to choose a meaningful membership function for $\mu_{A\to B}(x, y)$. A possible candidate could be derived from the logical equivalence $(p \to q) \Leftrightarrow (\neg p \vee q)$ by using a T-conorm:

$$\mu_{A\to B}(x, y) = \mu_{\bar{A}\cup_s B}(x, y) = s\big(1 - \mu_A(x), \mu_B(y)\big) \tag{9.15}$$

But this choice would cause the same problem as mentioned in Sect. 9.2.1.3, that is, yielding nonzero results although the antecedent $\mu_A(x)$ is zero. Diverging from the standard definition of implication in propositional logic, a T-norm evaluates to zero if any of its arguments is zero. The monotonicity of T-norms assures that the higher the degrees of truth of its arguments are, the higher the degree of truth of the result is. Thus, a T-norm can be seen as an extension of an implication operation which preserves cause and effect, and therefore it is reasonable to choose one as fuzzy implication operator. Using the two most common T-norms minimum and product, one could derive the following membership functions for the fuzzy implication:

$$\begin{aligned} \mu_{A\to B}(x, y) &= \min\big\{\mu_A(x), \mu_B(y)\big\} \\ \mu_{A\to B}(x, y) &= \mu_A(x) \cdot \mu_B(y) \end{aligned} \tag{9.16}$$

Until now, we only considered two-valued (fuzzy) relations, which only allow implications from a single antecedent to a single consequent. The extension to multiple antecedents is straightforward by replacing the single element x by a vector of elements $\bar{x} \in X_1 \times X_2 \times \cdots \times X_n$. Equation (9.13) then extends to

$$\begin{aligned}\mu_{B^*}(y) &= \sup\{\mu_{\bar{A}^*}(\bar{x}) \star \mu_{\bar{A}\to B}(\bar{x}, y) \mid \bar{x} \in X_1 \times \cdots \times X_n\} \\ &= \sup\{\mu_{A_1^* \times \cdots \times A_n^*}(\bar{x}) \star \mu_{A_1 \times \cdots \times A_n \to B}(\bar{x}, y) \\ &\qquad \mid \bar{x} \in X_1 \times \cdots \times X_n\}\end{aligned} \tag{9.17}$$

where $\mu_{A_1^* \times \cdots \times A_n^*}(\bar{x})$ is the cartesian product realized by combining $\mu_{A_1^*}(x_1)$, $\ldots, \mu_{A_n^*}(x_n)$ by T-norms to $\mu_{A_1^*}(x_1) \star \cdots \star \mu_{A_n^*}(x_n)$ leading to

$$\begin{aligned}\mu_{B^*}(y) = \sup&\left\{ \bigcap_{i=1}^{n}{}_t\, \mu_{A_i^*}(x_i) \star \bigcap_{i=1}^{n}{}_t\, \mu_{A_i}(x_i) \star \mu_B(y) \right. \\ &\left. \mid (x_1, \ldots, x_n) \in X_1 \times \cdots \times X_n \right\} \\ = \sup&\{\mu_{A_1^*}(x_1) \star \cdots \star \mu_{A_n^*}(x_n) \star \mu_{A_1}(x_1) \star \cdots \star \mu_{A_n}(x_n) \star \mu_B(y) \\ &\mid (x_1, \ldots, x_n) \in X_1 \times \cdots \times X_n\}\end{aligned} \tag{9.18}$$

which in case of singleton fuzzification reduces to

$$\begin{aligned}\mu_{B^*}(y) &= \mu_{\bar{A}\to B}(\bar{x}', y) = \bigcap_{i=1}^{n}{}_t\, \mu_{A_i}(x_i') \star \mu_B(y) \\ &= \mu_{A_1}(x_1') \star \cdots \star \mu_{A_n}(x_n') \star \mu_B(y)\end{aligned} \tag{9.19}$$

Keep in mind that this image of the fuzzy implication relation $\mu_{A\to B}(\bar{x}, y)$ is itself a fuzzy set with Y as its universe of discourse. Equation (9.18) or its reduced form (9.19) deduce a valid conclusion from a given explicit premise $\bar{x}'$ backed by generalized modus ponens. Due to the choice of a T-norm as implication operator, the problem of an unsound argument as a result of a false premise disappears.

9.2.3.3 Fuzzy Logic Systems

We have seen how a fuzzy implication relation can be used to map a set of antecedents to a consequent, that is, how a conclusion can be derived from a set of premises using a single rule of the form *if* $\bar{x} \in \bar{A}$ *then* $y \in B$. In practice, one typically wants to derive a single result from a *set of rules* which define consequents for different (or each) combinations of antecedents.

If a conclusion should be derived from several rules, the individual results have to be combined appropriately. The image of a set of fuzzy implication relations is the disjunction of their individual images using a T-conorm:

$$\mu_{B^*}(y) = \bigcup_{j=1}^{m}{}_s\, \mu_{B_j^*}(y) = \bigcup_{j=1}^{m}{}_s\, \sup\{\mu_{\bar{A}_j^*}(\bar{x}) \star \mu_{\bar{A}_j \to B_j}(\bar{x}, y) \mid \bar{x} \in X\} \tag{9.20}$$

which can be further reduced when using *singleton fuzzification* to

$$\mu_{B^*}(y) = \bigcup_{j=1}^{m}{}_{s}\, \mu_{B_j^*}(y) = \bigcup_{j=1}^{m}{}_{s} \left(\bigcap_{i=1}^{n}{}_{t}\, \mu_{A_{j,i}}(x_i') \star \mu_{B_j}(y) \right) \tag{9.21}$$

Definition 9.7 A *fuzzy inference engine* is a function $X_1 \times \cdots \times X_n \to F(Y)$ which maps a vector of crisp inputs $\bar{x} \in X_1 \times \cdots \times X_n$ to a conclusion fuzzy set $\mu_{B^*}(y) \in F(Y)$. The mapping is specified by a fuzzifier, a set of rules (fuzzy rule base)

$$R_j : IF\ x_1\ is\ A_{j,1}\ AND\ x_2\ is\ A_{j,2}\ and\ \ldots\ and\ x_{n_j}\ is\ A_{j,n_j}\ THEN\ y\ is\ B_j$$

where n_j is the number of premises of rule j and $A_{j,i}$ specifies the fuzzy set of premise i in rule j, and the definition of a T-norm and T-conorm for conjunction and disjunction operations.

The most common combinations of disjunction and conjunction are (bounded) sum-product and max-min:

$$\begin{aligned} \mu_{B^*}(y) &= \min\left\{ \sum_{j}^{m} \left(\mu_{B_j}(y) \cdot \prod_{i}^{n_j} \mu_{A_{j,i}}(x_i') \right), 1 \right\} \\ \mu_{B^*}(y) &= \max_{j \in \{1,\ldots,m\}} \left\{ \min\left\{ \mu_{A_{j,1}}(x_1'), \ldots, \mu_{A_{j,n_j}}(x_{n_j}'), \mu_{B_j}(y) \right\} \right\} \end{aligned} \tag{9.22}$$

A fuzzy inference engine maps a (crisp) vector $\bar{x}'$ of premises to a conclusion fuzzy set $\mu_{B^*}(y)$ which assigns a degree of truth to every possible $y \in Y$. Often, one is interested in a single (crisp) representative $\bar{y}$ from the set of possible solutions Y, which should correspond to some kind of "most typical" solution. The process of deriving such a single, crisp value is called *defuzzification*.

Definition 9.8 A *defuzzifier* is a function $F(Y) \to Y$ which maps a fuzzy set $\mu \in F(Y)$ to a single, crisp value $\bar{y} \in Y$.

As for fuzzification, again there exist a wide variety of different *defuzzification* strategies. We will introduce only two of them, the *centroid defuzzifier* and the *height defuzzifier*.

The *centroid defuzzifier* computes the centroid (center of gravity) $\bar{y}$ of the conclusion fuzzy set μ_{B^*}:

$$\bar{y} = \frac{\int_Y y \cdot \mu_{B^*}(y)\, dy}{\int_Y \mu_{B^*}(y)\, dy} \tag{9.23}$$

Using $\bar{y}$ as the representative of the conclusion is quite intuitive. It can be seen as the mean or expected conclusion. However, it is quite expensive to compute. The *height defuzzifier* uses the centroids of each individual rule conclusion to determine the defuzzified value of the full result:

$$\bar{y} = \frac{\sum_{j=1}^{m} \bar{y}_j \cdot \mu_{B_j^*}(\bar{y}_j)}{\sum_{j=1}^{m} \mu_{B_j^*}(\bar{y}_j)} \tag{9.24}$$

Notice that when using *height defuzzification* the single rules do not need to be combined to $\mu_{B^*}(y)$ using a T-conorm, as the defuzzified value can be computed directly from the individual rule conclusions. Typically, the center of gravity $\bar{y}_j$ of each single conclusion is known before, so the main computational effort reduces to the evaluation of each $\mu_{B_j^*}(\bar{y}_j)$. From (9.19) (or more generally (9.18)) and the definition of T-norms, it is obvious that the center of gravity of any $\mu_{B_j^*}$ equals the center of gravity of μ_{B_j} of the according succedent fuzzy set and thus is not influenced by the parameters $\bar{x}'$.

Definition 9.9 A *fuzzy logic system* (FLS, also called fuzzy logic controller) is a function $f : X_1 \times \cdots \times X_n \rightarrow Y$ which maps a vector of crisp input values to a crisp output via a fuzzy inference engine (including a fuzzifier) and a defuzzifier.

The fuzzy inference engine is defined by a set of if-then rules. Each crisp input x_i is discretized by the fuzzy sets used in the premises of the rule set. The evaluation of the inference engine for a given input $\bar{x}'$ yields a fuzzy set $\mu_{B^*}(y)$ which is then defuzzified to a single crisp output $\bar{y}$ (Fig. 9.5).

A fuzzy logic system can be represented as a single formula. Using the product as T-norm and height defuzzification, the according fuzzy logic system is defined as:

$$\bar{y} = f(\bar{x}) = \frac{\sum_{j=1}^{m} \bar{y}_j \cdot \prod_{i}^{n_j} \mu_{A_{j,i}}(x_i')}{\sum_{j=1}^{m} \prod_{i}^{n_j} \mu_{A_{j,i}}(x_i')} \tag{9.25}$$

As stated in the introduction, the main idea of our Petri net with fuzzy logic approach is to allow a discrete, natural language description of objects and a rule based definition of processes. It was shown how this can be achieved using the mathematical framework of fuzzy sets and fuzzy logic based reasoning. We have motivated the fuzzy discretization of states and derived the validity of fuzzy reasoning from basic assumptions of propositional logic. Finally, the whole process of discretization and reasoning was condensed to a simple, single function—a fuzzy logic system.

9.2.4 Application to Petri Nets

Petri nets with their elements (places, transitions and arcs) are used for the visualization and definition of models of biological systems. Objects within such systems, biological entities like proteins, genes, RNA or metabolites but also more abstract

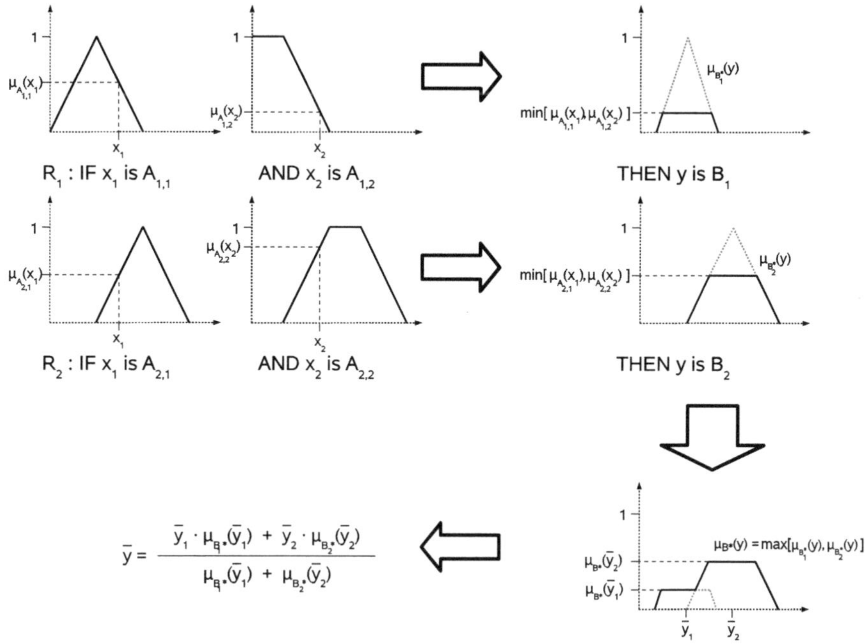

Fig. 9.5 A schematic representation of a fuzzy logic system. Crisp inputs x_1 and x_2 are fuzzified by the premises of the given rules. For each rule, a single conclusion fuzzy set $\mu_{B_1^*}$ and $\mu_{B_2^*}$ is derived by a minimum T-norm. Those conclusions are then combined to the final concluding fuzzy set μ_{B^*} using a max T-conorm. As a last step, μ_{B^*} is defuzzified using the height defuzzification. The centers of gravity $\bar{y}_1$ and $\bar{y}_2$ are those from the conclusion fuzzy sets $\mu_{B_1^*}$ and $\mu_{B_2^*}$, that is, known beforehand. Notice that μ_{B^*} does not need to be computed explicitly when using height defuzzification. Here, it is shown for the sake of completeness

entities like the accessibility of a promoter region, external influences etc., are visualized as places within the Petri net. Possible processes within the system, like creation, transport, conversion, degradation of objects or, generally, state-changes of biological or abstract objects are realized by transitions [271].

Every entity can be described using linguistic terms, like *highly concentrated*, *rather active* or *fully accessible*. Such terms can be formalized using appropriately defined fuzzy sets and the state of an entity can be described by defining degrees of similarity to these fuzzy sets. Using the same approach, uncertain measurements can be described by mapping their (crisp) value to fuzzy sets (fuzzification). A defuzzification technique can then be utilized to derive continuous, real-valued numbers as the most typical representatives of the fuzzy sets, which in turn are then stored on according places in the Petri net. Fuzzification and a subsequent defuzzification can be seen as a type of filtering for uncertain measurements.

The defuzzified values are representatives of the inexact linguistic terms used to describe knowledge and measurements and one should keep in mind that they are not exact, definite results.

The real-valued numbers stored in places are used as (crisp) inputs to fuzzy logic systems which in turn generate (crisp) real-valued outputs. Each fuzzy logic system discretizes the crisp inputs from places according to its definition, that is, according to the premises of its rule base.

Definition 9.10 A *Petri net with fuzzy logic* (PNFL) is an instance of a hybrid functional Petri net defined as a 6-tuple $PN = (P, T, A, F, W, M_0)$ where

$P = \{p_1, p_2, \dots, p_m\}$	is a finite set of places
$T = \{t_1, t_2, \dots, t_n\}$	is a finite set of transitions
$A \subseteq P \times T \cup T \times P$	is a set of arcs
$F = \{f_0, f_1, \dots, f_s\}$	is a finite set of functions $f_i : \mathbb{R}^* \to \mathbb{R}$ including the zero function $f_0 : \mathbb{R}^* \to 0$
$W : A \to F$	is an assignment of a function $f_i \in F$ to an arc
$M_0 : P \to \mathbb{R}$	is the initial marking

with $P \cap T = \emptyset$. For a general definition of a Petri net, see Chap. 3.

The values of all places connected as inputs to a transition can be used as arguments for the functions assigned to the arcs. A function assigned to an output arc ($t_j \to p_i$) defines the amount added to p_i whenever transition t_j fires, while a function assigned to an input arc ($p_i \to t_j$) defines the consumption from p_i. In most cases, these functions are fuzzy logic systems, but to keep a PNFL as flexible and user-friendly as possible, we allow the assignment of arbitrary functions to arcs. Where appropriate, the zero function $f_0 \equiv 0$ can be assigned to input arcs, i.e. the content of the adjacent place is not changed during firing, while its value can still be used in premises of other fuzzy logic systems. Such arcs are often called *read arcs* or *test arcs*. Read arcs reflect the fact that many biological processes are influenced by (biological) entities which are themselves not affected by the process, for example, the pH-value (seen as an abstract entity) of the cellular environment of an enzyme may affect its conversion rate while it may not be affected by the enzymatic reaction. The order of firing transitions is determined by firing rules. A PNFL does not impose any restriction to the type of such firing rules, that is, they can be chosen freely according to the modeling intention. Examples for firing rules are the random firing rule (transitions are randomly chosen) or stochastic firing rules based on the Gillespie algorithm [130], which preferably fire transitions according to the attributes of adjacent places. A PNFL for the feed-forward motif is presented in Fig. 9.6, specifying the according definition of the 6-tuple $PN = (P, T, A, F, W, M_0)$.

9.3 Results

9.3.1 Design Principles

In this section, the design process leading to a PNFL model is highlighted on the example of the already introduced feed-forward motif. The four noisy simulated

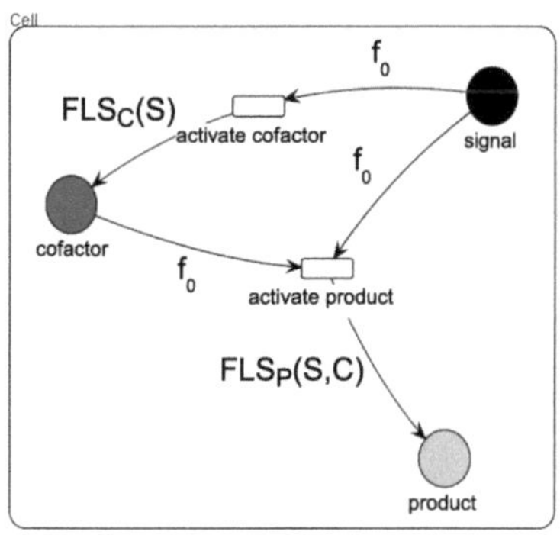

P = {Signal, Cofactor, Product }

T = {activateCofactor, activateProduct }

A = {(Signal, activateCofactor), (activateCofactor, Cofactor), (Signal, activateProduct), (Cofactor, activateProduct), (activateProduct, Product) }

F = {f_0, $FLS_C(S)$, $FLS_P(S,C)$ }

W :
f_0 → (Signal, activateCofactor)
f_0 → (Signal, activateProduct)
f_0 → (Signal, activateProduct)
$FLS_C(S)$ → (activateCofactor, Cofactor)
$FLS_P(S,C)$ → (activateProduct, Product)

$\mathbf{M_0}$: <1.0 low, 0.0 high> → Signal
<1.0 low, 0.0 high> → Cofactor
<1.0 low, 0.0 high> → Product

$FLS_C(S)$ rule system,
calculates the new Cofactor state:
IF signal is *high* **THEN** cofactor is *high.*
IF signal is *low* **THEN** cofactor is *low.*

$FLS_P(S,C)$ rule system,
calculates the new Product state:
IF signal is *high* **AND** cofactor is *high* **THEN** product is *high.*
IF signal is *high* **AND** cofactor is *low* **THEN** product is *low.*
IF signal is *low* **AND** cofactor is *high* **THEN** product is *low.*
IF signal is *low* **AND** cofactor is *low* **THEN** product is *low.*

Fig. 9.6 An example Petri net with fuzzy logic (PNFL) representing the feed-forward motif. The two fuzzy logic systems $FLS_C(S)$ and $FLS_P(S, C)$ calculate new states for C and P whenever the according transitions fire. Neither signal nor cofactor are consumed or modified by the activation reactions, still they are directly influencing them. Thus, the input arcs of this system are read arcs, that is, the zero-function f_0 is assigned to them. Initially, all concentrations of this system are low and the model is in a steady state. To simulate the reaction of the feed-forward loop on signal molecule concentrations changes, one assigns experimental measurements to the signal place and then fires the transitions. Examples for resulting time courses are given in Figs. 9.7 and 9.8

time courses of the concentration of the signal are used as inputs to the system. The time courses of concentrations of cofactor and product are to be deduced by the PNFL model. In practice, there are two basic design principles for a PNFL which specify the roles of transitions in the model.

9.3.1.1 Semi-Discrete Modeling

The first is similar to boolean networks (or, generally, discrete multi-valued logic networks). In boolean networks, starting from the current states of its entities at time point t_i, the following states at time point t_{i+1} are deduced using rule bases (truth tables), that is, the current state $S_A(t_i)$ of an entity A is replaced by a state $S_A(t_{i+1})$ which might be independent from the old state. Whenever applied, the functions of this system define the *new* state of an entity.

Following this definition, the firing of a transition in a semi-discrete PNLF should lead to a replacement of the marking of a place, that is, define a new state. The state

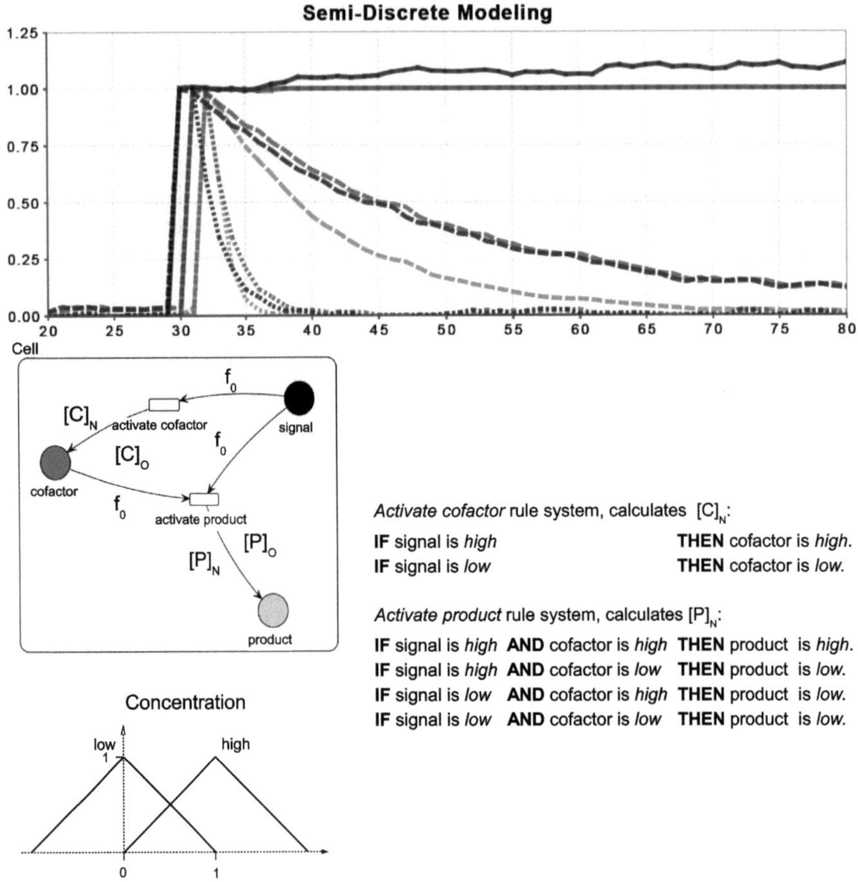

Fig. 9.7 Semi-discrete modeling. $[P]_O$ and $[C]_O$ indicate functions that completely remove the markings from the according places. $[P]_N$ and $[C]_N$ are fuzzy logic systems that define the new markings. f_0 is the null function, that is, the marking is not changed (read or test arcs). The very simple rule sets lead to a behavior that is qualitatively similar to the experimental measurements in Fig. 9.1. *Signal* (*black*) is used as input, *cofactor* and *product* are simulated (*gray* and *light-gray*). Root mean square deviation (RMSD) is about 0.09

of an entity in PNFL is described by a fuzzy discretization, that is, the degrees of membership to several fuzzy sets (Fig. 9.7).

To keep the rule bases as simple as possible, all concentrations were fuzzified using two triangular fuzzy sets *low* and *high* only. Using the boolean network like design principle, we obtain two fuzzy logic systems, one for each transition in the network. The induction of the cofactor is influenced by the concentration of the signaling molecule only and can therefore be described by two rules, while the induction of the product is influenced by both cofactor and signaling molecule and is described by four rules. Due to the simplicity of this model, the rules can easily be adjusted manually. The conclusion of a single rule is either *low* or *high*; the conclusion of a rule base is therefore a fuzzy set combined from (weighted) fuzzy sets *low* and *high*. The new state of an entity is then defined as the defuzzified value of this conclusion.

9.3.1.2 Semi-Continuous Modeling

The second design principle is similar to systems of ordinary differential equations (ODEs). ODEs define the change $\Delta[A]$ of an entity A during a time interval. The new state (value) $[A](t_i + \Delta t) = [A](t_i) + \Delta[A]$ of the variable hereby depends on the old state. Whenever applied, the functions of this system define a *change* of the old state of an entity (Fig. 9.8).

Following this definition, the firing of a transition in a semi-continuous PNFL changes the marking of a place by adding a positive or negative number.

The premises of the rule bases are still based on the two-valued discretization (*low* and *high*) of the current concentrations. However, the conclusions now define concentration changes and therefore do not correspond to the fuzzy sets *low* and *high*. Instead, they can be chosen from several singleton fuzzy sets with centers of gravity distributed across $[-1, 1]$ in 0.1 valued steps. As we use height defuzzification, the shapes of the conclusion fuzzy sets are unimportant, as only their centers of gravity are used in the calculations. In this setting, the concentration of an entity is defined by two fuzzy logic systems. One defines the increase in concentration caused by high levels of effector molecules, the other causes a self-degradation depending on the current concentration level. To derive those conclusions minimizing the root mean square deviation an optimization technique can be applied,for example, a genetic algorithm mutating the conclusions. Keep in mind that the choice of number, shape and location of fuzzy sets is user and application dependent. We just state some exemplary values here.

The semi-continuous modeling approach for a PNFl can be compared to the hybrid modeling approach as described in Chaps. 6, 13. There, the functions assigned to input and output arcs define the modifications of tokens. One could, for example, assign differential equations to the arcs and thereby approximate or mimic an ODE system. Keep in mind that a PNFL can be seen as an instance of a hybrid Petri net where fuzzy logic systems are used to define transition effects. For more details on continuous modeling approaches, see Chap. 8.

9.3.2 PNMA

The Petri Net Modeling Application (PNMA) is a software platform developed in Java for modeling, simulation and parameters estimation of Petri nets. It is possible to graphically create Petri nets in a rich user interface, which incorporates all the functionalities, the user is nowadays accustomed to. Models, together with their simulations and parameter optimization procedures, as well as results can be organized in a project management system.

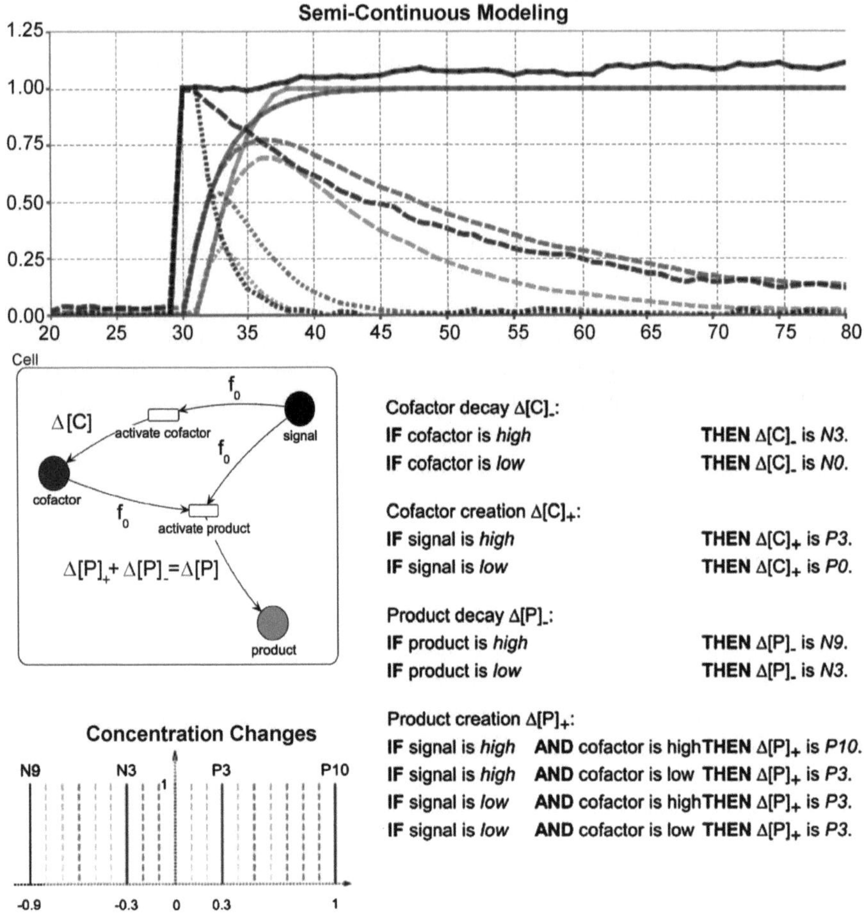

Fig. 9.8 Semi-continuous modeling. $\Delta[C]$ and $\Delta[P]$ are the sum of two fuzzy logic systems each and define the change of markings. Due to the definitions of the conclusion singleton fuzzy sets these changes are in the interval $[-1, 1]$. Only the two fuzzy logic systems defining $\Delta[P]$ are shown. The resulting time courses are very similar to the experimental with an RMSD of about 0.06

PNMA's user interface provides sophisticated graphical editors for Petri nets that allow to customize the appearance of places, transitions, cells etc., to define the arc functions and to even visualize the progress of simulations directly on the Petri net. Assistants can be used to intuitively create Petri nets built of modules. Tokens of the Petri net can be visualized in various way, for example, in tabular form, directly on the Petri net view or in time course plots. Also, the fuzzy logic components such as fuzzy sets with their membership functions or fuzzy rules can graphically be edited.

PNMA implements Petri net with fuzzy logic (PNFL) in an extended way: Instead of only allowing real numbers as markings of places, objects of any Java class are allowed. Accordingly, any Java method can be inscribed on an arc to manipulate markings when firing a transition. For simulation, PNMA exploits the Java compiler to generate an efficient bytecode version of each transition. Once the fuzzy sets and rule system is defined in PNMA's user interface, fuzzy functions are compiled as well and can thus directly be used on arcs.

The order of firing transitions is determined by various available firing rules. The most basic firing rule determines all enabled transitions at each step and fires them in a random order, skipping disabled transitions. The *simultaneous firing rule* used by PNFL fires all enabled transitions at each step, without writing results to output places, that is, all transitions fire under the same condition in a single step. After each transition has calculated its results, they are written to output places and conflicts are resolved appropriately.

Stochastic firing rules, for example, based on the Gillespie algorithm [130], provide another prominent way of simulating Petri nets. They assume a well-mixed system of molecules in which many nonreactive collisions of the molecules occur until eventually some colliding molecules react. Molecular numbers are stored as marking on places and a reactive collision corresponds to a firing transition. In such Petri nets (originally termed *reaction networks*), the Gillespie algorithm creates trajectories of molecular populations according to laws of stochastic chemical kinetics. Tau-leaping [56], hybrid methods [312] and other variants overcome certain drawbacks of the original Gillespie algorithm that is quite inefficient in models with large populations of molecules, high numbers of transitions or multiscale models. The Java library FERN [112] implements the state-of-the-art algorithms and is integrated in PNMA.

In contrast to PNFL, simulation of stochastic chemical kinetics is a very fine grained way of modeling biological system, since basically each molecule is represented as a token. The proposed functional hybrid Petri nets are able to model both PNFL and stochastic chemical kinetics and it is in principle even possible to create models of mixed granularities. In addition, the Petri net implementation of PNMA can also handle spatial models, where not only temporal dynamics are considered, but also the interactions of components across spatial dimensions.

9.3.3 *Parameter Estimation*

In most cases, modeling of biological networks is a reverse engineering problem: Experimental data determine the desired simulation results and the task is to find the model and its parameterization that is able to reproduce these data. In the case of PNFL, these parameters are fuzzy rule definitions. This reverse engineering can be reduced to an optimization problem: Given a similarity function for simulation results and experimental data, the task is to minimize this function with respect to the model parameters. Such a function is, for instance, the root mean square deviation (RMSD):

$$R_X(Y) = \frac{1}{N}\sum_{i=1}^{N}(X_i - Y_i)^2 \qquad (9.26)$$

where X is the list of the N experimental data points and Y is the simulated data and as such determined by the model and its parameters. Each parameter combination P results in a vector Y_P of simulated data and thus in a similarity value $R(Y_P)$. The task is to find the best that is, minimal value of $R(Y_P)$ for all parameter combinations P. In PNFL, P is for instance a choice of conclusions for predefined fuzzy rules for each transition. Note that in this case, the parameter search space (i.e., the set of all possible parameter combinations) is finite, which is not the case if for instance rate constants are used instead of fuzzy logic systems. However, the search space can still be very large: Consider a transcription factor regulatory network of 10 genes and assume, for simplicity, that each gene is regulated by exactly two of the other genes and only the two fuzzy sets *inhibition* and *activation* are allowed as conclusions. If each transcription factor concentration is only modeled as fuzzy sets *low* and *high*, 2^4 parameter combinations are possible for a single fuzzy function. Thus, for the network's ten fuzzy functions, there are 2^{40} possible combinations, which prohibits a full enumeration of the parameter search space to find the best $R(Y_P)$.

As indicated, minimization of $R(Y_P)$ with respect to P is a hard problem and is usually done using heuristic techniques like genetic algorithms or simulated annealing [136, 192]. Briefly, these heuristics perform walks on the parameter search space attempting to walk towards the global minimum. At each step (also called move), the objective function is evaluated to assess the usefulness of the last move. The methods vary in many ways, for example, which moves are allowed and how to handle bad moves. The most promising approaches combine elements of different techniques in order to find this minimum more quickly and reliably. Furthermore, it is of great importance to integrate as much available data as possible into the optimization procedure to keep the parameter search space tractable.

Thus, a general parameter estimation framework has to fulfill high standards in flexibility and efficiency. PNMA provides a very powerful system for optimization that allows the user to include various types of data and to customize the optimization procedure itself. The optimization workflow can be specified using a Petri net. Hence, PNMA does not only use its Petri net implementation for modeling biological systems, but also to allow the user to graphically compose full-fledged optimization programs. Several building blocks are provided to create for example, a simulated annealing procedure in a straight forward way (see Fig. 9.9) and, using the plugin system, additional modules for this graphical programming can easily be integrated.

9.3.4 Application

The DREAM challenge [242] is a scientific initiative to assess the power of different methods to infer biological meaningful networks from experimental data. In each challenge, data is provided and scientists are invited to infer the underlying networks with their favorite method. So far, four challenges have taken place, each consisting of three to five subchallenges.

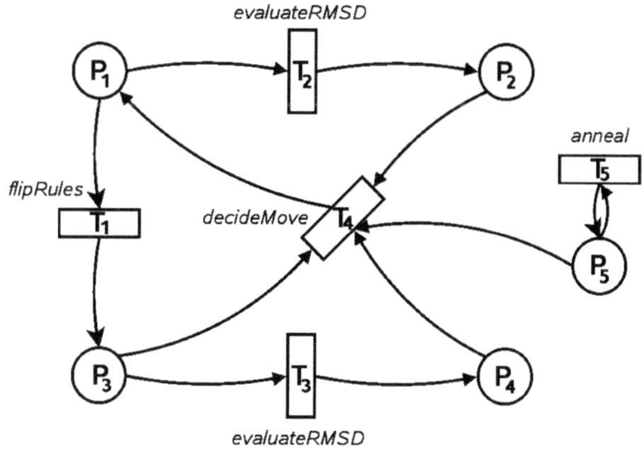

Fig. 9.9 A simple optimization program. Conclusions of the rule system of fuzzy function *activateProduct* of the feedforward motif (see Fig. 9.7) are estimated in a simulated annealing-like manner. Place P_1 always contains the currently best estimate for all conclusions of *activateProduct* and P_2 its RMSD. P_3 contains the next guess for the conclusions and P_4 its RMSD. Transition T_1 generates a next guess by randomly flipping the conclusions of some rules (e.g., from fuzzy set *low* to *high*). T_2 and T_3 evaluate $R(Y_{P_1})$ and $R(Y_{P_3})$, respectively, by starting a simulation of the model Petri net and comparing its results to experimental data. T_4 compares the scores and decides if the token on P_1 remains unchanged or gets replaced by the token from P_3. A classic simulated annealing would be: If $P_4 > P_2 \rightarrow$ replace, otherwise replace with probability $\exp(-\frac{P_4-P_2}{T})$. T is the temperature on place P_5 and is decreased during the optimization by transition T_5 to disfavor bad moves in the late steps of the optimization

The DREAM4 In Silico Network Challenge goal is to reverse-engineer regulatory networks of 10 genes and 100 genes, respectively. For each of these networks, that are known to be subnetworks of gene regulation in *E. coli* or *S. cerevisiae*, four kinds of data sets have been generated in-silico. Only the generated data is published and neither the network topology nor the exact parameterization of the in-silico network. The goal is to infer the network as a list of confidence ranked interactions between the genes. The data sets are:

1. Knock-down data: Steady state levels of wild-type and heterozygous knock down strains for each gene are given. A knock-down strain is simulated by cutting the expression rate of the knocked-down gene in half.
2. Knock-out data: Similarly to the knock-down, steady state levels are given for strains with completely disabled expression for each gene.
3. Time series data: 21 measurements of the network's simulation are given after recovering from an external perturbation. It is unknown, which genes have been perturbed. For the networks of size 10, size 50 and size 100, 4, 23 and 46 time courses of different perturbations are provided.
4. Multifactorial data: Some of the genes' kinetic parameters are slightly altered and steady state levels are given.

Random noise has been added to all data types to mimic experimental variances and inexact measurements.

Since the data provided is not collected by experiment but simulated in-silico, it is obvious that a computational model consisting of the given number of genes (10 or 100) actually exists that can reproduce all given data (disregarding the noise). In real wet-lab applications, this assumption does not hold true, as the number of involved genes is usually not really known and side effects are possible. But also when considering in-silico experiments, most probably there is not a single model that can explain the data [241] but a multitude of them.

Of course, the network and its parameters are unknown. An optimization program can be used to infer the network, which typically involves repeated simulation with varying parameters and comparing the outcome to the reference. This can be a time consuming task even for the size 10 network. For both, efficient parameter estimation and meaningful predictive power, a simple model is required. A lot of parameters are hard to estimate and tend to overfit given data. However, the model must be complex enough to represent the key aspects of the real system. We propose that fuzzy logic as a basis of modeling biological systems with Petri nets can help to walk the fine line between oversimplification and data overfitting.

Using semi-quantitative methods like fuzzy logic to handle the dynamics and introducing prior knowledge are potent means to restrict the search space, most notably, fuzzy modeling often allows to completely dispose continuous parameters, as indicated above. Our method for network reconstruction is to simulate all given experimental conditions (knock-down, knock-out, perturbations, multifactorial) with PNFL and derive a total score (e.g., by evaluating the RMSD) by comparison to the reference data. Using a heuristic optimization procedure based on a combination of genetic algorithms and simulated annealing, the optimal network topology and its parameters are inferred.

The results of the DREAM4 challenge give an impression of the power of our approach: All of the PNFL derived models we sent in scored well above our competitors. How well PNFL works for experimentally derived data must still be elucidated but our success in the DREAM4 challenge strengthens our point that a Petri net with fuzzy logic is an appropriate model for biological systems.

9.4 Related Work

9.4.1 Common Network Motifs and Cell Cycle Model

Bosl [42] showed the suitability of fuzzy logic based models to describe well known and typical biological motifs by examining several small networks. These simulated networks contained representatives of positive and negative feedback loops, feed-forward motifs, single input motifs as well as classical chemical reaction kinetics. All dynamics were derived from natural language descriptions of those motifs, as they could for example be found in publications. In addition to the simple motifs, an oscillating model of the mammalian cell cycle was built which demonstrates the ability to reproduce complex behavior. Cell growth is integrated into this model seamlessly using the same fuzzy logic framework as used for protein dynamics. This shows the flexibility of the framework and its ability to model dynamics on several levels.

9.4.2 Comparison of Fuzzy Logic to Multi-Class Discrete Logic

A comparison of fuzzy logic to discrete (ordinary) logic based modeling was done by Aldridge et al. [7] on a model of a signaling pathway downstream of TNF, EGF and insulin receptors in human colon carcinoma cells. They predicted concentration changes of downstream molecules as a response to different stimuli, that is, a treatment with several different combinations of concentrations of TNF, EGF and insulin. Time series data of downstream molecules were collected experimentally and compared to the fuzzy logic and to the discrete logic based simulations. The set-based discretizations and rule bases had been manually adjusted and the same rule systems were used for fuzzy logic as well as for discrete logic based modeling. Keep in mind that ordinary sets can be used as premises and conclusions in fuzzy logic systems, as they are special cases of fuzzy sets. The fuzzy logic system showed an improved fitness which is caused by the soft transitions between states and the ability to adopt and stay in intermediate states. Additionally, Aldridge et al. reduced both the fuzzy logic as well as the discrete logic to two-states models (the latter is similar but not identical to a boolean model) and optimized the parameters of the system using a global nonlinear least squares regression. Both models were then compared to a true boolean model. Again, in this setup the two-state fuzzy logic performed better than its two-state discrete model counterpart, while both models had a better fitness than the boolean model.

9.4.3 Classification of Gene Expression Data

There is a wide range of applications for fuzzy logic in computational biology. It is not only suited for modeling and simulation of networks but for example also used for applications like classification of gene expression data. Schaefer et al. [340] use FLSs to classify three independent data sets of cancer patient biopsy samples. For each data set, the top fifty significant genes were considered for training and testing of the fuzzy classifier. This example is suited to elucidate the *curse of dimensionality* every rule based system suffers from. Even if only two fuzzy sets would be used to describe the expression of a gene (*low* and *high* expression) a single FLS using 50 genes as premises contains a rule base consisting of $2^{50} \approx 10^{15}$ different rules. The evaluation of such a FLS is clearly prohibitive due to computational requirements. This problem could be solved by realizing the classifier as a combination of $\binom{50}{2} = 1225$ FLSs with two premises each only, that is, a FLS is created for each pair of genes. Each of these FLS classifies a sample based on two genes only and the classification of the sample in total is therefore based on $\binom{50}{2}$ single classifications. $\binom{50}{2}$ fuzzy logic systems with two premises discretized by two fuzzy sets lead to $4 \cdot 1225 = 4900 \ll 2^{50}$ rules which are computationally tractable. It is quite intuitive that not all of the 1225 gene pair expressions levels are suited to distinguish between the considered classes. Therefore, the number of FLS can be reduced further without decreasing the classification accuracy significantly. The fuzzy classifier performs well with accuracy levels comparable to a nearest neighbor classifier, where the latter is well suited for gene expression classification.

9.5 Summary

The adaption of fuzzy sets for representing states of entities and fuzzy logic based reasoning for describing processes can be used to model biological systems. Fuzzy sets capture the typically inexact, qualitative knowledge about biological entities and are well suited to represent limited and/or incomplete knowledge, inexact measurements as well as error prone data. Due to the fact that they can stand for arbitrary properties, it is possible to uniformly represent all types of external and internal factors influencing a system. Fuzzy sets can be designed freely by a user according to his needs. Fuzzy logic systems allow the formulation of biological processes using simple yet powerful rule systems, which can be formulated using natural language. Therefore, hypotheses concerning the behavior of entities or influences between entities can be translated directly into executable systems. The representation using Petri nets clearly visualizes entities, processes and dependencies within a biological system. A Petri net and fuzzy logic based system can easily be outlined in a pen-and-paper style by creating drafts of entities and their dependencies and describing the desired properties and effects of dependencies and influences in natural language.

9.6 Problems

9.1 Fuzzification

Two fuzzy sets *inactive* and *active* are specified by a set of point coordinates each:

$$\textit{inactive} \cdots \big[(-1,0),(0,1),(1,0)\big] \tag{9.27}$$

$$\textit{active} \cdots \big[(0,0),(1,1),(2,1),(3,0)\big] \tag{9.28}$$

Draw a sketch of these fuzzy sets in a coordinate system. Fuzzify the crisp numbers 0.5, 0.75, 1, 1.5. Notice that in some cases the crisp numbers differ while their fuzzy discretizations can not be distinguished.

9.2 Defuzzification

Defuzzify the fuzzy discretizations derived in Exercise 9.1 using the height defuzzification. Notice that the defuzzified numbers may differ from the original numbers.

9.3 Implication

An input vector $\bar{x} = (0.5, 0.75)$ and a rule system are given:

$$\text{IF } x_1 \text{ is inactive AND } x_2 \text{ is inactive THEN } y \text{ is inactive} \quad (9.29)$$

$$\text{IF } x_1 \text{is inactive AND } x_2 \text{ is active THEN } y \text{ is active} \quad (9.30)$$

$$\text{IF } x_1 \text{ is active AND } x_2 \text{ is inactive THEN } y \text{ is active} \quad (9.31)$$

$$\text{IF } x_1 \text{ is active AND } x_2 \text{ is active THEN } y \text{ is active} \quad (9.32)$$

Sketch the conclusion fuzzy sets as in Fig. 9.5. Derive the formula for the fuzzy logic system using a product t-norm, sum t-conorm and height defuzzification. Calculate $\bar{y}$.

9.4 Create a PNFL

Create a complete Petri net with fuzzy logic based on a very short description of a small, well-known biological motif (negative-feedback loop):

"The basal transcription of X is completely inhibited by the active form of Z. X itself is a transcription factor that is necessary for the creation of Y. The inhibitor Z can only be found in its active state in the presence of Y."

To create the PNFL, first draw a Petri net representing the dependencies between the given entities. Then define fuzzy sets that discretize the possible states of the entities and convert the given description to collections of if-then rules. Choose t-norm, t-conorm and a defuzzification strategy. Derive the formulas for the resulting fuzzy logic systems.

Part III
Biochemical Applications

Chapter 10
Topological Analysis of Metabolic and Regulatory Networks

Stefan Schuster and Björn H. Junker

Abstract The theoretical apparatus of Petri nets has been widely used for visualizing metabolic and regulatory networks and for describing their properties and behavior in a quantitative way. In this chapter, the theoretical basis, algorithmic issues and biological applications of using Petri nets in that field are reviewed, in particular, in view of topological (structural) analyses. Several useful notions such as T-invariants, P-invariants, and Maximal common transition sets are explained. The correspondence between several of these concepts and similar concepts in traditional biochemical modeling, such as between minimal T-invariants and elementary flux modes, is discussed. The presentation is illustrated by several hypothetical and biochemical examples. A larger running example is taken from sucrose metabolism in plants. For this, an important difference in functioning between monocotyledon and dicotyledon plants is explained. Algorithms and software tools for determining structural properties of Petri nets are briefly reviewed.

10.1 Introduction

The theoretical analysis, modeling and computer simulation of metabolic and regulatory networks is an integral part of Systems Biology (cf. [157, 205, 290, 414]). These approaches have manifold applications in biotechnology, medicine and other fields. As described in other chapters in this book, various properties of, and phenomena occurring in, such networks are worth being studied, such as the dynamic

S. Schuster (✉)
Biologisch-Pharmazeutische Fakultät, Lehrstuhl für Bioinformatik, Friedrich-Schiller-Universität Jena, Ernst-Abbe-Platz 2, 07743 Jena, Germany
e-mail: stefan.schu@uni-jena.de

B.H. Junker
Leibniz Institute of Plant Genetics and Crop Plant Research (IPK), Corrensstr. 3, 06466 Gatersleben, Germany
e-mail: junker@ipk-gatersleben.de

I. Koch et al. (eds.), *Modeling in Systems Biology*, Computational Biology 16,
DOI 10.1007/978-1-84996-474-6_10,

behavior, optimality properties, stability and robustness. For many of these studies, a rather detailed knowledge of parameters is necessary. For example, to perform dynamic simulations (e.g., of oscillations), usually many kinetic parameters must be known. This information is not often available completely, all the more when these parameters change in time, such as enzyme concentrations. For signal transduction and gene regulatory networks, measuring kinetic parameters is particularly difficult. Therefore, it is of interest to restrict the analysis, at least at the outset, to phenomena that are independent of imperfectly known parameters. And even if they were known, it is worthwhile starting the analysis by studying the invariants of the system because this provides a basic understanding, which is very helpful at later stages of the analysis [157, 291, 316]. Moreover, certain properties such as basic routes in signaling systems can only be derived by appropriate tools, which differ from dynamic simulation [327, 351].

In metabolic and regulatory systems, the structural (network) properties are practically invariant since these are given by the stoichiometry of the reactions involved, which is constant in most cases. Sometimes, variable stoichiometries are observed, such as the number of protons needed to produce one mole of ATP by H^+-ATPase [414]. However, this results from a higher-level description of that enzyme. If the elementary steps are considered, stoichiometry is constant.

About two decades ago, the search for methods to analyze invariants of intracellular networks led to the development of an entire field of research—the topological analysis, also called structural analysis or network analysis [157, 170, 290]. A subfield is called metabolic pathway analysis or constraint-based modeling [291, 355, 397], in which the structure of pathways (routes) going through the system is detected and/or optimal flux distributions are calculated. Network analyses of other systems, such as sociological networks and electrical circuits, have a long tradition. They are still flourishing and rapidly growing fields, for example, with respect to social networks [118], the Internet [421], and car traffic [394]. From a relatively scarce amount of input information, multitudinous, far-reaching conclusions can be drawn. For example, the robustness of networks against failure of, or attacks on, particular constituents can be predicted [33, 170, 375]. In the networks under study in this chapter, this concerns the robustness against knockout mutations or enzyme deficiencies.

Since metabolic systems can be regarded as networks (as can be seen on the well-known Boehringer charts decorating many biochemistry labs), it is tempting to describe them as graphs in the sense of graph theory. Graphs consist of nodes and edges, which might represent metabolites and reactions, respectively. However, this is difficult due to the presence of bi- or multimolecular reactions, which would require that one arc could link three or more nodes. As a simple, yet instructive biochemical example, consider a system describing the pathway by which sucrose is converted to starch in higher plants (Fig. 10.1). The topological properties of simpler versions of that network have been analyzed earlier for the case of the potato tuber [211, 322]. Note that besides the ubiquitous ATP-dependent phosphofructokinase (PFK) in plants there additionally is a pyrophosphate-dependent phosphofructokinase (PFP) that uses pyrophosphate rather than ATP as phosphate donor [259].

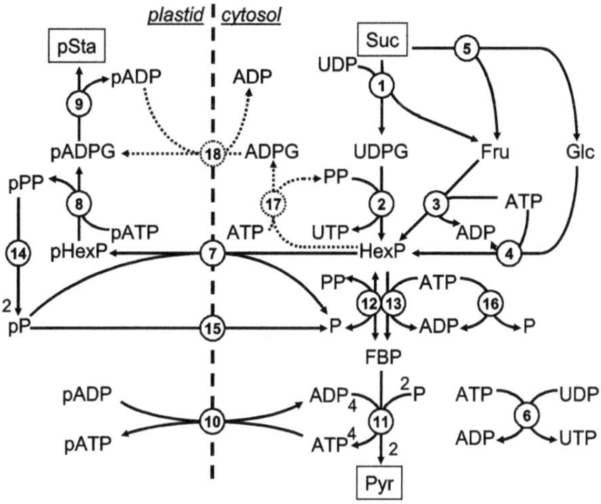

Fig. 10.1 Reaction scheme representing the pathway of sucrose-to-starch conversion in storage tissue of monocotyledon and dicotyledon plants. For clarity, most reversible reactions are shown in their predominant direction only. Metabolites treated as external (for definition, see Sect. 10.2) are boxed. The representation somewhat differs from that normally used for Petri nets in that reaction numbers are encircled rather than boxed. Reactions 17 and 18 (*dashed lines*) are present in the extended model (monocotyledons) only (cf. [82]). Enzyme numbers with names and abbreviations: 1, sucrose synthase (SuSy); 2, UDP-glucose pyrophosphorylase (UGPase); 3, fructokinase (FK); 4, hexokinase (HK); 5, invertase (Inv); 6, nucleoside diphosphokinase (NDPK); 7, hexose phosphate transporter (HexPT); 8, plastidial ADP-glucose pyrophosphorylase (pAGPase); 9, plastidial starch synthase (pStSy); 10, ATP/ADP translocator (AATL); 11, lumped reaction of lower glycolysis (Glyc); 12, pyrophosphate-dependent phosphofructokinase (PFP); 13, ATP-dependent phosphofructokinase (PFK); 14, plastidial pyrophosphatase (pPPase); 15, phosphate transport (PT); 16, lumped reaction of ATP consumption (ATPcons); 17, cytosolic ADP-glucose pyrophosphorylase (AGPase); 18, ADP-glucose transporter (ADPGT). Abbreviations of metabolites: ADP, adenosine diphosphate; ADPG, adenosine diphosphate glucose; ATP, adenosine triphosphate; FBP, fructose-1,6-bisphosphate; Fru, fructose; Glc, glucose; HexP, hexose phosphates; P, phosphate; PP, pyrophosphate; Pyr, pyruvate; Sta, starch; Suc, sucrose; UDP, uridine diphosphate; UDPG, uridine diphosphate glucose; UTP, uridine triphosphate. The small letter "p" in front of a metabolite name indicates the plastidial pool of this metabolite

Examples of bimolecular reactions in that scheme are sucrose synthase and invertase. By a simple graph-theoretical analysis, one can easily deduce that there is a route from sucrose to pyruvate. However, the exact overall stoichiometry (net reaction equation) is difficult to derive because the fate of the byproducts of the bimolecular reactions (e.g., fructokinase), which may give rise to products being co-substrates further down in the pathway must be taken into account.

There are even cases where connected routes do not imply a steady-state mass flow. An illustrative example is the conversion of even-chain fatty acids into sugar [89, 412]. It is well known that sugar in the human diet can be converted into fat. It is not so easy to decide, by inspecting the biochemical network, whether the opposite transformation is feasible, at least for the major constituents of lipids, even-chain fatty acids. Due to the irreversibility of some reactions (e.g., pyruvate dehydrogenase), the pathway exactly reverting the conversion pathway from sugar to fatty acids is infeasible. However, there is another connected route running via the tricarboxylic acid cycle (cf. [378]). Importantly, it can be shown mathematically that, in most animals including humans, this route can not carry a steady-state flux [89, 412] because not every metabolite (e.g., oxaloacetate) is balanced with respect to production and consumption although there are producing and consuming reactions for each of these. In contrast, green plants, fungi, nematodes and many bacteria are capable of converting fatty acids into sugar, via the glyoxylate shunt, which is absent from most animals (cf. [378]). This example shows that simple graphs are generally insufficient to describe stationary flux distributions (cf. [199]). Nevertheless, for other investigations, for example, in analyzing the flow of radioactive tracer, methods based on simple graphs are appropriate.

An appropriate way of coping with mass balance in metabolic networks is provided by Petri nets (for a definition, see the introductory chapter on Petri nets in this book). They are bipartite graphs with two types of nodes, called places and transitions. In metabolic systems, places represent metabolites and transitions stand for reactions. Thus, bimolecular reactions can be described properly (cf. [63, 208]). Petri nets involve the stoichiometry by weighting the arcs by the stoichiometric coefficients. Early papers applying Petri nets to metabolic systems were published by Hofestädt [162] and Reddy et al. [315]. Later, many other papers (some of them reviewed in this chapter) followed. Sometimes, bipartite graphs are used without calling them Petri nets (e.g., [170, 426]).

In the field of Petri net theory, the interest in structural properties started relatively early. Notably the definition of T-invariants and P-invariants [219] (cf. [370]) was a remarkable milestone (*see Chap.* 3: *P- and T-invariants*). Intuitively speaking, T-invariants are flux distributions that allow the network to be at a stationary state. P-invariants are linear combinations of variables (e.g. substance concentrations) that remain constant in time. Note that this constancy property of P-invariants holds in all dynamic regimes (oscillations etc.) and is trivially fulfilled when the system is at steady state. Later, other invariants in Petri nets have been defined and investigated, such as the Maximal common transition sets (MCT-sets, [327], *see Chap.* 4). In this chapter, these structural properties will be reviewed and relevant biological applications will be discussed.

10.2 T-invariants

10.2.1 Modelling Steady States

The time behavior of biochemical and regulatory networks can, under certain simplifying conditions such as spatial homogeneity, be described by the ordinary differential equation system

$$\frac{\mathrm{d}X}{\mathrm{d}t} = \underline{N}V \tag{10.1}$$

where X and V denote the vectors of concentrations and net fluxes, respectively, $\underline{N}$ stands for the incidence matrix of the net and t denotes time. The fluxes correspond to the flow of tokens per time in Petri nets. In addition to the usual places in Petri nets, input and output places can be defined, in which the token reservoir is inexhaustible and, thus, assumed to be unaffected by token flow. Therefore, these places do not enter the equation system (10.1). In metabolic networks, they correspond to the *external metabolites*. In contrast, internal metabolites (places) do enter (10.1).

Metabolic networks often attain, after some initial transient period, a stationary state. Equation (10.1) can then be simplified to the algebraic equation system:

$$\underline{N}V = 0 \tag{10.2}$$

because the concentrations do not then change in time. This equation is isomorphic to (4.3) *in Sect.* 4.4 ("*Transition invariants*"), which defines the T-invariants in Petri nets.

A T-invariant (transition invariant) is a vector with the property that if each transition fires as many times as the corresponding entry of the vector indicates, the original marking is regained. These vectors are the solutions of (10.2). Thus, T-invariants correspond to flux distributions in reaction networks at steady state [354, 407].

Since Petri nets usually include irreversible transitions only, all components of a T-invariant must be non-negative. Such invariants are called true T-invariants. Very often, the net direction of all biochemical reactions in a given system is known, that is, whether the forward or reverse reaction step is faster. This knowledge is available, for example, when the reaction is irreversible or fulfills a defined biochemical function. In this case, the orientation of all reactions can be chosen in such a way that their (net) fluxes are nonnegative. Then only steady-state flux distributions corresponding to true T-invariants are relevant. There are many cases, though, where the net direction is not fixed for all reactions. For example, the net flux though isomerase reactions can change direction upon changing physiological conditions (cf. [378]). Usually, for each of these reversible reactions, a forward transition and backward transition are included separately into the Petri net.

10.2.2 Minimal T-invariants and Elementary Modes

The set of all T-invariants of a network forms a linear vector space. If I_1 and I_2 are invariants, also the linear combination $\lambda_1 I_1 + \lambda_2 I_2$ (with λ_1, λ_2 being real numbers) are invariants of the net (cf. [318]). The solution space of the algebraic equation (10.2) is infinite, which is a motivation to search for a unique set of basic vectors in that space. For obvious reasons, it is convenient to find a set of vectors that are as simple as possible. An analogy to Fourier analysis can be drawn. Complicated sounds can be decomposed into pure harmonics. Likewise, all T-invariants can be written as a linear combination of minimal T-invariants [219] (*for their definition, see Sect.* 4.4 "*Transition invariants*"). However, in contrast to Fourier analysis, the coefficients in the linear combination must be nonnegative in order not to violate the irreversibility constraint for transitions.

Independently of the developments in Petri net theory, the concept of elementary flux modes was introduced in metabolic network analysis [346, 349], based on earlier attempts in theoretical chemistry by Clarke [78]. Intuitively speaking, an elementary flux mode is a minimal set of enzymes that could operate at steady state and complies with the directionality of irreversible reactions. That means that no other flux distribution at steady state is a proper subset of an elementary mode.

To decompose flux distributions into such simplest relevant routes helps one better understand the behavior of biochemical systems. The concept has been applied to numerous biochemical systems that are relevant in biotechnology and medicine. For example, elementary modes in carbon fixation in green plant seeds [356], in methylotroph bacteria [403], in cyanophycin production by recombinant strains of bacteria [92] and in the conversion of fatty acids into sugar [89] have been analyzed. There is a close similarity between minimal T-invariants and elementary modes [354, 407]. The concept of elementary modes is somewhat more general because reversible reactions need not be decomposed. This saves the effort of discarding cyclic minimal T-invariants within one and the same reaction. Algorithms for computing minimal T-invariants and elementary flux modes have first been developed in parallel. Interestingly, they show some similarities (*see Sect.* 10.3).

We now return to our biochemical example (Fig. 10.1). The pathway by which sucrose is metabolized to starch in the potato tuber has been reported to yield 12 minimal T-invariants [211]. This number results from a few sub-pathways that can be present in different combinations in the T-invariants. To concentrate on the main metabolic process mediated by this network, that is, the conversion of sucrose to starch, we simplified the system by (a) combining the hexose phosphates (glucose 1-phosphate, glucose 6-phosphate, and fructose 6-phosphate) into one pool (as has been done earlier by Rohwer and Botha [322]), (b) omitting sucrose phosphate synthase and -phosphatase, and (c) treating sucrose synthase irreversible in the direction of sucrose cleavage. As a result, no futile cycles can be detected. They have been studied in-depth earlier [211, 322]. The remaining, relative simple metabolic network is complicated enough that the pathways across the system can not easily be seen by inspection. Sucrose, pyruvate and starch are here considered as external metabolites. It is often not easy to decide whether cofactors should be treated to be internal or external. The rule of thumb is that they should be set to external status if they participate not only in the reactions included in the model but also in non-negligible other reactions, because then, their mass balance can not be described by the system equations of the model. In contrast, if the network under study is large enough to balance them, they can be treated to be internal. This is the case here for ATP, ADP and pyrophosphate. As starch is a polymer of glucose, we count the mole number of starch in hexose units.

Computation of the T-invariants with the program METATOOL [406] (http://pinguin.biologie.uni-jena.de/bioinformatik/) reveals that the simplified network as

Table 10.1 Minimal T-invariants of the model describing sucrose metabolism in plants[a]

Number	List of enzymes	Overall stoichiometry
1	Inv, FK, PFK, Glyc, HK, HexPT, pAGPase, pStSy, PT, pPPase, AATL	Suc = 2Pyr + pSta
2	4*SuSy, 4*UGPase 4*FK, 5*HexPT, 5*pAGPase, 5*pStSy, 5*AATL, 5*pPPase, 5*PT, 7*PFK, 4*PFP reserve, 3*Glyc, 4*NDPK	4Suc = 6Pyr + 5pSta
3	5*Inv, 5*FK, 5*HK, 2 HexPT, 2*pAGPase, 4*AGPase, 4*ADPGT, 6*ptStSy, 2*AATL, 2*pPPase, 2*PT, 4*PFP, 4*Glyc	5Suc = 8Pyr + 6pSta
4	2*SuSy, 2*UGPase, 2FK, HexPT, pAGPase, 2*AGPase, 2*ADPGT, 3*pStSy, AATL, pPPase, PT, PFK, Glyc, 2*NDPK	2Suc = 2Pyr + 3pSta
5	5*SuSy, 5*UGPase, 5*FK, HexPT, pAGPase, 7*AGPase, 7*ADPGT, 8*pStSy, AATL, pPPase, PT, 2*PFP, 2*Glyc, 5*NDPK	5Suc = 4Pyr + 8pSta

[a]The minimal T-invariants numbered 1 and 2 apply to dicotyledons. In the extended model applicable to monocotyledons, all five minimal T-invariants are valid. For abbreviations, see legend to Fig. 10.1

depicted in Fig. 10.1 yields two minimal T-invariants only, given in the upper part of Table 10.1. Both can be easily interpreted in biochemical terms: they describe the formation of starch from sucrose and differ in that the first uses the enzyme invertase to cleave sucrose, while the second uses sucrose synthase for that purpose. A consistency check can be made by counting carbons in the overall stoichiometries. By taking into account that sucrose, pyruvate and (one monomer of) starch include 12, three and six carbons, respectively, the results are indeed consistent.

The elementary modes (minimal T-invariants) can be used to determine the maximal molar yield (product : substrat ratio). While T-invariant 1 generates only one mole of starch (counted as hexose monomer) per mole of sucrose, T-invariant 2 is slightly more efficient by producing 5 moles of hexose units in starch from 4 moles of sucrose. It can be shown mathematically that the flux distribution allowing the maximum yield with respect to a given substrate-product pair always coincides with an elementary mode [349]. Thus, $5 : 4 = 1.25$ is the maximum yield possible by this network. The ability to use sucrose synthase (in the reverse direction) to cleave sucrose is an advantage of plants over many other organisms such as animals and yeasts, which exclusively use invertase to cleave sucrose into fructose and glucose. Sucrose synthase, in contrast, gives fructose plus glucose-phosphate, so that one ATP for phosphorylation is saved. Thus, green plants show both a higher redundancy and a better ATP economy in sucrose degradation than other organisms. The pathways described above for the potato tuber are characteristic for one of the two large groups of flowering plants, the dicotyledons (their seedlings typically have two cotyledons, that is, leafs of the seedling). The other group, the monocotyledons, exhibit an alternative route of starch synthesis: they possess a cytosolic form of the enzyme ADP-glucose pyrophosphorylase, and an ADP-glucose transporter in the plastidial membrane (cf. [82]). When these two reactions are included in the metabolic network described in Fig. 10.1, the system gives rise to five minimal T-invariants (Table 10.1). Apart from the two T-invariants of the basic model, there are three T-invariants making use of the two additional reactions. While in T-invariants 4 and 5, sucrose synthase is used to cleave sucrose, the way via invertase is taken in T-invariant 3. The yield of starch per sucrose in the new T-invariants is 1.2, 1.5, and 1.6 for T-invariants 3, 4, and 5, respectively. Thus, the latter two are higher than with the plastidial AGPase alone. This effect can be explained by the fact that the energy-rich compound pyrophosphate produced by the cytosolic AGPase can be directly recycled in the UGPase reaction, while the pyrophosphate in the plastid can only be cleaved by pyrophosphatase and transported back to the cytosol, thereby loosing its function as a donor of chemical energy. Interestingly, the cytosolic AGPase always appears together with the plastidial AGPase, but never alone, in a flux ratio of 2:1 (T-invariants 3 and 4) or 7:1 (T-invariants 5). This is an effect of stoichiometric constraints that arise from balancing the cofactors. At first sight, this might not seem biologically significant as many other cofactor-metabolizing reactions are not included in the model. However, there is evidence from which it might be concluded that these ratios are close to those observed *in vivo*. First, the activity ratio of cytosolic versus plastidial AGPases has been shown to be in the range of 5:1 to 20:1 [82]. Furthermore, mutations in the cytosolic route had a major impact on starch synthesis (summarized in [82]). Unfortunately, there are no experimental flux ratios between cytosolic and plastidial AGPases available to date. The network shown in Fig. 10.1 is worth being analyzed further in the future. For example, it would be of interest to compute the ATP balance in each pathway and compare it to experimental values.

10.2.3 Clustering of T-invariants

As the number of minimal T-invariants grows rapidly (usually exponentially) with network size, the problem arises how to handle and interpret such large sets. Independently of each other, two groups in France and Germany have, therefore, proposed methods for clustering elementary modes and minimal T-invariants, respectively [142, 143, 297, 298]. In both approaches, the similarity between invariants is analyzed. Grafahrend-Belau et al. [142, 143] first translate each minimal T-invariant into its binary pattern (support vector). Such patterns had also been used and called activity sets by Nuño et al. [285]. Then the Tanimoto similarity coefficient and, from that, a distance between any two given minimal T-invariants are computed. This allows one to derive clusters of minimal T-invariants, the T-clusters (*see also Chap.* 4).

Pérès [297] took into account the signs of the components in the elementary modes. In her method, all elementary modes are compared with each other. Two integer threshold values have to be prescribed, s and τ. All elementary modes that share more than s common reactions of the same direction (called common motifs) are grouped into one cluster. Thereafter, clusters are merged if their common motifs share more than τ common reactions. Thus, the clustering is performed without using a distance measure. The method was applied to part of mitochondrial metabolism, and the clusters could indeed be interpreted in biochemical terms [298].

10.3 P-invariants

10.3.1 Definition and Biochemical Interpretation

P-invariants (place invariants) are vectors, Y, with the property that multiplication of these vectors with any marking that can be reached from a given initial marking gives the same result. With M_0 being the initial marking and M some arbitrary marking, the relation $Y^T \cdot M = Y^T \cdot M_0$ defines a P-invariant. Algebraically, these vectors are solutions of the equation

$$Y^T \underline{N} = 0 \tag{10.3}$$

(*see* (4.1) *in Sect.* 4.3, "*Place invariants*").

In biochemical networks, P-invariants describe conservation relations for metabolites. For example, the biochemical network N_1 *in Sect.* 4.3 "*Place invariants*" involves the P-invariants $ATP + ADP = const.$, $NADP^+ + NADPH = const.$ and $2GSSG + GSH = const.$ (cf. [378]). Then, also any superposition of these, say $2ATP + 2ADP + 3NADP^+ + 3NADPH = const.$, is a P-invariant. Usually, invariants are chosen so as to involve the smallest integer coefficients, and they are decomposed into the minimal terms (such as $ATP + ADP = const.$).

In order to describe the conservation of atom groups within molecules, the coefficients in the conservation relations must be nonnegative. Thus, only nonnegative conservation relations are relevant (also called semi-positive since at least one coefficient must be positive) [345, 347]. In a closed system, that is, a system without external substances, always one strictly positive conservation relation holds, in which the conservation quantity represents total mass. By "perturbing" the coefficients of this relation in all directions of the null-space of $\underline{N}$, it can be shown that the number of linearly independent semi-positive conservation relations in any closed system equals the dimension of that null-space [345].

When there is a positive linear combination that permanently increases or decreases in time, the system is called superconservative or subconservative, respectively [111]. The terms conservative, superconservative and subconservative are also used for Petri nets. The program INA developed by Starke and coworkers in Berlin (http://www2.informatik.hu-berlin.de/~starke/ina.html) determines whether a Petri net has one of these properties. It is worth noting that these three cases do not cover all networks (cf. [345]). In fact, biochemical systems are usually open systems with

Table 10.2 P-invariants of the network shown in Fig. 10.1. For the dicotyledon (a) and the monocotyledon (b) models, all P-invariants are the same, except for P-invariant 4[a]

Number	List of metabolites (sum = constant)
1	UTP + UDP + UDPG
2	pADPG + pATP + pADP
3	UTP + HexP + pHexP + P + pP + 2 PP + 2 pPP + 2 FBP + pATP + ATP
4a	−UTP − HexP − pHexP − P − pP − 2PP − 2pPP − 2FBP − pATP + ADP
4b	−UTP − HexP − pHexP − P − pP − 2PP − 2pPP − 2FBP − pATB + ADP + ADPG
3 + 4a	ATP + ADP
3 + 4b	ATP + ADP + ADPG

[a]For abbreviations, see legend to Fig. 10.1

an input and output of mass. This mass transfer is described by a flux between external metabolites. Therefore, these systems may, depending on conditions, have a positive or negative mass balance.

Many regulatory networks, by contrast, can be written such that they are conservative. Examples for superconservativity are provided by systems of prebiotic evolution, in which nucleic acids accumulated more and more. Interestingly, the chemical organizations [93] allow both for conservativity and superconservativity.

The test for conservativity [345] has attracted renewed interest with the advent of online databases of reconstructed metabolic networks. Often, information in such databases is prone to errors. To check the stoichiometric consistency and completeness of a network, it is worthwhile running a test for conservativity [126]. This show whether, in principle, the network complies with the condition of mass conservation.

A complete set of linearly independent P-invariants of the metabolic network shown in Fig. 10.1 are listed in Table 10.2. The sum of all uridine-containing metabolites is constant (P-invariant 1) because synthesis or degradation of the uridine residue is not included in the model. A similar reasoning can be applied to the plastidial adenosine residues in the metabolites of P-invariant 2. P-invariant 3 is a sum of phosphate groups. However, this is not the sum of all phosphate groups in the system but only of those that can be cleaved off from metabolites. Therefore, ATP enters that P-invariant with the factor of unity although it contains three phosphate groups. FBP and PP are counted twice as they can release two phosphate groups each.

Interestingly, P-invariant 4 (versions a and b corresponding to the dicotyledon and monocotyledon models, respectively) involves negative coefficients. This is because the program METATOOL computes the P-invariants only by (10.3) without taking into account non-negativity conditions. In the system under study, it is easy to convert the P-invariants so that those conditions are satisfied. We can sum up P-invariants 3 and 4 (Table 10.2). The resulting invariants are easy to interpret as conservation of the adenosine moiety in the cytosol. In more complex networks, such linear combinations are not so easy to perform. It is then sensible to use specific algorithms for computing semi-positive P-invariants (see Sect. 10.3.2).

Table 10.3 Concepts from Petri net theory and their counterparts in the modeling of biochemical systems

PN theory	Biochemical modeling
T-invariants	Steady-state flux distribution, flux modes
True T-invariants	Non-negative steady state flux distribution
Minimal T-invariants	Elementary flux modes
P-invariants	Conservation relations
True P-invariants, semi-positive P-invariants	Semi-positive conservation relations
Conservative, subconservative and Su(pe)rconservative Petri nets	Conservative, subconservative and Superconservative systems
Maximal common transition sets	Sets of partially coupled reactions
Deadlock	False equilibrium

As outlined in Sects. 10.2 and 10.3, several concepts in Petri net theory have counterparts in "traditional" biochemical modeling. Table 10.3 gives an overview of the parallel concepts (see also [354, 427]). There are a number of other concepts such as trap, siphon, deadlock, and liveness, which have not been considered intensely in biochemical modeling so far. Some preliminary ideas of how to use them for modeling biochemical networks have been presented [427]. It is certainly worth using those concepts more extensively.

10.3.2 Algorithms and Software Tools for Computing Minimal Invariants

All the algorithms for both, P- and T-invariants, have to solve linear algebraic equation systems using Integer Linear Programming (ILP) or related methods, for which many algorithms exist. For Petri nets, one algorithm is that for computing minimal P-invariants (without originally dealing with biological networks) presented by Colom and Silva [80]. It involves row operations on the incidence matrix augmented with an identity matrix, thus computing consecutive tableaux like in the Gauss-Jordan method. In the course of the algorithm, care has to be taken to eliminate duplicate and non-minimal T-invariants. Colom and Silva [80] proposed two alternative tests to do so. By transposition of the incidence matrix, that algorithm can be used also for calculating minimal T-invariants.

A number of software tools for computing structural properties of Petri nets such as minimal P-invariants and T-invariants have been developed. Examples are INA ([371], http://www2.informatik.hu-berlin.de/~starke/ina.html), CPN-AMI (http://move.lip6.fr/software/CPNAMI/) and the Petri Net Toolbox of MatLab (The MathWorksTM). A list (including web links) of these and several other packages has been compiled in the "Petri Nets World" archive (http://www.informatik.uni-hamburg.de/TGI/PetriNets/).

An algorithm for computing elementary flux modes based on methods from convex analysis was presented by ourselves [350] and implemented in the program METATOOL [302], independently of Petri net theory. It shares several features with the Colom-Silva algorithm, for example, the row operations on the incidence matrix augmented with an identity matrix. However, elementary modes are allowed to involve reversible reactions. This is considered by partitioning the stoichiometry matrix into "reversible" and "irreversible" parts. Thus, the elementary-flux modes algorithm saves much memory space during the computation, because it uses reversible reactions explicitly. Moreover, the test for eliminating non-minimal and duplicate T-invariants (flux modes) is different (for a more detailed comparison, see [351]). METATOOL also involves routines for determining other properties such as the left and right null-spaces of the stoichiometry matrix and the enzyme subsets (see Sect. 10.4).

For the entire algorithm for computing elementary modes, an alternative has been proposed [402]. This computes a basis of the nullspace first and then combines these vectors to give the remaining elementary modes. This algorithm empirically shows a higher performance on metabolic networks, although it is unclear so far whether this holds for all types of networks. The current version METATOOL 5.1 [406] involves the Urbanczik–Wagner algorithm [402] and a very efficient minimality test based on determining the rank of a submatrix of the stoichiometry matrix, as devised by Klamt et al. [197].

There are a number of alternative programs for computing elementary modes, such as EFMTool [387] and CellNetAnalyzer [198]. With the latter, also semi-positive conservation relations and many other properties can be determined, both for metabolic and signaling networks.

10.4 Other Topological Properties

Sackmann et al. [327] introduced the concept of Maximal common transition sets (MCT-sets). An MCT-set represents a maximal set of transitions that always operate together. In other words, two transitions belong to the same MCT-set, if and only if they participate in exactly the same minimal T-invariants (*see Chap.* 4). Each Petri net can be decomposed into MCT-sets, that is, disjoint subpathways, which can be interpreted as building blocks of the net. Earlier, the concept of enzyme subsets had been introduced [302]. Their definition is stricter: Two reactions in a biochemical pathway belong to the same enzyme subset if they always operate together and do so in fixed flux proportions. Importantly, neither the transitions belonging to one MCT-set nor those belonging to one enzyme subset need to be adjacent to each other, that is, these sets need not be connected [147, 302]. For all steady-state analyses (for example, for detecting routes in the form of minimal T-invariants), all transitions belonging to the same enzyme subset can be lumped into one overall transition. This usually reduces both the numbers of transitions and of places. Moreover, finding such subsets can help make predictions about genetic regulation because the genes

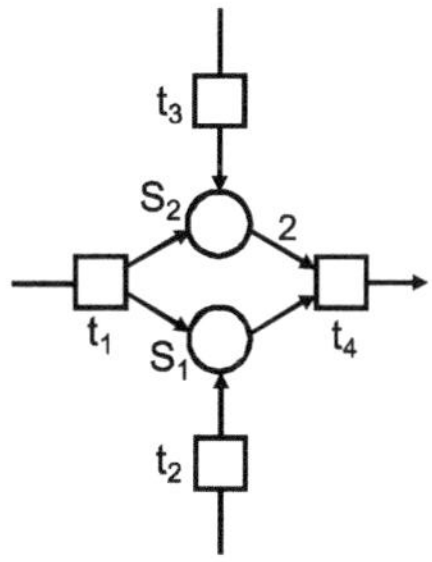

Fig. 10.2 Example network in which an MCT-set is not an enzyme subset. Note that the arrow leading from S_2 to t_4 is weighted by the factor 2

corresponding to the enzymes within one enzyme subset (or likewise one MCT-set) are likely to be regulated coherently [317, 352].

A constant ratio of fluxes is not required for MCT-sets. Thus, all enzyme subsets correspond to MCT-sets, while the opposite is not true. An example is shown in Fig. 10.2. Note that in that example system, two molecules of S_2 are consumed when transition t_4 fires. Thus, transitions t_1 and t_4 alone can not maintain a steady state. Transition t_3 must fire as well to refill S_2. In addition, transition t_2 may be operative. Its flux must then have a counterpart as part of the flux through transition t_3 to guarantee a steady state. Thus, transitions t_3 and t_4 form an MCT-set because they are always operative together. However, their fluxes need not be proportional to each other. In fact, in this system, each reaction constitutes an enzyme subset on its own. The flux vector can be written as

$$V = \begin{pmatrix} a \\ b \\ a+2b \\ a+b \end{pmatrix} \tag{10.4}$$

with a and b being positive values depending on the kinetic parameters. Burgard et al. [53] presented an approach called "Flux Coupling Analysis", which can be considered as a generalization of the concept of enzyme subset. Within that approach, "partially coupled reactions" are defined such that if one of these carries a steady-state flux, also the other do so, yet without a fixed flux ratio. Thus, an MCT set is equivalent to a set of partially coupled reactions. In the terminology of Flux Coupling Analysis, an enzyme subset is a set of fully coupled reactions.

10.5 Related Work with Concrete Biological Examples

Extensive work on application of Petri net theory to biochemical and regulatory systems has been done by Ina Koch and Monika Heiner and their groups. For example, Heiner et al. [153] applied that theory for the modeling of apoptosis (programmed cell death). The model gives rise to 10 minimal T-invariants, which can be classified according to the biochemical stimulus. Koch et al. [211] modeled sucrose utilization in the potato tuber. The net under study, which is related to the one shown in

Fig. 10.1, gives rise to 12 non-trivial minimal T-invariants. In three of these, sucrose is cleaved by sucrose synthase (used in the reverse direction) and in the other 12, by invertase. In both cases, the resulting hexoses can go into glycolysis and/or starch synthesis. Several of the invariants involve futile cycles, which permanently hydrolyze ATP and may play a role in regulation.

Sackmann et al. [327] used a Petri net approach to study the mating pheromone response pathway in *Saccharomyces cerevisiae* as a case study (see also [143]). The study had a strong focus on the analysis of invariants. Some of the computed P-invariants represent the switching behavior of compounds between their activated and inactivated forms. Moreover, MCT-sets (which were defined in that paper) and minimal T-invariants were computed. The latter correspond to the known signal flows through the net. Sackmann et al. [328] analyzed the homeostasis of iron in humans. The 85 feasible minimal T-invariants computed were grouped into ten clusters by the UPGMA method. Grunwald et al. [147] modeled gene regulatory processes relevant for the Duchenne muscular dystrophy. Simão et al. [362] studied tryptophan synthesis in *E. coli*. They took into account two types of regulatory feedbacks: the direct inhibition of the first enzyme of the tryptophan pathway by its final product, and the transcriptional inhibition of the Trp operon by the Trp-repressor complex.

Recently, Petri net approaches have been used to analyze the regulation of alternative splicing. RNA splicing is a process by which parts of mRNA molecules (introns) are sliced out and the remaining parts (exons) are sealed together (cf. [178]). The term alternative splicing refers to the situation where different sets of introns, or introns of different length, are clipped out depending on developmental stage, tissue, disease etc. Splicing is performed by a large ribonucleoprotein complex, called the spliceosome. This complex is assembled in a cascade of binding and modification processes (cf. [321]). Kielbassa et al. [190] analyzed the assembly of the U1 snRNP subunit of the spliceosome. The Petri net model is covered by 19 minimal T-invariants. One of them can be considered as predominant, it describes virtually the entire assembly pathway. For that network, also the MCT-sets have been computed. Their number is 42, with 14 of them containing more than one transition. The results point to the importance of the stability of complexes during the maturation pathway. It was demonstrated that complexes that dissociate too fast, hinder the formation of the complete U1 snRNP. In consequence, it was concluded that only a long contact time of spliceosomal factors ensures the formation of the complete U1 snRNP complex, in accordance with Rino et al. [321].

The assembly of an entire spliceosome has recently been analyzed by Bortfeldt et al. [41]. The network consists of 161 transitions and 140 places. All reactions are part of at least one of the 71 resulting minimal T-invariants. These define pathways, which are in good agreement with the current knowledge and known hypotheses on reaction sequences during spliceosome assembly. The complexity of the network was further increased by two switches, which introduce alternative routes during formation of the *A-complex* in early spliceosome assembly and upon transition from the B-complex to the C-complex. By compiling known reactions into a complete network, the combinatorial nature of invariant computation leads to pathways that have previously not been described as connected routes, although their constituents were known. T-clusters divide the network into modules, which were interpreted as building blocks in spliceosome maturation [41].

10.6 Conclusions

Metabolic and regulatory networks play an enormous role in living cells. They form a bridge between the genotype and phenotype. They are more complicated than graphs (in the sense of graph theory) in which reactions would correspond to arcs and metabolites would correspond to vertices. Due to the presence of bimolecular and higher-molecular reactions, there should be arcs connecting more than two vertices. Mathematically, metabolic networks are a sort of hypergraphs (cf. [199]). Another option is to describe them by bipartite graphs, in which the two types of nodes correspond to reactions and metabolites. This is realized by Petri nets (cf. [63, 208]). Analyzing topological properties of such nets allows one to derive far-reaching conclusions from a rather limited set of input information (stoichiometry and reversibility).

However, Petri nets are not the only tool used for representing, simulating and analyzing biochemical networks. Many authors use graphical representations as usual in biochemistry textbooks and ordinary differential equations for dynamic simulation. When the analysis is restricted to steady states, the differential equations can be simplified to algebraic equations. These can be analyzed by various methods, for example, from matrix algebra. An advantage of using Petri nets is the visualization of the flow of tokens by appropriate simulators (see the "Petri Nets World" archive). To some extent, it can be considered a matter of taste whether Petri nets or other appropriate methods are used. The computation of T-invariants or, alternatively, of elementary modes, correctly takes into account the condition of mass balance. It is important that connectedness of a network does not necessarily imply a steady-state flow. To make a distinction between (a) paths (connected routes) in the sense of graph theory and (b) routes that are capable of carrying a net flux at stationary states, a distinction in terminology is useful. The term "pathway" should be used for (b) only (cf. [304]).

In regulatory networks, the steady-state condition is not as important as in metabolic networks. Often, signals are transmitted by pulses along certain routes, that is, by nonstationary processes [158]. On the other hand, averaged over longer periods, also regulatory networks are at steady state because they must regenerate before they can transmit the next signal (cf. [32]). Another difference to metabolic networks is that information flow is more important than mass flow. Although mass balance must be satisfied as well, this is not as useful a criterion as in metabolic networks to derive interesting properties (cf. [428]). Therefore, the observation that minimal T-invariants often describe interesting signaling routes [147, 327, 328] can be considered a proof of principle while it is not completely understood theoretically so far. Klamt et al. [195] have translated signaling systems into interaction graphs, that is, directed graphs with signed arcs. Signalling paths in such graphs can then be determined by computing elementary modes, although classical graph theoretical algorithms (e.g., depth-first search) perform that task faster. Recently, we have shown that a sound theoretical basis can be given for the special case of enzyme cascades [32].

In summary, we have shown here that the topological analysis of Petri nets is extremely useful in analyzing metabolic and regulatory networks. By a running example taken from sugar metabolism in plants, we have demonstrated that important properties such as optimal yields, redundancy and conserved moieties can be derived by analyzing T-invariants and P-invariants. By an overview of the literature, we have shown that these techniques are widely used for a large variety of biological systems.

10.7 Problems

10.1 Some metabolite is connected to n irreversible reactions (transitions) that either produce that metabolite from external sources or consume it, leading to external sinks. What is the maximum number of minimal T-invariants in this net upon variation of the number of transitions that produce that metabolite (with n fixed and the remaining transitions consuming the metabolite)?

10.2 Determine the minimal T-invariants to

1. the net shown in Fig. 10.2 (to that end, consider which flux distributions are feasible at steady state, or by which linear combinations the flux vector given in Equation (10.4) can be obtained).
2. the net shown in Fig. 10.2 after replacing the factor 2 attached to the arrow leading from S_2 to t_4 by 1.

10.3 Assume that in the extended model describing sucrose metabolism in monocotyledons (Fig. 10.1),

1. the enzyme phosphofructokinase (PFK)
2. the enzyme hexokinase (HK)

has been knocked out.

1. In which of the two cases more minimal T-invariants would drop out?
2. Assume that the system responds by using that minimal T-invariant still allowing the highest starch-over-sucrose yield. In which case the yield would be higher, and what would its value be?

Acknowledgement The authors would like to thank Monika Heiner, Steffen Klamt, Ina Koch, Sabine Pérès, and Andrea Sackmann for very helpful discussions.

Chapter 11
Analysis of Dynamical Models of Signaling Networks with Petri Nets and Dynamic Graphs

Simon Hardy and Ravi Iyengar

Abstract The static representation of biological interaction networks can be misleading. All interactions do not occur simultaneously. On the other hand, differential equations can represent a dynamical system, but the topology of the interactions is not explicitly accessible from the calculations of system dynamics. To have a graph representation of a dynamical system, we have developed the dynamic graph. We used the Petri net representation of an ODE system and invariant analysis to identify the main components of a signaling network and thus bridge the two formalisms. The result is a method that can be used to analyze the dynamics of the network topology. Its main feature is the highlighting of the function and interactions of regulatory motifs in the emergence of a complex biological behavior. The example used here is the Bhalla–Iyengar model of the MAPK/PKC signaling pathway in fibroblasts. A property of this pathway is the ability to operate both in a monostable or bistable regime. We show with dynamic graphs that both the topology and the kinetics of this model are responsible for this behavior.

11.1 Introduction

The cell is in constant interaction with its environment. It senses extracellular substances through molecular receptors which often triggers biochemical or morphological responses. The mechanisms relaying extracellular information to specific control elements of the cell are grouped under the term signal transduction. The signaling network of the cell contains numerous components and many pathways formed by successive biochemical interactions and activation events

S. Hardy (✉) · R. Iyengar
Department of Pharmacology and Systems Therapeutics, Mount Sinai School of Medicine, One Gustave L. Levy Place, New York, NY 10029-6574, USA
e-mail: simon.hardy@mssm.edu

R. Iyengar
e-mail: ravi.iyengar@mssm.edu

I. Koch et al. (eds.), *Modeling in Systems Biology*, Computational Biology 16,
DOI 10.1007/978-1-84996-474-6_11, © Springer-Verlag London Limited 2011

of biomolecules. An example of a signaling pathway is Ligand → Receptor → Gprotein → Kinases → Transcription factor → DNA. Other effectors include metabolic enzymes to alter metabolism, ion channels to alter cellular excitability and cytoskeletal proteins to alter cell shape or movement. Along the way, small intracellular signaling molecules called second messengers, such as cyclic adenosine monophosphate (cAMP), calcium ions and diacylglycerol (DAG), can be produced and then rapidly diffuse and transmit the signal. Signaling proteins usually have active and inactive states and they can be activated by binding to other molecules or by posttranslational modifications like phosphorylation.

First viewed as isolated cellular signaling wires, intracellular signaling pathways are rarely simple unidirectional paths from the membrane receptors to their targets [413]. Signaling pathways include bifurcation and integrator proteins that spread or combine signals. Signaling loops are also present to provide dynamic adaptation to stimuli. Several of these topological features have been linked to regulatory properties. The complex cellular signaling networks discovered by biologists led to the hypothesis that they have not only transmittal functions, but also information processing capabilities [238]. Hypothetically, the transformation of biochemical information into cellular responses is achieved by small recurring circuits in the network named regulatory motifs. Examples of motifs are positive and negative feedback loops, feedforward loops, bifans, gates and switches [12]. These motifs were identified because their occurrence is statistically greater in biological networks than in randomly built networks. This was first demonstrated in gene regulation networks [264, 361] and later confirmed for signaling networks [239].

Signaling networks are suitable systems for Petri net modeling. Like other biochemical reaction systems signaling networks are intrinsically bipartite and binding and enzymatic reactions have defined Petri net representations. Also, signaling networks are formed by sequences of signaling activation events and by the concurrent actions of interconnected pathways. It is easy to represent the chain of activation events—for example, an upstream active enzyme relaying the signal by activating the next enzyme—with a graphical formalism that is appealing to biologists. Furthermore, models of signaling networks can be analyzed qualitatively and quantitatively using Petri net theory methods presented in earlier chapters of this book.

The usual Petri net modeling methodology is to start with a qualitative model and then iteratively perform analyses and add quantitative information [154, 211]. The first step of this process is to extract and build a model from the biology literature. Second, qualitative analyses are done using the mathematical tools of the Petri net theory. This provides a formal validation of the model, the analyses results can be used to generate a structural portrait of the system and give hints to key quantitative parameters [230, 327]. Third, stochastic and/or deterministic information about quantities and events can be added to the model to transform the qualitative model into a quantitative one. This time-dependent model can be represented with different Petri net extensions or with other mathematical formalisms and be analytically solved or numerically simulated. This type of model is the basis for a quantitative analysis of the behavior of biological system models.

The work presented in this chapter does not fit in this general framework. Rather than using Petri nets to build a qualitative or a quantitative representation from experimental data, we present a computational method that uses Petri nets to analyze an existing quantitative model. We will take the mathematical model of a signal transduction network specified with ordinary differential equations (ODE), convert it into a discrete Petri net model and perform invariant analyses to extract structural properties that are biologically meaningful for cellular signaling. Then, these properties will serve to reconstruct the directed interaction graph of the model, also known as the influence graph. By doing so, we bridge two important formalisms for biological modeling: the dynamic representation of ODEs and the static representation of graph theory. Once this bridge is established, it is possible to use numerical data from a simulation run of the ODEs to create a dynamic representation of the interaction graph. The result is a dynamic graph. We refer to Chap. 8 from Ackermann for more information about ODE modeling of biological systems.

The main objective of this computational method is to analyze the dynamic behavior of biological systems at a higher level of abstraction. Instead of looking at concentration variations over time, the focus of the analysis becomes the dynamic of the regulatory motifs. As we have seen, the motifs are the building blocks of signaling networks. Models with a complex behavior are bound to comprise many motifs, stacked and/or nested. Understanding the interrelated dynamics of these motifs is a key to understanding the systemic properties emerging from the interactions between hundreds of individual molecular components. In this chapter, the running case study to illustrate the Petri net-based method for regulatory motif analysis is the published ODE model of the mitogen-activated protein kinase (MAPK) 1,2/protein kinase C (PKC) system, a signaling pathway activated by the fibroblast growth factor (FGF) and parameterized with experiments done with mouse NIH 3T3 fibroblasts [34]. In this chapter, we refer to it as the Bhalla–Iyengar model. This model was developed to study the interaction between a positive feedback loop (MAPK-PLA_2-PKC) and a negative feedback loop (MAPK-MKP1) and how a network that can operate both in a monostable or bistable regime emerges from this special topology.

11.2 Methods and Concepts

The motivation to develop a computational method connecting ODE models and interaction graphs comes from the need to have a network view of a dynamic system. The combination of the two formalisms gives rise to a new type of investigation: the analysis of the dynamic of regulatory motifs with dynamic graphs. Many steps are necessary to achieve this integration of graph theory and dynamical systems using Petri nets. This approach is different from hybrid Petri net modeling as described in the chapters by Miyano and Matsuno because two goals of the dynamic graph

$$d[A]/dt = -k_1[A][B] + k_2[AB] \quad (11.1)$$

$$d[B]/dt = -k_1[A][B] + k_2[AB] \quad (11.2)$$

$$d[AB]/dt = k_1[A][B] - k_2[AB] \quad (11.3)$$

$$d[C]/dt = -\frac{k_{cat}[AB][C]}{k_M + [C]} \quad (11.4)$$

$$d[D]/dt = \frac{k_{cat}[AB][C]}{k_M + [C]} \quad (11.5)$$

$$\Longrightarrow$$

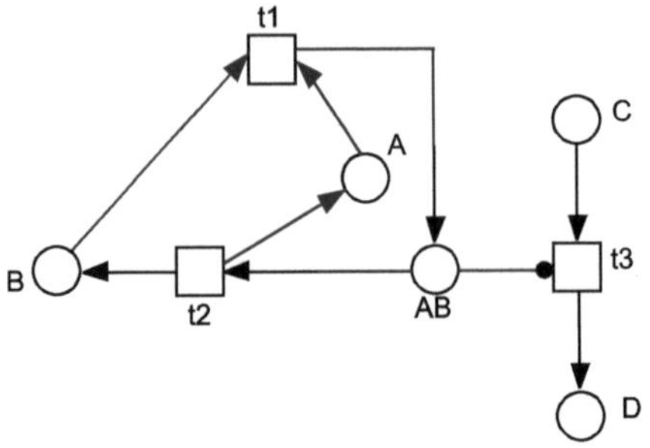

Fig. 11.1 Conversion of five ordinary differential equations representing a molecular binding reaction and an enzymatic reaction into a discrete Petri net model

method are the reduction of the number of variables and the creation of a monopartite representation. In this section, we describe each step, from ODEs to Petri nets to graphs.

11.2.1 From Ordinary Differential Equations to a Petri Net Model

The relationship between ODEs and Petri nets is documented [49, 129, 248]. In biological applications, the usual process is to use a validated Petri net model to set up the structure of an ODE model or of a continuous Petri net. We use the same relationship between the qualitative and quantitative representations, but to do a reverse-engineering of a dynamic model. The variables on the left side of the differential equations become places, and every unique rate term on the right side becomes a transition. Figure 11.1 is an example of this conversion. This system of five equations corresponds to one mass action reaction and one Michaelis–Menten enzymatic reaction. Each variable representing the molecular specie A, B, AB, C and D corresponds to a place. Each unique term, $k_1[A][B]$, $k_2[AB]$ and $\frac{k_{cat}[AB][C]}{k_M+[C]}$, corresponds to transitions t_1, t_2 and t_3 respectively. If the term is negative in the rate equation of a variable, then an arc from the associated place to the transition representing this term is added. If the term is positive, then the arc goes from the transition to the associated place. Read arcs represent the action of an enzyme if the Michaelis–Menten formulation is used. Read arcs represent a reading relation which denotes that the firing of transition will read the current value of the place but without modifying its content. The arc connecting place AB and transition t_3 in Fig. 11.1 is a read arc. The interpretation of this representation is that the enzyme AB participates in the reaction t_3 as a catalyst but it is not modified by it. If enzymatic reactions are modeled as two-step mass action reactions, then the Petri net model should be free of read arcs.

The Petri net software Snoopy [155] has a conversion feature for importing SBML model and automatically generating Petri net models (discrete or continuous). The Systems Biology Markup Language is a generic format for various formalisms, not just ODE models. Thus, some information can be software tool specific

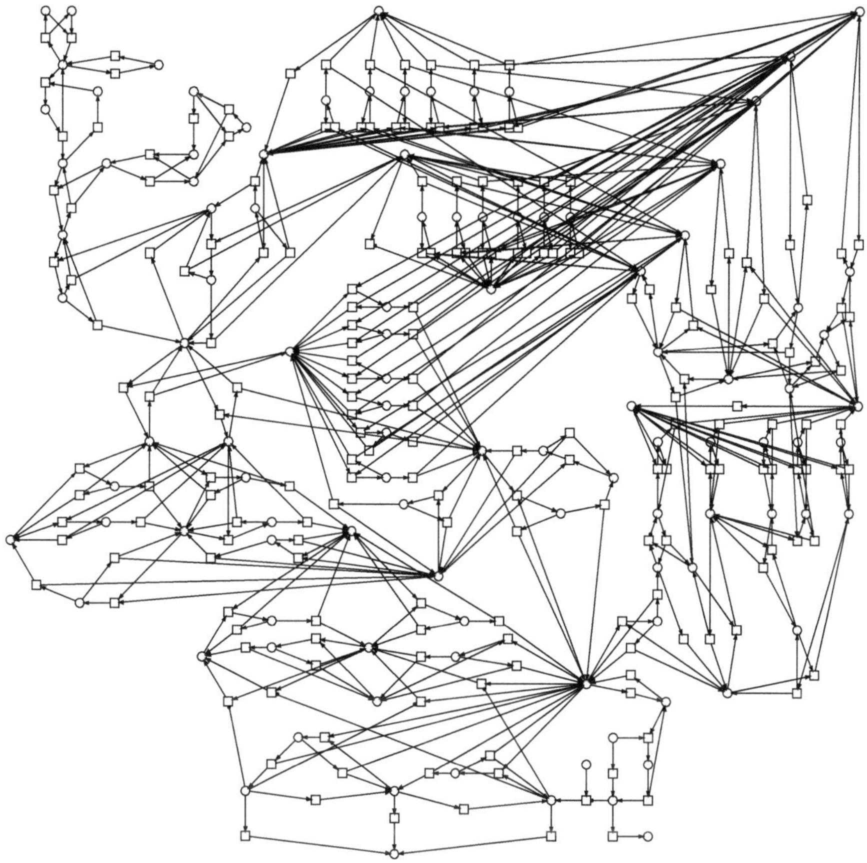

Fig. 11.2 Petri net model generated from the ordinary differential equations of the Bhalla-Iynegar model of the MAPK/PKC pathway

and be misinterpreted by other tools, despite their SBML compatibility. For that reason, the result of the import of a SBML model with Snoopy should be inspected and validated. We used Snoopy to import the SBML file of the Bhalla–Iyengar model and the generated Petri net model was manually curated. In this model, most enzymatic reactions are modeled as two-step mass action reactions, but some are modeled with the Michaelis–Menten formalism, hence there are a few read arcs in the Petri net of the Bhalla–Iyengar model shown in Fig. 11.2.

Several biological models using differential equations are not pure ODE models. Some models might include logical statements. It is also usual to clamp some variables to a certain value for modeling simplicity. This is almost always the case for adenosine triphosphate (ATP) in signal transduction models, the major source of energy in the cell. The underlying assumption is that the metabolic processes of the cell are maintaining ATP at a constant level. These modeling implications must be individually dealt with for an appropriate conversion into a Petri net model. We give more details on these special situations in Sect. 11.2.2.

11.2.2 From a Petri Net Model to an Interaction Graph

11.2.2.1 Interaction Graphs and Conservative Components

An interaction graph $G = (V, E)$ consists of a finite set V of vertices and a finite set $E \subseteq V \times V$ of edges. The edge $(x, y) \in E$ is a link between vertices x and y. The vertices correspond to molecular species and an edge connects two vertices if an interaction involves both species. The meaning of an edge in protein-protein interaction graphs is that there is an experimentally confirmed direct interaction between the two linked molecular species. Using graph theory and bioinformatic data mining methods, structural information and unexpected distal relationships can be extracted from large networks built from the literature or based on data from high-throughput experiments. See [25, 237, 246] for recent reviews on this subject. For our purpose, the interaction graph is an intermediary step to build the influence graph of an ODE model. The first Petri net theory tool that we use to build the interaction graph of the MAPK/PKC signaling pathway is the conservative invariant analysis (also known as P-invariant analysis).

A conservative invariant analysis identifies the sets of places of a Petri net model that respect a conservation property. Theses sets are called conservative components. The conservation property states that the markings of such a set of places form a linear combination that is always constant no matter what is the state of the system (see definition in the chapter by Reisig). This Petri net conservation property can be related to a modeling conservation property that is very common in signal transduction models. It is a widespread practice to model the total concentration of groups of signaling molecules as constant quantities. For most molecular species, only the distribution between the different states of the molecular species changes to reflect fluctuating levels of activity and not the total amount. For example, MAPK can be sequentially phosphorylated by a kinase at two different amino acid residues. This means that there are three different phosphorylation states for MAPK: unphosphorylated, singly phosphorylated or dually phosphorylated. The concentration distribution (or token distribution in the case of a Petri net model) between the three variables changes during cellular signaling, but the sum of the three concentrations remains constant because the synthesis and degradation of this protein are not regulated by the signal and are more or less constant. Considering this assumption, one can predict the result of the conservative invariant analysis of a signal transduction model: every molecular species with a constant total concentration are conservation components and the different activity states are identified as the places of these components. The relationship between conservative invariants and biochemical modeling concepts has already been discussed for metabolic models [407, 427], signaling models [327]. Furthermore, combined networks of gene regulation and signal transduction have been explored using P-invariants [147].

11.2.2.2 Building the Interaction Graph of the Bhalla–Iyengar Model

In the Petri net model of the MAPK/PKC pathway, a conservative invariant analysis identified 19 P-invariants.[1] This model is not completely covered by P-invariants. This incomplete coverage means that some places are not part of any conservative invariant. It can be explained by the fact that in many signaling models, not all molecular species have a conservation rule. This is the case for molecules with a regulated production (or release) and degradation (or restoring) such as metabolites, second messengers, ribonucleic acids (RNA) or particular enzymes. In the Bhalla–Iyengar model, two molecular species have no conservation rule: MAPK phosphatase 1 (MKP1) and MKP1 mRNA. The expression of the enzyme MKP1 is controlled by the level of activity of MAPK. The elevation of MAPK in the nucleus promotes the transcription of MKP1 mRNA, which eventually leads to an increased synthesis of this phosphatase. For ODE modeling, it is sufficient to consider mRNA transcription and protein synthesis as processes creating entities out of nothing because it does not affect the dynamic of the system. It is outside the boundaries of the model. However, this does not respect the mass conservation law underlying every biochemical systems. mRNAs are made of nucleotides and proteins are synthesized from amino acids. The same reasoning applies to destruction processes. Macromolecules do not disappear into the void; they are degraded into smaller parts. For a P-invariant analysis, it is thus desirable to take the mass conservation law into account and slightly modify the model. In the Bhalla–Iyengar model, the addition of three places for nucleotides, amino acids, and degraded mRNAs is enough to completely cover the Petri net model with P-invariants. Note that this alteration is only intended for invariant analysis. The stoichiometric relationship between the smaller molecules and the macromolecule is not considered. Therefore, the modified model can not be used with tokens for state space analysis without further modifications.

The modified model of the MAPK/PKC pathway has 21 P-invariants P', where P' is a set of places. Each P-invariant is a vertex in the interaction graph shown in Fig. 11.3. Edges between vertices are determined by the transition connectivity in the Petri net model. Two conditions are used to evaluate the connectivity. An edge links the vertices x and y if

1. $P'_x \cap P'_y \neq \emptyset$; the P-invariants P'_x and P'_y have at least one place in common, or
2. $\exists t : (W(p_1, t) > 0 \wedge W(t, p_2) > 0) \vee (W(p_2, t) > 0 \wedge W(t, p_1) > 0)$, where $p_1 \in P'_x$, $p_2 \in P'_y$, $t \in T$ and $W : ((P \times T) \cup (T \times P) \rightarrow \mathbb{N})$; the places p_1 and p_2 from the P-invariants P'_x and P'_y are connected through transition t from the set of transitions T. W is the multiset of arcs of the Petri net.

The second condition is necessary only if the Petri net model has read arcs. The interaction graph of Fig. 11.3 contains all the edges that can be detected in the MAPK/PKC model with these two conditions. The interactions between P-invariants mean that (1) the two molecular species can form a complex or that

[1]Conservative invariant analysis was performed with the software Charlie, a companion tool of Snoopy.

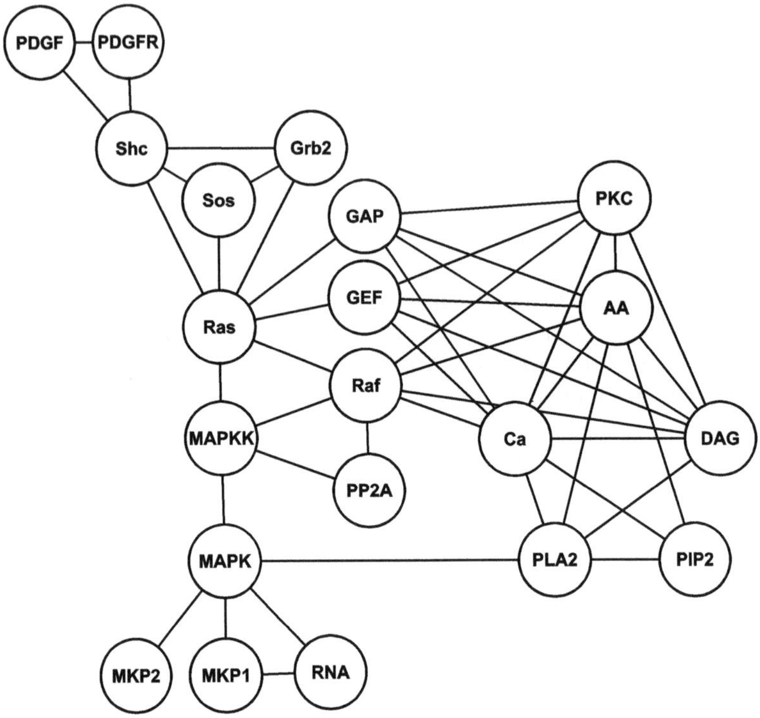

Fig. 11.3 Interaction graph of the Bhalla–Iyengar model of the MAPK/PKC pathway

(2) one species is a catalyst for a reaction involving the second species. The interaction graph, an intermediary representation of the structural information in the ODEs, is now built. In the next section, it is transformed into an influence graph to include the directionality of signal propagation.

11.2.3 From an Interaction Graph to an Influence Graph

11.2.3.1 Influence Graphs and Repetitive Components

An influence graph $I = (V, A, T)$ consists of a finite set V of vertices and a finite set $A \subseteq V \times V$ of arcs. The arcs $(u, v) \in A$ are directed edges going from the vertex u to the vertex v. An arc is associated to an interaction type i from a set of types I $(T : A \rightarrow I)$. The biological influence graphs that are built with the Petri net-based computational method presented in this chapter have two types of interaction: activation and inhibition. In biological terms, these interactions represent signaling activation or inactivation of one molecular species by another. The underlying biological process can vary—phosphorylation, binding, etc.—but the perspective is always the same: is the interaction promoting or impeding signal propagation. To assign directionality and interaction type to the edges of the interaction graph, a repetitive component analysis (or T-invariant analysis) is done, not on the entire Petri net model, but on subgraphs. These subgraphs are graphs formed by the places of P-invariants and their connected transitions; they are specific to one molecular species. This analysis identifies sets of transitions that are involved in an equal flow property at steady state. This reveals the signaling segments of the model. A repetitive component is a set of transitions that if fired sequentially cause a return to the initial state. This sequence may therefore be repeated. It is a cycle. In Petri net models of signaling networks, repetitive components have been used to identify functional units [327] and transduction activation components [230].

11.2.3.2 Building the Influence Graph of the Bhalla–Iyengar Model

There are 21 subgraphs to analyze in the Bhalla–Iyengar model. We exemplify the process of identifying signaling segments with the subgraph of Ras, shown in Fig. 11.4. In the Ras subgraph, there are 17 minimal T-invariants (including the trivial ones). To identify signaling segments, all T-invariants having one or more transitions in common are grouped together. For the Ras subgraph, this operation generates 7 signaling segments. $\{t_1, t_2, t_3, t_4, t_5, t_6, t_7, t_9, t_{10}, t_{11}\}$, $\{t_{12}, t_{13}\}$ and $\{t_{22}, t_{23}$, and $t_{24}\}$ are among these segments. The interesting property of these segments is that at steady state, the cumulative flow is null for each segment independently. Because kinetic rates and initial concentration values have not been considered so far, the identification of signaling segments does not find the steady state solutions. What it does provide is the minimal grouping of the reactions that have rates canceling each other out at steady state.

With the signaling segments, it is now possible to follow the propagation of a signal in the pathway. The starting point is the source of the signal. It is usually a single variable in the ODE model that has an input function, for example a pulse, a step increase or an oscillation. The source of the signal is also part of a P-invariant, which is a vertex in the interaction graph. This perturbation causes changes throughout the network revealing its dynamic behavior. In the model, the source of the signal is the free ligand variable, associated with the PDGF vertex. This is the growth factor that is added to starved cells in experiments. The procedure to create the influence graph is to follow the signal in the Petri net model and mark the signal propagation onto the interaction graph. Here is an example. The place of the free ligand is connected to one signaling segment and the output place of this segment is part of the receptor P-invariant (PDGFR). This propagation is depicted as an arc from the PDGF vertex to the PDGFR vertex in the influence graph (see Fig. 11.5). From the place of the receptor-ligand complex, the signal is transmitted to two other signaling segments. One segment leads to the internalized receptor state. It is a signaling deadend because it is not connected to any other P-invariant. The other segment is connected to the Shc P-invariant. This propagation is represented as an arc from the PDGFR vertex to the Shc vertex. Note that it is impossible to go back to a segment that has already been explored.

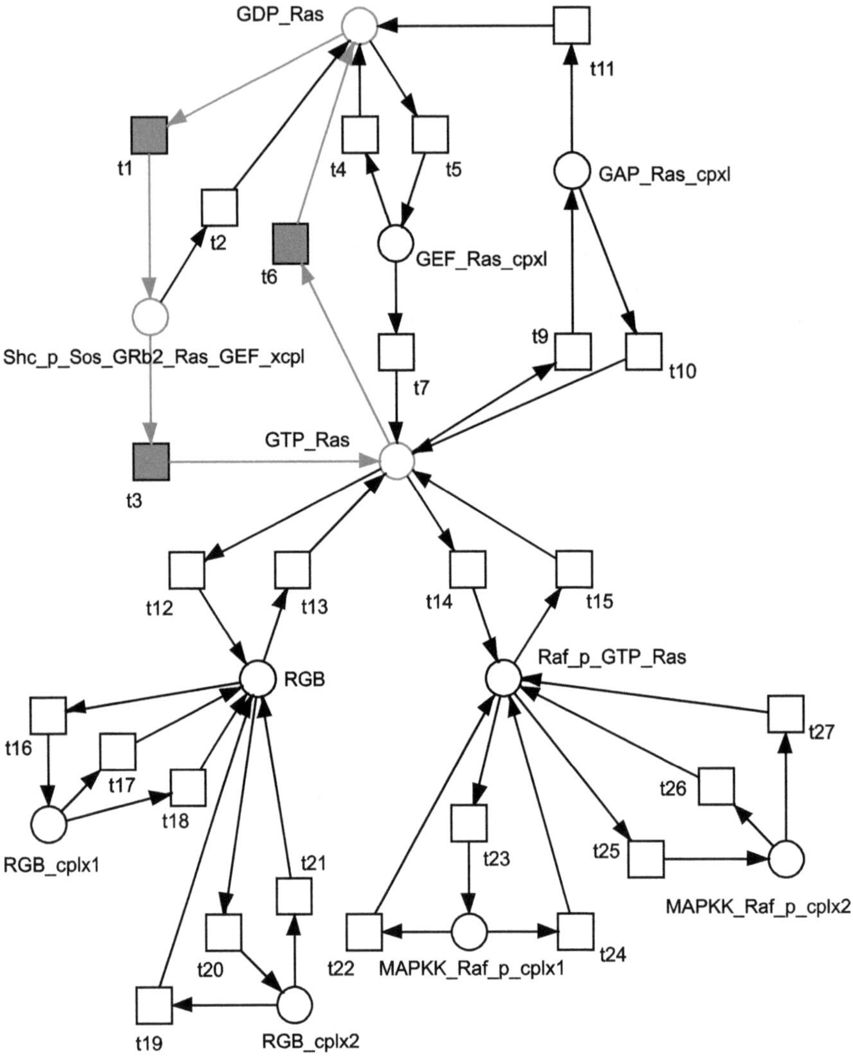

Fig. 11.4 Subgraph of the Ras conservative component of the Petri net of the Bhalla–Iyengar Petri model of the MAPK/PKC pathway

It is interesting to observe that some edges in the interaction graph are not represented in the influence graph: for example, the edge between the PDGF and Shc vertices. This edge is present in the interaction graph because PDGF and Shc can be in the same trimeric complex with PDGFR. However, this is a consequence of the linear composition of two successive events: the binding of the ligand to the receptor and then the phosphorylation of the kinase by the activated receptor. The tracking of the signal propagation eliminates these interactions sharing a common interactor.

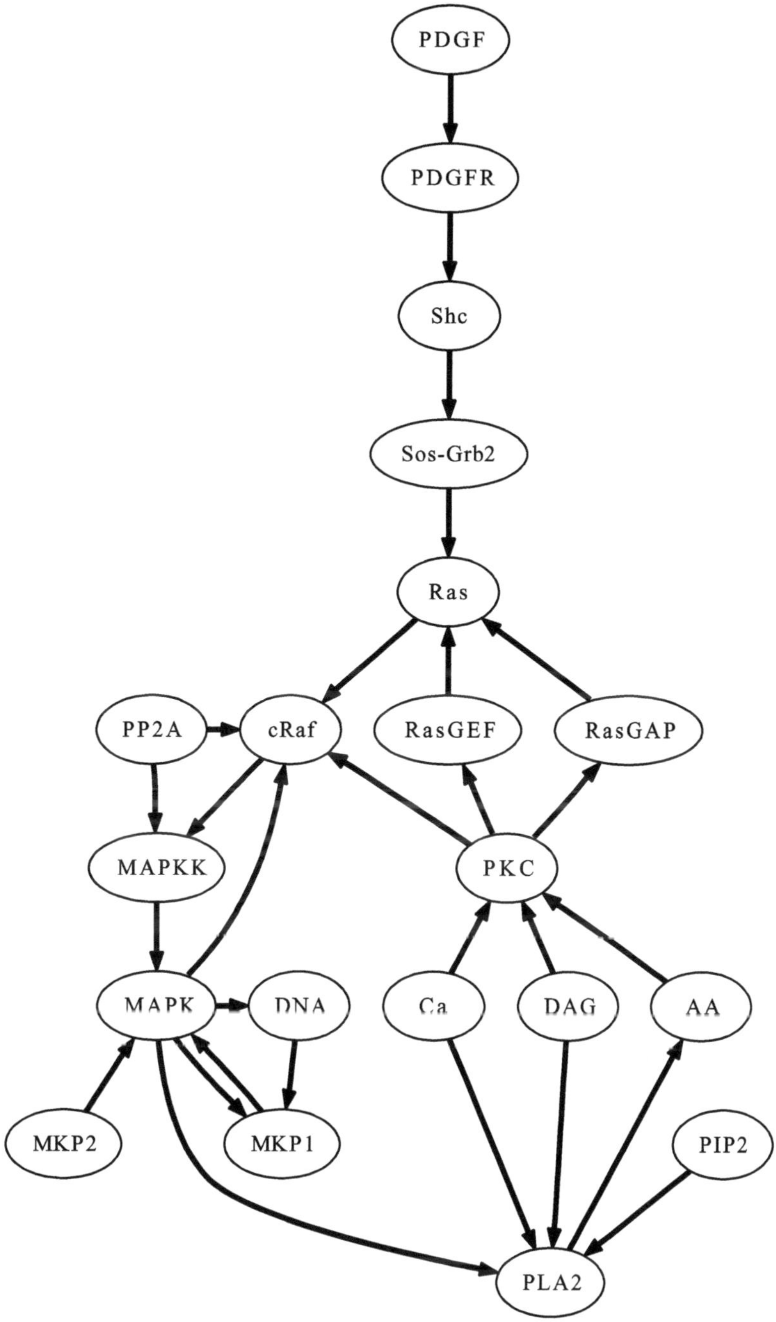

Fig. 11.5 Influence graph of the Bhalla–Iyengar model of the MAPK/PKC pathway in which PDGF is the source of the signal

For the first interactions considered so far, we have assumed that the influence type was activation. The signal could always propagate further down in the pathway. However, when interactions are in opposition, for example, when the signal is propagating in an opposite direction on a signaling segment, their interaction type must be different. This is the case for the vertices GAP and GEF. The former is inactivating Ras and the latter is activating it. These vertices are connected to the Ras vertex through the same signaling segment, but they have opposite effects. For that reason, one arc is of the activation type and the other is of the inactivation type.

Some vertices, like unregulated proteins, are left out of the propagation of the signal. To include them in the influence graph, one has to verify the vertices from the interaction graph that have not been visited by the signal. An example of this kind of node is the unregulated phosphatase MKP2 in the Bhalla–Iyengar model. It is possible to connect this vertex to the influence graph because of its connectivity through a signaling segment in the Petri net model. The effect of MKP2 is to inactivate MAPK by dephosphorylation. Since the signal is activating MAPK, the arc from MKP2 to MAPK is of the inactivation type.

The influence graph of the Bhalla–Iyengar model is shown in Fig. 11.5. In this subsection, we explained how to transform an interaction graph into an influence graph by using structural information extracted from the ODEs using Petri net theory. In the next, we introduce heuristic visualization rules to map simulation data onto the influence graph in order to get a dynamic graph.

11.2.4 From an Influence Graph to Dynamic Graphs

Biologists are used to seeing a network of biochemical interactions as an influence graph. However, the static aspect of this depiction can be misleading. These interactions are dynamic. The activation of the different signaling components does not occur exactly at the same time. Complex signaling networks usually have complex dynamic behavior. With the dynamic graph, it is possible to represent the temporal activity of the biochemical system on its topology by mapping the simulation data onto the influence graph. The dynamic graph of a biochemical network is a graph in which the color of vertices and edges represents the relative value of concentration and flow at a specific simulation time step. The intensity of a color represents the level of activity of a molecular species or of a reaction flow. It is thus possible to analyze the signal propagation and the activity of the regulatory motifs. This is what we call analysis of the topology dynamic.

The influence graph of a biological network is built from the Petri net model, thus the vertices and edges of the graph are derived from the places and transitions of the model. This same Petri net model was first generated from the ODE model, thus the places and transitions are associated to the variables and rates of the mathematical representation. By this linear association, the vertices and edges of the influence graph can be traced back to the variables and rates of the ODE model. We exploit these associations to transform an influence graph into a dynamic graph with simulation data. Each vertex and edge of the dynamic graph is assigned a heuristic rule that converts simulation data into colors.

11.2.4.1 Activation Ratio Using P-invariant to Display Activity of Signaling Components

For molecular species, a good indicator of the level of activity is the ratio between the quantity of molecules in an active conformation and the total quantity for this molecular species. A heuristic rule can be based on this ratio. For example, the molecular species of the small GTPase Ras is inactive when it is bound to GDP and is active when it is bound to GTP. In Fig. 11.4, the marking of all the places of the Ras P-invariant that contain the active GTP-bound form (*GTP_Ras*, *RGR*, *RGR_cplx1*, *RGR_cplx2*, *Raf_p_GTP_Ras*, *MAPKK_Raf_p_cplx1*, *MAPKK_Raf_p_cplx2*) can be summed and then divided by the value of the P-invariant. The result of this operation is always between 0 and 1 and gives the relative activation value at any given time step. Equation (11.1) is the general formula to compute the relative value of activation of a molecular species with its P-invariant.

$$r_t = \frac{\sum_{X_j \in A} w_j \cdot m(X_{j,t})}{\sum_{X_j \in PI} w_j \cdot m(X_{j,t})} \tag{11.1}$$

where X_j is a place of the Petri net, A is the set of places identified as the active conformations of a molecular species, PI is the set of places of the conservative component, w_j is the weight of place X_j in the P-invariant vector and $m(X_{j,t})$ is the marking of place X_j at time t. The next step is to convert this numerical value into a color using a scale (i.e., white for 0, black for 1). For better visualization results, it is also possible to adapt the color scaling to a smaller range if preferred.

11.2.4.2 Reaction Flow Ratio Using Maximum to Display Strength of Interaction Between Signaling Components

For molecular reactions, a comparable ratio can be calculated. The sum of the flows of the signaling segments linking two P-invariants is divided by the maximal value reached by this sum during the simulation run. In this sum, flows in opposite directions have opposite signs. This ratio can be negative when the system is returning to basal state. For example, consider the flow between the vertices Raf and MEK in the influence graph. Raf is activated when it is bound to Ras and in turn, the complex activates MEK by phosphorylation. Raf can also be phosphorylated by PKC, thus becoming a more potent activator of MEK because of a lower dissociation constant with Ras. In the model, unphosphorylated and phosphorylated Raf can form a complex with GTP-Ras. Consequently, the two molecular complexes *RGR* (unphosphorylated Raf) and *Raf_p_GTP_Ras* (phosphorylated Raf) of the Ras subgraph in Fig. 11.4 can activate MEK by phosphorylating two of its residues. As a result, there are four signaling segments between the Raf and MEK invariants (2 complexes each phosphorylating 2 residues). The sum of the rates of the transitions t_{18}, t_{19}, t_{24}, and t_{27} correspond to the global phosphorylation of MEK by Raf. This sum divided by the maximum that this sum reaches during the simulation period is an example of a heuristic rule for a flow ratio.

The kinetic parameters and the initial conditions of the ODE models have been ignored until the generation of a dynamic graph. This is why there is only one influence graph by pair of ODE model and source of signal. Two different influence graph can be generated from the same ODE model by considering two different sources of signal. However, the dynamic of this model depends not only on its structure, but also on the set of parameters. Each set of parameters has its own dynamic graph. Furthermore, each set of heuristic rules will create a different dynamic graph. The latter flexibility and tailoring are useful to compare the systemic effects of different parameters such as different amplitude of signals or parameter variations. For example, the denominator of the heuristic rule of an edge can be set to the same value for different dynamic graphs to enable comparison between them.

A single frame of a dynamic graph of the MAPK/PKC signaling pathway is shown in Fig. 11.6. At that time step, the source of the signal returned back to zero, but some downstream motifs are still active. The dynamic graphs of two simulations are available as supplemental material. Use the control buttons under the figures to play the animations (http://pnbook.uni-frankfurt.de/). This concludes the description of the Petri net-based methodology to generate dynamic graphs from ODE models of signaling pathways (see Fig. 11.7 for summary). In the next section, we analyze the signal propagation and the topology dynamic of the model.

11.2.5 Tool Support

The Bhalla–Iyengar model is available in the DOQCS database [363] at http://doqcs.ncbs.res.in, accession number 4, in GENESIS format [404], Matlab format and SBML format. This model was imported and simulated with Virtual Cell [234]. The Virtual Cell model, *MAPK_Feedback_Bhalla*, is available in the public domain at http://www.vcell.org under the shared username *sihar*.

11.3 Results

With the Bhalla–Iyengar model and experiments, Bhalla, Ram and Iyengar identified a cellular mechanism that controls the transition from a bistable behavior to a monostable behavior and that is dependent on the previous history of the cell [34]. A saturating stimulation of the cells with PDGF for 5 minutes induces a MAPK response that is sustained for at least 45 minutes (bistable regime). However, if the same cells are restimulated one hour later with a second saturating PDGF stimulation, the activity of MAPK is transient and returns to basal levels in less than 30 minutes (monostable regime). The duration of the second response is dependent on the dose of the first stimulation.

The Bhalla–Iyengar study contains many simulations. In this section, we analyze two of them that capture the main features of the dynamic behavior of the

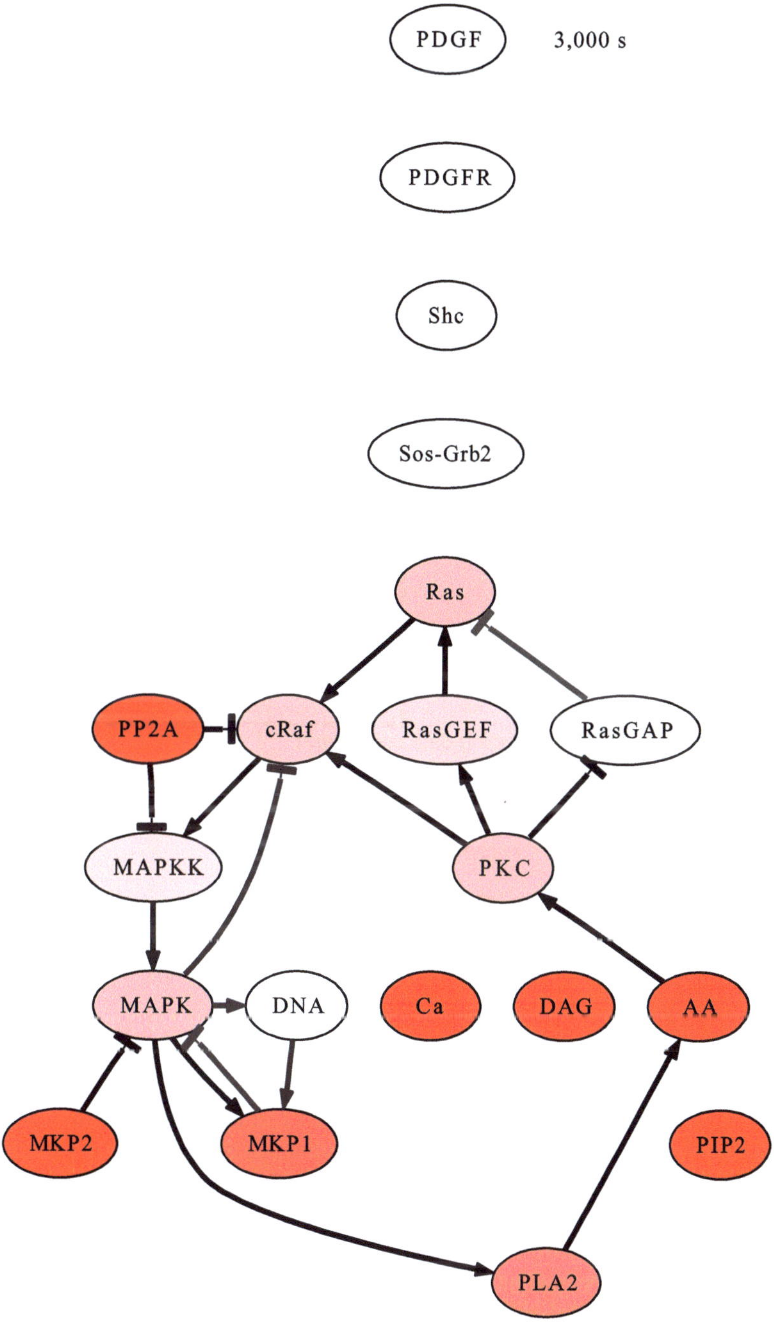

Fig. 11.6 Frame of one dynamic graph of the Bhalla–Iyengar model of the MAPK/PKC pathway 50 minutes after stimulation

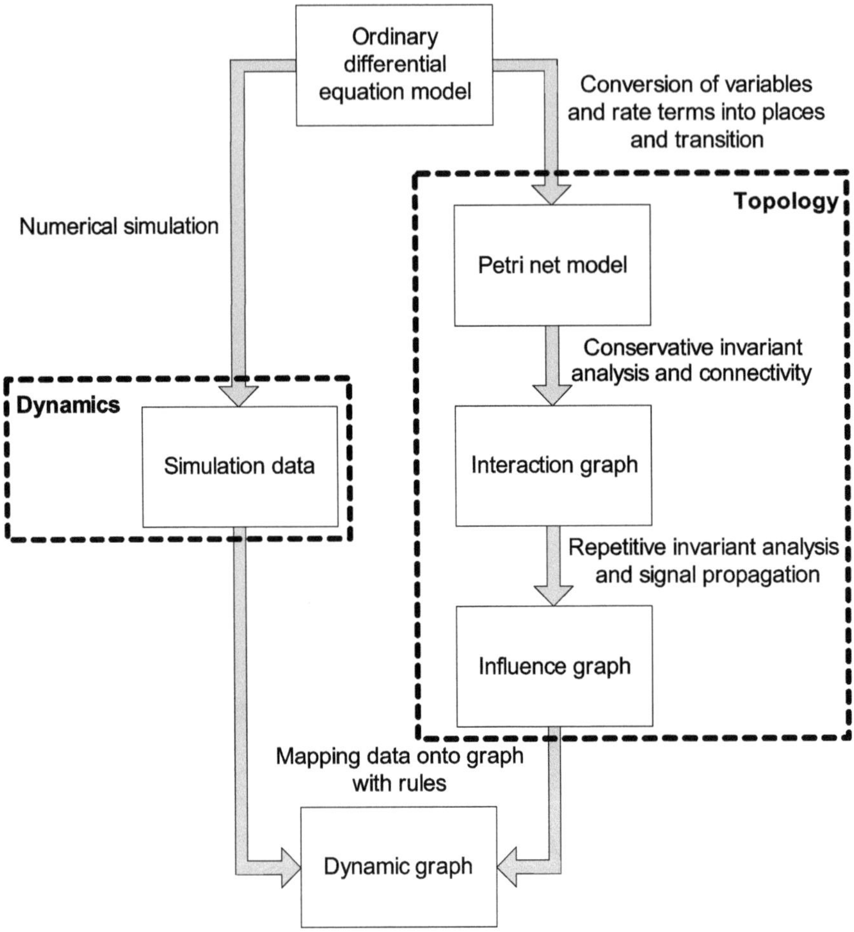

Fig. 11.7 Process of building a dynamic graph from an ODE model using Petri net theory

MAPK/PKC system: a first simulation with a single saturating stimulation and a second simulation with two saturating stimulations. We created the dynamic graphs for the two simulations (see supplemental Figs. 1 and 2 at http://pnbook.uni-frankfurt.de/). These representations show clearly the dynamics of this model as described in the paper. A brief saturating stimulation lasting 5 minutes is enough to initiate the PKC positive feedback loop that sustains the signal long after the stimulation is removed. In this positive loop, a MKP1 negative feedback loop is nested. This topology could have created an oscillatory system. Instead, the output of this topology is a sustained signal that is deactivated when the negative loop reaches a certain threshold. The reason for this delayed inactivation rather than an oscillation is that the loops are operating at two different time scales: the positive loop is made of fast, binding and enzymatic reactions while the negative loop is mostly dependent

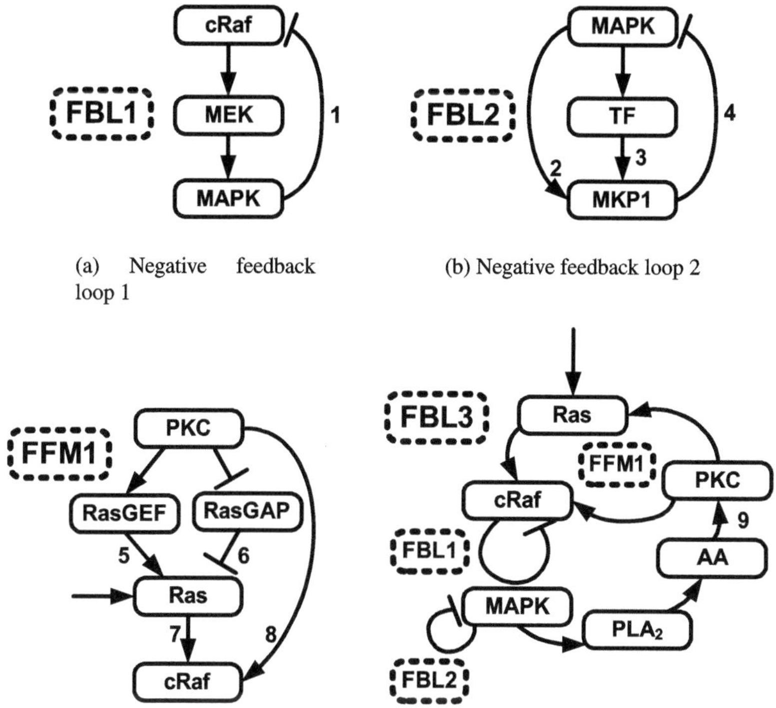

(a) Negative feedback loop 1

(b) Negative feedback loop 2

(c) Nested feedforward motif 1

(d) Positive feedback loop 3

Fig. 11.8 Regulatory motifs of the Bhalla–Iyengar model of the MAPK/PKC pathway

on the slow processes of gene expression and protein synthesis. A second stimulation, an hour after the first one, transiently activates MAPK because MKP1 levels are already up. In other words, the negative regulators are already synthesized and for that reason the same stimulation does not produce a prolonged response.

A network analysis of the influence graph of the Bhalla–Iyengar model reveals more regulatory motifs than the two loops previously studied. Some of these new motifs also play a role in the dynamic behavior of the model. We divided the motifs into four modules: a negative feedback loop between MAPK and cRaf (FBL1, Fig. 11.8(a)), a feedforward motif (FFM) nested in the negative feedback loop between MKP1 and MAPK (FBL2, Fig. 11.8(b)), a 3-branch feedforward motif between PKC and cRaf (FFM1, Fig. 11.8(c)) and the PKC positive feedback loop where the three other motif modules are nested (FBL3, Fig. 11.8(d)). Guided by the dynamic graphs, we methodically analyzed the dynamic of these motifs when the system is stimulated once. Then, we studied the motif lifetimes to analyze the model dynamic with two spaced stimulations.

11.3.1 Dynamic and Function of the Regulatory Motifs

The regulatory motifs of cellular signaling networks have information processing capabilities. The analysis of their dynamic is essential to understand how a specific input signal can be transformed into a completely different output signal. Even though they form an integrated system, the four motif modules of the Bhalla–Iyengar model have distinct signal processing functions that we describe bellow. The reaction flows and concentrations analyzed in this section were identified by our methodology and they correspond to the values of the edges and nodes of the dynamic graph.

11.3.1.1 Negative Feedback Loop 1: MAPK Inhibition of Upstream Activator cRaf is Inconsequential

The motif FBL1 is a simple negative feedback loop: the output MAPK is inhibiting the input cRaf [401]. This type of motif usually has a purpose of adaptation to maintain the output response in a physiologically desired range. To measure the activity of feedback motifs, a first place to look is the value of the rate of the feedback flow. In the motif FBL1, it is the inhibitory edge going from MAPK to cRaf. It is identified as Flow 1 (flow numbers are indicated in Fig. 11.8). As expected, this flow follows the activation pattern of MAPK by reaching its maximum around 20 minutes (Figs. 11.9(a) and 11.10(c)). A reasonable hypothesis is that the motif FBL1 is causing the observed adaptation of active MAPK: after reaching its maximum, the level of active MAPK starts declining. This hypothesis is invalidated by the comparison of the activation of cRaf with and without FBL1 (Fig. 11.9(b)). The difference is negligible and consequently, the motif FBL1 seems inoperative. This is explained by the fact that the phosphorylation of cRaf by MAPK is equally balanced by its constant dephosphorylation by the phosphatase PP2A. This motif might play a role for the MAPK pathway in other cell types or when it is activated by other receptors—it can theoretically induce oscillations [187]—but this motif is not involved in the cell memory property of mouse fibroblasts stimulated by PDGF according to the Bhalla–Iyengar model.

11.3.1.2 Negative Feedback Loop 2: Phosphorylation of MKP1 and Increase of Its Expression by MAPK Are Both Needed to Halt MAPK Signaling

The motif FBL2 is a negative feedback loop with a feedforward motif nested inside it: MAPK can activate MKP1 through two paths and then, MKP1 inhibits MAPK. To understand the function of a FFM, one has to consider the dynamic of the converging branches and the mechanism of signal integration. For example, both inputs of the FFM might be necessary to activate the output; this type of FFM thus has a logical AND function. Its opposite is a FFM with a logical OR function in which

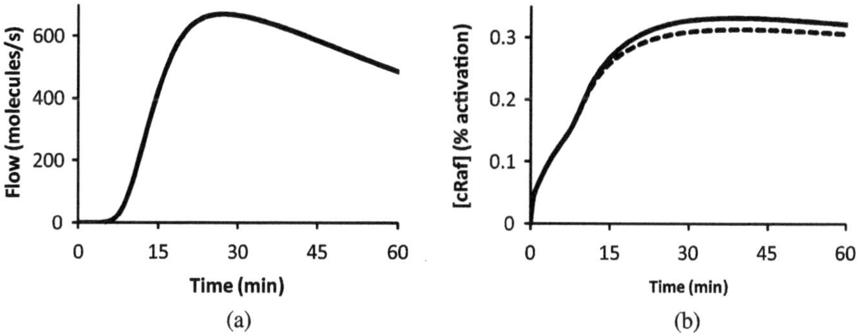

Fig. 11.9 Temporal activity of the negative feedback loop 1 (FBL1) between cRaf and MAPK. (**a**) Flow 1: Phosphorylation of cRaf by MAPK; (**b**) cRaf activity with (*dashed*) and without MAPK feedback (*solid*)

one path or the other can activate the output in parallel. There are also other FFM architectures where the inputs are incoherent; one is activating the target and the other is inhibiting it. Mangan and Alon studied the possibilities of this type of motif [240]. In the motif FBL2 of the Bhalla–Iyengar model, the FFM has a logical OR function. MAPK directly phosphorylates MKP1, which reduces its ubiquitination and degradation [50]. The expression of MKP1 is also transcriptionally regulated by MAPK [381]. Consequently, both inputs cause an increase of the levels of the phosphatase independently. However, their action is at different time scales. The phosphorylation is much faster than gene transcription and protein synthesis. There is a first phase of activation of MKP1 with a peak around 20 minutes corresponding to the phosphorylation of existing MKP1 by MAPK (Flow 2) and then a second phase of activation due to the late synthesis of MKP1 (Flow 3) and its subsequent phoshorylation (Fig. 11.10(a)). The slow steady increase of the dephosphorylation of MAPK by MKP1 (unphosphorylated and phosphorylated MKP1 can equally deactivate MAPK) is due to the large amount of newly synthesized MKP1 pouring into the cell during the first hour following the stimulation (Fig. 11.10(b)). Since the FFM nested in the motif FBL2 is an OR gate and because the feedback flow 4 follows the temporal pattern of MKP1 synthesis, an early conclusion would be that flow 2 is not required for the function of this motif. However, a selective disabling of flows 2 and 3 shows that MAPK is permanently activated if the FFM is not completely operational (Fig. 11.10(c)). The upregulation of the transcription and synthesis and the downregulation of the degradation are both necessary for the feedback loop to turn off the bistability. This might seem paradoxical: both branches of the OR gate nested in the motif FBL2 must function for the overcoming of the positive feedback by the negative feedback. This illustrates the additive amplitude increase of some FFMs. The flow 2 or the flow 3 alone is insufficient to let the negative feedback reach the threshold where MAPK is irreversibly inactivated. This is an important point in network biology: concurrent paths seem redundant, but in this situation, their additive effect is needed to achieve to motif function.

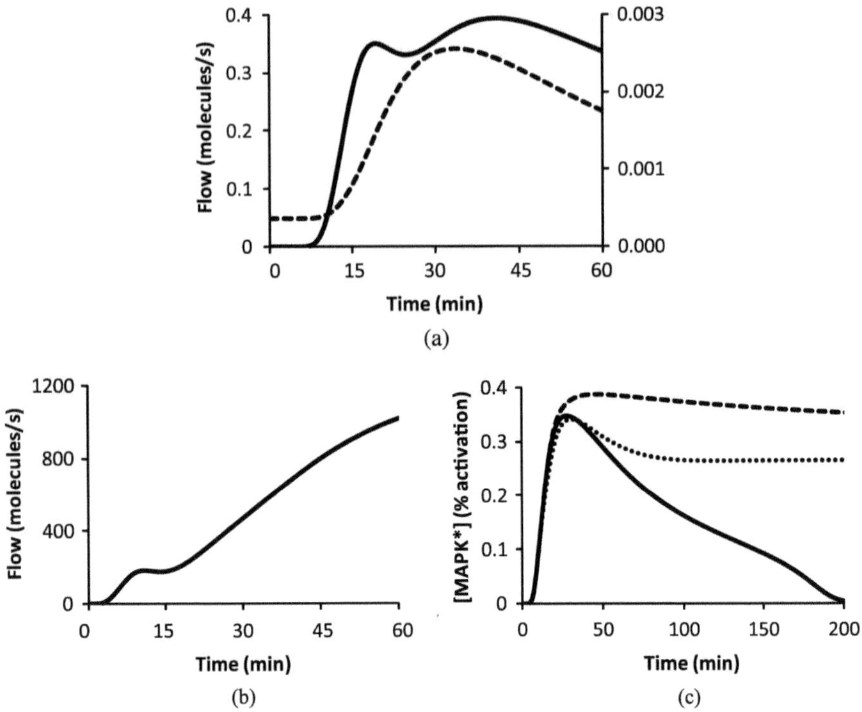

Fig. 11.10 Temporal activity of negative feedback loop 2 (FBL2) between MAPK and MKP1. (**a**) Reaction flows of the two paths of the nested feedforward motif from MAPK to MKP1. Flow 2: Phosphorylation of MKP1 by MAPK (*solid*, *left axis*), Flow 3: Synthesis of MKP1 (*dashed*, *right axis*); (**b**) Reaction flow of the inhibition of MAPK by MKP1. Flow 4: Dephosphorylation of MAPK by MKP1; (**c**) Control of the MAPK bistability duration in three different conditions: with full feedforward motif (*solid*), without the flow 3 for when the IEG regulated transcription of MKP1 is inhibited (*dashed*) and without the flow 2 for when MAPK can not regulate MKP1 degradation (*dotted*)

11.3.1.3 Nested Feedforward Motif 1: The Three Paths of the Motif Participates in a Cumulative Amplification of the Signal

The motif FFM1 is a nested feedforward motif with three paths: PKC can activate (1) cRaf directly by phosphorylation and indirectly through the activation of Ras by (2) the upregulation of RasGEF and (3) the downregulation of RasGAP. The first FFM with Ras as the output is an OR gate. RasGAP and RasGEF have independent opposing effects. However, their coherent regulation by PKC increases the amplitude of the output. Approximately 10 minutes after stimulation, the GTP exchange by RasGEF (Flow 5) leaves its basal level and the rate of GTP hydrolysis by RasGAP (Flow 6) decreases (Fig. 11.11(a)). This is when PKC is activated by arachidonic acid (AA, Fig. 11.12(a)). This second Ras activation occurs after the first activation by the stimulation signal that is coming from the receptor bound GEF Sos. These two phases of activation are observed on the

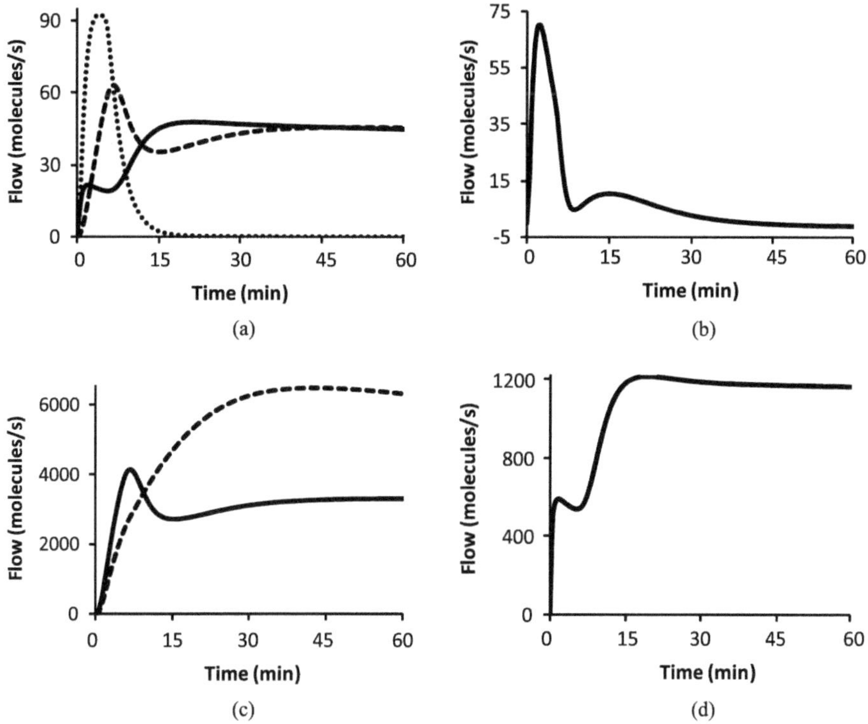

Fig. 11.11 Temporal activity of the nested feedforward motif 1 (FFM1) from PKC to cRaf. (**a**) Reaction flows of the three reactions controlling the levels of active Ras. Flow 5: Ras GTP exchange catalyzed by RasGEF (*solid*), Flow 6: Ras GTP hydrolysis catalyzed by RasGAP (*dashed*) and Stimulation signal: Ras GTP exchange catalyzed by Sos (*dotted*); (**b**) Reaction flow of the activation of cRaf by Ras. Flow 7: Binding of Ras and cRaf; (**c**) Decomposition of the cRaf activation into its two components. Forward binding of Ras with cRaf (*solid*) and with phosphorylated cRaf (*dashed*); (**d**) Reaction flow of the third path of the FFM1. Flow 8: Phosphorylation of cRaf by PKC

binding rate of Ras with cRaf (Flow 7, Fig. 11.11(b)). Flow 7 is the cumulative binding flow of Ras to unphosphorylated and phosphorylated cRaf. A closer look at two components of this flow—the forward binding of Ras with the two cRaf conformations—confirms that the latter activation is caused by the phosphorylation of cRaf by PKC (Flow 8, Figs. 11.11(c) and 11.11(d)). As previously noted, this modification of cRaf increases its activity because the dissociation constant of phosphorylated cRaf with Ras is lower than for its unphosphorylated counterpart. This biophysical integration mechanism creates an amplifying FFM. The full amplification of FFM1 is needed to sustain the activity of the positive feedback loop: MAPK is back to its basal level if either of the flow 5, 6 or 8 is disabled (data not shown).

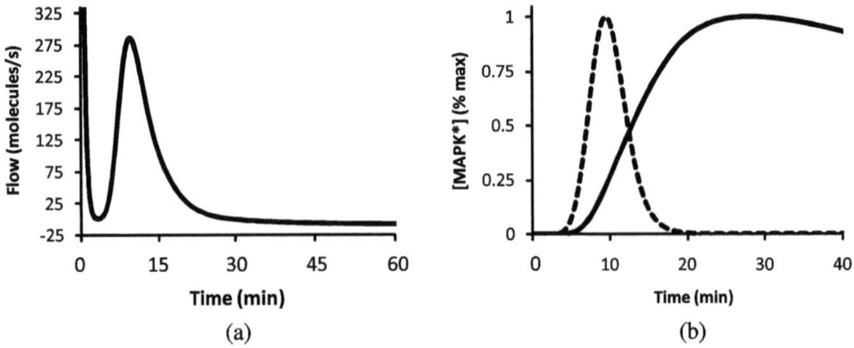

Fig. 11.12 Temporal activity of the positive feedback loop 3 (FBL3) between MAPK and PKC. (**a**) Flow 9: Binding of AA and PKC; (**b**) MAPK activity with PKC activity and the positive feedback (*solid*) and without (*dashed*). Traces were independently scaled

11.3.1.4 Positive Feedback Loop 3: MAPK and PKC Form a Timed Bistable Switch

The last motif module of the Bhalla–Iyengar model is the motif FBL3. It is a positive feedback loop between PKC and MAPK. The three other motif modules are embedded inside it. This loop functions as a bistable switch that can transform a brief stimulation into a sustained response. Flow 9 shows that this loop becomes active approximately 5 minutes after the PDGF stimulation; it reaches an active quasi-steady state around 25–30 minutes; and slowly decreases thereafter (Fig. 11.12(a)). Flow 9 is a binding flow: its value is the difference between the forward and backward rates of the binding reaction. This means that if the flow value is 0, it is a steady state, not necessarily an inactive state. The inhibition of a molecular species of the loop, like PKC, breaks the loop and impedes the persistent activation of MAPK (Fig. 11.12(b)). As discussed previously, the roles of the different nested motif modules inside the motif FBL3 vary. The motif FBL1 has a negligible effect. The motif FBL2 is an antagonist of the positive feedback loop and the strength of FBL2 will determine the outcome of the bistable switch. The motif FFM1 is an amplifier of the positive feedback that is necessary to achieve bistability. Connecting the modules together creates a signaling network that has the necessary complexity to reproduce the experimental observations. A reduced dynamic graph, showing only the key components and flows of the system is shown in the supplementary material.

The modularization of the Bhalla–Iyengar model in terms of regulatory motifs is needed to thoroughly dissect its complex behavior and analyze the dynamic of its topology. Next, we take a step further with this approach and define the lifetime of motifs. This measure will highlight the history-dependency feature of the model.

11.3.2 Lifetime of the Regulatory Motifs

Now that we gained an understanding of the inner mechanisms of the Bhalla–Iyengar model from a modular analysis of its regulatory motifs based on dynamic graphs, we will apply this knowledge to understand the dynamic behavior of this system when subjected to two saturating stimulations each lasting 5 minutes, one hour apart. Up to the second stimulation, the dynamic response of the simulated system is identical to the one stimulation experiment as expected. The stimulation sets the positive feedback loop into action with the help of the signal amplification of a nested feedforward motif. The positive feedback sustains the signal long after the initial stimulus is removed. This long MAPK activity initiates a slower negative feedback loop driven by the transcription and the synthesis of the negative regulator MKP and reduces its degradation by phosphorylation. When the second saturating dose of PDGF is added, both the positive and negative feedback loops are already active. This activity of the loops induces, after an identical stimulation, a different response from the cell. This is why this system is history-dependent.

The reduction of the system and its behavior to its regulatory motifs provides the main components for the analysis of the dynamics of the system. The level of activation at these signaling components is an indication of the activity of the regulatory motifs. Fig. 11.13 shows the output of the four regulatory motifs of the MAPK/PKC model: MAPK* (FBL1), MKP1 (FBL2), cRaf*-RasGTP (FFM1) and PKC* (FBL3). The MAPK activity profile shows the double activation (Fig. 11.13(a)); the second stimulation induces a shorter response than the initial stimulation. A new maximum of activity is reached for FFM1 with the second FGF stimulation (Fig. 11.13(c)) because cRaf is simultaneously stimulated by the FGF signal and the positive feedback loop signal coming from PKC. However, this higher activity is not transmitted downstream, because MAPK is already saturated. The second stimulation has also a minimal effect on FBL3 (Fig 11.13(d)): PKC* is below its maximum by 5% when the second stimulating signal reaches this motif. The most significant state difference for a regulatory motif is in FBL2 (Fig. 11.13(b)). At the moment of the second stimulation, MKP1 synthesis is in progress, the levels of MKP1 are already high and it has started to downregulate MAPK. The negative regulation is overpowering the positive regulation and for that reason, the second stimulation induces only a transient response. See the reduced dynamic graph in the supplemental material to observe this behavior.

The regulatory motifs are functional modules useful to decompose a signaling network and understand its complex dynamic behavior. It is a level of organization above individual signaling components. Once the functions of the regulatory motifs are understood, it is possible to represent the behavior of the system just by their activity. We developed the concept of motif lifetime to display this information. When the signal of the output signaling component of a motif is above 25% of its usual maximal level, the motif is considered active. With this simple metric, the period of activity of motifs can be determined. The lifetime of the four regulatory motifs of the Bhalla–Iyengar model is displayed in Fig. 11.14, where one can see that the second stimulation occurs when the MKP1 loop (FBL2) is still active. A single

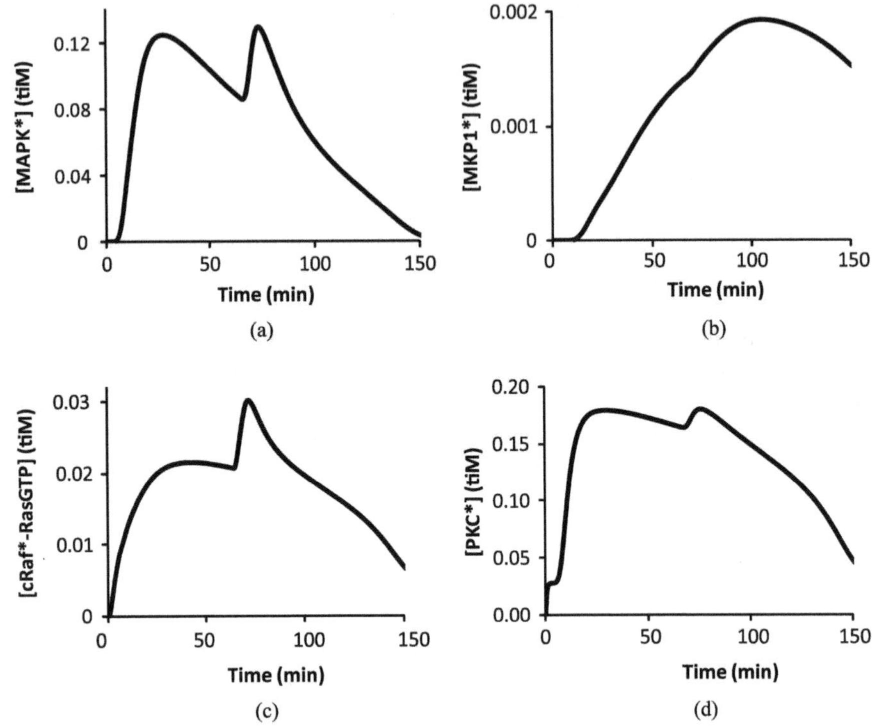

Fig. 11.13 Outputs of the four regulatory motif modules with a first 5 minute stimulation at 0 min and a second one at 65 min. Concentrations of (**a**) MAPK* (FBL1); (**b**) MPK1 (FBL2); (**c**) cRaf*-RasGTP (FFM1) and (**d**) PKC (FBL3)

Fig. 11.14 Lifetime of the four regulatory motifs of the Bhalla–Iyengar model. Stimulations indicated by *short black rectangles*

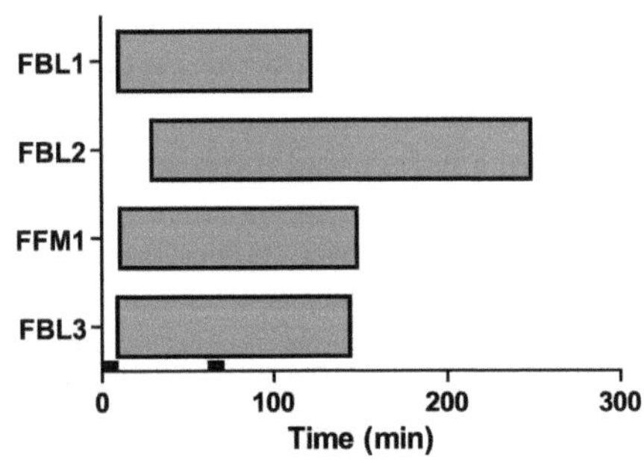

PDGF stimulation produces a MAPK response that is above the 25% threshold for approximately 120 minutes. The second stimulation induces a shorter response of 40 minutes. From the lifetime of the motif FBL1, it is possible to see that the model predicts that the monostable regime of these cells could last up to four hours after the initial stimulation.

11.4 Related Work

The relationship between the dynamics of a network and its topological organization has been subject to debate. Studies using graph theory established profiles of the signaling processing capabilities of network configurations [239] and characterizations of the robustness of a system [310] without any kinetic information. However, Ingram et al. studied the dynamics of the bifan regulatory motif with mathematical modeling and concluded that it is difficult to get insights into biological function just by considering topology [169]. The theoretical analysis of the bifan motif and of its filtering and synchronization properties was extended by Lipshtat et al. [233], again using mathematical modeling. They also concluded that a thorough study of the range of possible behaviors of a motif needed more than nodes and links alone, emphasizing this statement for a network of coupled motifs. Dynamic graphs, based on the simulation data of a mathematical model, are a step further in that direction. Kinetic dynamic and topology are both needed to fully understand the behavior observed in distant input-output relationships between components of cellular signaling networks.

A previous attempt to connect differential equation modeling and graph theory by Fages and Soliman demonstrated in [113] that an influence graph could be derived from a system of ODEs using the Jacobian matrix, but dynamics were not considered.

11.5 Summary

In this chapter, we described a new type of representation for dynamical systems: the Dynamic Graph. Dynamic Graphs bridge two formalisms commonly used in computational biology—differential equations and graph theory—with Petri net theory. The transformation of an ODE model into an animated influence graph is done in several steps. First, the mathematical model is converted into a Petri net model. Second, an interaction graph is built with the results of a conservative component analysis of this model. Third, the interaction graph is modified into an influence graph with the results of a repetitive component analysis of subgraphs of the Petri net model. Finally, the definition of heuristic visualization rules allows the mapping of the simulation data, generated by the numerical simulation of the ODEs, onto the influence graph. The result is a Dynamic Graph. We demonstrated the building process of a dynamic graph with the Bhalla–Iyengar model and analyzed the dynamics of its topology. The model was divided into four regulatory modules and functions were assigned to each module. This model successfully explains the transformation from a bistable to a monostable system. We have also completed a more detailed analysis of this system with the dynamic graph approach and described the role of the motifs that gives to this system its history-dependent characteristic.

Dynamic Graphs are an intuitive representation of the activity of a signaling network. They are powerful tools to analyze the propagation of a molecular signal inside the cell and to grasp the formation of regulatory motifs as the signaling information is processed by the cell's pathways. They also enable a functional analysis of complex cellular behaviors that are based on regulatory motifs instead of individual signaling components. As biomodels are likely to become more complex, Dynamic Graphs and dynamic topology analysis will become powerful tools for computational biologists.

11.6 Problems

11.1 Transform the following ordinary differential equations into a Petri net model

$$\begin{aligned}
d[Ligand]/dt &= -k_1[Ligand][Receptor] + k_2[Receptor.act] \\
d[Receptor]/dt &= -k_1[Ligand][Receptor] + k_2[Receptor.act] \\
d[Receptor.act]/dt &= k_1[Ligand][Receptor] - k_2[Receptor.act] \\
&\quad - k_3[Receptor.act][Kinase_1] + k_4[R.K_1.complex] \\
d[Kinase_1]/dt &= -k_3[Receptor.act][Kinase_1] + k_4[R.K_1.complex] \\
d[R.K_1.complex]/dt &= k_3[Receptor.act][Kinase_1] - k_4[R.K_1.complex] \\
d[Kinase_2]/dt &= k_{\mathrm{cat1}}[R.K_1.complex][Kinase_2]/\big(K_{m1} + [Kinase_2]\big) \\
&\quad - k_{\mathrm{cat2}}[Phosphatase]\big[Kinase_2^*\big]/\big(K_{m2} + \big[Kinase_2^*\big]\big) \\
d\big[Kinase_2^*\big]/dt &= -k_{\mathrm{cat1}}[R.K_1.complex][Kinase_2]/\big(K_{m1} + [Kinase_2]\big) \\
&\quad + k_{\mathrm{cat2}}[Phosphatase]\big[Kinase_2^*\big]/\big(K_{m2} + \big[Kinase_2^*\big]\big)
\end{aligned}$$

11.2 It is possible to distinguish two types of edge in the interaction graphs generated from a Petri net model even if this information is not used in the method presented in this chapter.

1. Which characteristics of the Petri net model connectivity allow for this distinction? What is the biological interpretation for each type of connectivity?
2. Generate the interaction graph of the Petri net model of Question 11.1 above. Divide the edges of this graph in the two types.

11.3

1. Some edges present in an interaction graph might not be in the influence graph derived from it. How is this possible?
2. In the interaction graph of Question 11.2, which edges do not become arcs in an interaction graph when the signal comes from the Ligand?

11.4 Can a Petri net with more than one P-invariant be covered by only one signaling segment? Why?

11.5 The following questions are based on the model of the JAK-STAT pathway [420]. The Virtual Cell version of this model, *JAK-STAT Yamada model*, can be found under the username *sihar*. A SBML file can be exported from the Virtual Cell model and then read with Snoopy.

1. How many P-invariants do the Petri net of this model have? What is their biological meaning?
2. Is the model covered with P-invariants? If not, what modifications would give it this property?
3. What is the interaction graph of this model?
4. Which edges were not determined with the connectivity condition using the intersection of conservative components?
5. How many T-invariants are there in the STAT subgraph? What are the signaling segments of the STAT subgraph?
6. What is the regulatory motif of the model of this signaling pathway?

Acknowledgements The work is supported by NIH grants GM072853 and P50GM071558 to R.I.S.H. holds NSERC Postdoctoral Fellowship BP-342902. The Virtual Cell is supported by NIH Grant Number P41RR013186 from the National Center for Research Resources.

Chapter 12
A Modular, Qualitative Modeling of Regulatory Networks Using Petri Nets

Claudine Chaouiya, Hanna Klaudel, and Franck Pommereau

Abstract Advances in high-throughput technologies have enabled the delineation of large networks of interactions that control cellular processes. To understand behavioral properties of these complex networks, mathematical and computational tools are required. The multi-valued logical formalism, initially defined by Thomas and coworkers, proved well adapted to account for the qualitative knowledge available on regulatory interactions, and also to perform analyses of their dynamical properties. In this context, we present two representations of logical models in terms of Petri nets. In a first step, we briefly show how logical models of regulatory networks can be transposed into standard (place/transition) Petri nets, and discuss the capabilities of such a representation. In the second part, we focus on logical regulatory modules and their composition, demonstrating that a high-level Petri net representation greatly facilitates the modeling of interconnected modules. Doing so, we introduce an explicit means to integrate signals from various interconnected modules, taking into account their spatial distribution. This provides a flexible modeling framework to handle regulatory networks that operate at both intra- and intercellular levels. As an illustration, we describe a simplified model of the segment-polarity module involved in the segmentation of the *Drosophila* embryo.

C. Chaouiya (✉)
Instituto Gulbenkian de Ciêencia, Rua da Quinta Grande, 6, 2780-156 Oeiras, Portugal
e-mail: chaouiya@igc.gulbenkian.pt

C. Chaouiya
TAGC Inserm U928, Campus de Luminy, case 928, 13288 Marseille, France

H. Klaudel · F. Pommereau
IBISC, Université d'Evry, 523 place des terrasses, 91000 Evry, France

H. Klaudel
e-mail: klaudel@ibisc.univ-evry.fr

F. Pommereau
e-mail: pommereau@ibisc.univ-evry.fr

I. Koch et al. (eds.), *Modeling in Systems Biology*, Computational Biology 16,
DOI 10.1007/978-1-84996-474-6_12, © Springer-Verlag London Limited 2011

12.1 Introduction

Great advances in molecular biology, genomics and functional genomics open the way to the understanding of regulatory mechanisms controlling essential biological processes. These mechanisms interplay and operate at diverse levels (transcription and translation of the genetic material, protein modifications, etc.). They define large and complex networks, which in turn constitute a relevant functional integrative framework to study the regulation of cellular processes. To assess the behaviors induced by such networks, dedicated mathematical and computational tools are very much required. In general, mathematical models for concrete regulatory networks are defined as a unique whole, considering networks of limited sizes (up to few dozens of components). This approach is not scalable and has to be modified as networks are increasing in size and complexity. One main purpose of this chapter is to present a compact, qualitative modeling framework to represent large regulatory networks and analyze them.

We rely on a qualitative discrete framework for the modeling of regulatory networks, namely the generalized logical formalism, initially proposed by Thomas in the 70s [389–392]. The logical formalism has been applied to a variety of regulatory networks comprising relatively large numbers of components (e.g., [330, 335]). To tackle the modeling of networks encompassing hundreds of nodes or interacting cells, we propose here to resort to modular modeling. In particular, in the case of patterning in developmental processes, one has to consider patches of communicating cells. In such processes, modularity clearly arises, each intra-cellular network defining a module. More precisely, in this chapter, we provide a convenient way to define the modeling of interacting regulatory modules.

After defining the semantics underlying regulatory interactions (as opposed to biochemical reactions that compose e.g., metabolic networks), the Sect. 12.2 gives the basis of the logical formalism. In [65–67, 374], a standard (i.e., P/T) Petri net representation of logical regulatory graphs have been proposed. This representation is summarized and discussed in Sect. 12.3.

The rest of the chapter is dedicated to the specification of a framework that addresses module composition in the context of patches of communicating cells. In Sect. 12.4, we show how, based on the logical framework, high-level Petri net representation provides a very compact and efficient means to compose regulatory modules.

Finally, to illustrate the modeling framework delineated in Sect. 12.4, we show the dynamical analyses (in particular expression pattern identification) for various composition scenarios of a simple module, and of the segment-polarity module involved in the segmentation of the *Drosophila* embryo.

12.2 Logical Modeling of Regulatory Networks

Regulation refers to the molecular mechanisms responsible for the changes in the concentration or activity of a functional product. Such mechanisms range from

DNA-RNA transcription to post-translational protein modifications. In regulatory networks, details on the precise molecular mechanisms that drive the regulation are often abstracted, the semantics associated to the interactions mostly reduces to activatory or inhibitory effects. Among the diversity of frameworks used to model such regulatory networks (see [175, 343]), one successful qualitative formalism is the *logical* approach, initially developed by R. Thomas and coworkers [389–392]. The logical formalism has been applied to model and analyze regulatory networks controlling a variety of cellular processes from pattern formation and cell differentiation (e.g., [258, 333–335]) to cell cycle (e.g., [115]). A software has been developed, GINsim, which enables the definition and analysis of logical models [132, 283] (see also Sect. 12.3.1). GINsim provides a number of exports of logical models among which several exports into Petri net formats.

When considering regulatory networks, the semantics associated with the interactions between components varies compared to that of, for example, reaction networks: levels of regulators do not change during the regulatory process. At the level of abstraction conveyed by the logical formalism, regulatory networks can be viewed as influence networks. In terms of PNs, to represent such interactions, *test arcs* provide a convenient solution. Moreover, in the case of an activation, the presence of an activator enhances the level of its target, but the absence of the activator may also have an effect on the target, decreasing its level (and the other way around for a repression). Such situations can be represented in PNs using *inhibitor arcs* that allow a test to zero. However, the analysis methods based on the matrix representation of PNs are no more valid when using inhibitor arcs. Opportunely, when places are bounded (their markings are limited), it is possible to replace inhibitor arcs by adding new complementary places. Section 12.3 relies on these principles to define a systematic translation of logical into Petri net models (see Part I, Chap. 3).

In the sequel, the definitions of logical regulatory graphs and their associated transition graphs are given, further details might be found in [64, 282].

12.2.1 Logical Regulatory Graphs (LRGs)

A Logical Regulatory Graph (LRG) is a graph, where each node represents a regulatory component, associated with a range of discrete functional levels (of expression or of activity). In most of the cases, the Boolean abstraction is sufficient (e.g., a gene is expressed or not, a protein is active or not), but there are situations where more than two levels are necessary to convey the functional role of a regulator that varies when the concentration of its product crosses different thresholds. In particular, this is the case when a product regulates several targets, these regulations possibly occurring at distinct thresholds. This leads to the arcs of the LRG that represent regulatory interactions. Finally, one needs to define the behaviors of the components submitted to regulatory interactions. This is done by setting up logical functions that define the target levels of the components (within the admissible ranges) for each possible combinations of incoming interactions. The formal definition of LRGs is provided below, and Fig. 12.1 p. 256 gives a simple example.

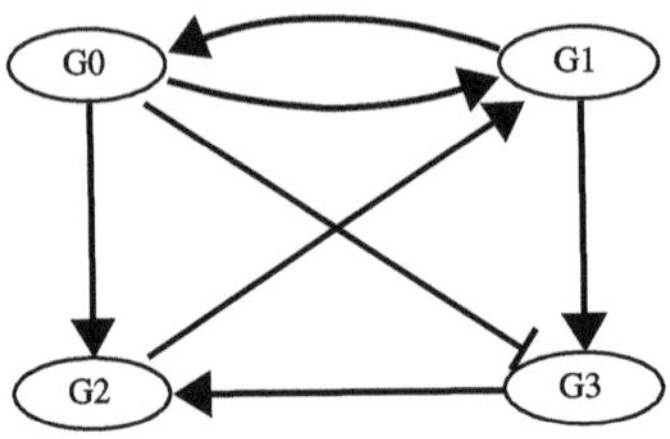

G0	G1	G2	G3	$\mathcal{K}_{G0}$
*	0	*	*	0
*	1–2	*	*	2

G0	G1	G2	G3	$\mathcal{K}_{G1}$
0	*	0	*	0
0	*	1	*	1
1–2	*	0	*	1
1–2	*	1	*	2

G0	G1	G2	G3	$\mathcal{K}_{G2}$
0	*	*	0	0
0	*	*	1	1
1–2	*	*	*	1

G0	G1	G2	G3	$\mathcal{K}_{G3}$
0–1	0–1	*	*	0
0–1	2	*	*	1
2	*	*	*	0

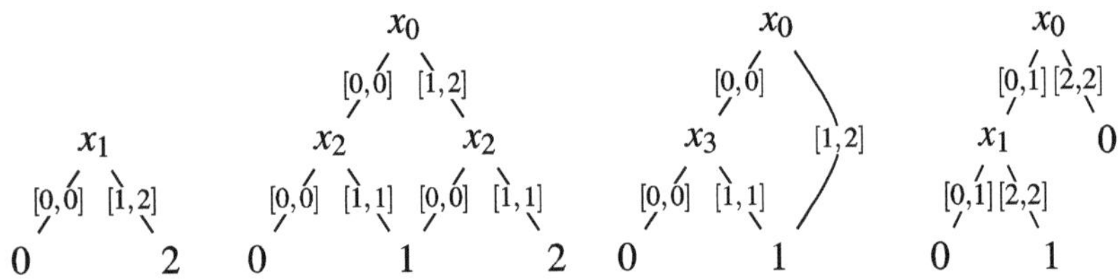

Fig. 12.1 Example of an LRG. The regulatory graph is displayed in the higher level, with the nodes denoting components and the arcs denoting interactions (*arcs with normal arrows* denote activations, whereas *arcs with blunt end* denote repressions). The logical functions are then given in the form of tables (one for each component), where each row corresponds to (a set of) state(s) with the corresponding function values ($*$ denotes one value among all possible values, 1–2 denotes value 1 or 2, etc.). For example, the *table on the left* defines the logical function $\mathcal{K}_{G0}$, which only depends on the levels of $G1$, the sole regulator of $G0$. The IDD representation of each logical function is given in the lower level (see Sect. 12.2.2 for explanations). For instance, $G3$ has two regulators ($G0$ and $G1$), its logical function $\mathcal{K}_{G3}$ specifies that, when both regulator levels are lower than 2, whatever the levels of the other components (which have indeed no effect on $G3$), the target level of $G3$ is 0 (first row of the table defining $\mathcal{K}_{G3}$). The IDD representing $\mathcal{K}_{G3}$ encompasses the decision variables x_0 and x_1 (the levels of $G3$ regulators) and the case just described is recovered by following the edges going out x_0 and x_1 labeled $[0,1]$, which leads to a leaf labeled 0 (for readability, this leaf has been duplicated)

Definition 12.1 A *logical regulatory graph* (LRG) is defined as a labeled directed multigraph[1] $\mathcal{R} = (\mathcal{G}, \mathcal{E}, \mathcal{K})$ where,

- $\mathcal{G} = \{g_1, \ldots, g_n\}$ is the set of nodes, representing *regulatory components*. Each $g_i \in \mathcal{G}$ is associated to its *maximum level* Max_i ($Max_i \in \mathbb{N}^*$), its *current level* being represented by the variable x_i ($x_i \in \{0, \ldots, Max_i\}$). We define $x =_{\text{def}} (x_1, \ldots, x_n)$ the current state, and $\mathcal{S} =_{\text{def}} \prod_{g_i \in \mathcal{G}} \{0, \ldots, Max_i\}$ the set of all possible states.

[1] A multigraph is a graph with possibly several edges between a pair of nodes.

- $\mathscr{K} = (\mathscr{K}_1, \ldots, \mathscr{K}_n)$ defines the *logical functions* attached to the nodes specifying their behaviors: $\mathscr{K}_j$ is a multi-valued logical function that gives the *target level* of g_j, depending on the state of the system: $\mathscr{K}_j : \mathscr{S} \to \{0, \ldots, Max_j\}$.
- $\mathscr{E}$ is the set of oriented edges (or arcs) representing *regulatory interactions*. An arc (g_i, g_j) specifies that g_i regulates g_j (when there is no possible confusion, i stands for g_i), that is, $\mathscr{K}_j$ varies with x_i. A regulatory graph may contain self-loops (an arc (i, i) represents a self-regulation of g_i).

  For each $g_j \in \mathscr{G}$, $Reg(j)$ denotes the set of its regulators: $i \in Reg(j)$ if and only if $(i, j) \in \mathscr{E}$.

Several remarks follow from Definition 12.1.

Remark 12.1 It is clear that, to determine the target level of a component, only the levels of its regulators are required (other components have no effect). In other words, $\mathscr{K}_j$ can be defined on the restricted domain $\prod_{g_i \in Reg(j)} \{0, \ldots, Max_i\}$. For example, in Fig. 12.1, since $G1$ is the sole regulator of $G0$, we could restrict the domain of $\mathscr{K}_{G0}$ to $\{0, \ldots, Max_{G1}\}$ (indeed, in the table defining $\mathscr{K}_{G0}$, we can verify that the values of $G0$, $G2$ and $G3$ do not matter).

Remark 12.2 If $g_i \in Reg(j)$ and $Max_i > 1$ (g_i regulates g_j and is multi-valued), g_i may have different effects onto a component g_j, depending on the current level of g_i, leading to the definition of a multi-arc between g_i and g_j. Such a situation typically happens when a component has a dual regulatory role, for example, activation at low and repression at high concentrations.

Here, to avoid the cumbersome notations resulting from such multi-arcs, we assume that all interactions are simple. However, it is straightforward to generalize all the definitions introduced in this chapter to LRGs encompassing multi-arcs (see, e.g., [282]).

Remark 12.3 The biologists often associate signs to the regulatory interactions, distinguishing between positive effect (activation or enhancing) and negative effect (repression or silencing). However, the actual effect of an interaction on its target may depend on the presence of cofactors; its sign may even change depending on the context. In any case, the signs of interactions can be derived from the logical functions $\mathscr{K}$s. Moreover, a *threshold* θ associated to an interaction from g_j to g_i with $1 \leq \theta \leq Max_j$ indicates for which level of g_j the interaction is *active* (i.e., when $x_j \geq \theta$, (g_j, g_i) is active). This threshold can also be recovered from the function $\mathscr{K}_i$: θ is the value for which there exists $x \in \mathscr{S}$ such that for x' defined as $x'_k =_{\text{def}} x_k, \forall k \neq j$ and $x'_j =_{\text{def}} x_j + 1 = \theta$, we have $\mathscr{K}_i(x) \neq \mathscr{K}_i(x')$. For example, in Fig. 12.1 the interaction from $G0$ to $G1$ is associated with a threshold 1 and a positive sign; this is visible by comparing the first and third rows of the table defining $\mathscr{K}_{G1}$ (also the second and the fourth rows), where, for fixed values of the other components, changing $G0$ level from 0 to 1 leads to a change in the target value of $G1$ from 0 to 1. Whereas the interaction from $G0$ to $G3$ is associated with a threshold 2 with a negative sign (determined by comparing second and third lines of $\mathscr{K}_{G3}$ table).

Finally, it is worth noting that a set of logical functions $\mathcal{K}_i, i = \{1, \ldots, n\}$, fully defines an LRG encompassing n regulatory components. In particular, for each $i \in \{1, \ldots, n\}$, its maximum level is given by the maximum value of $\mathcal{K}_i$.

12.2.2 Logical Functions Representation Based on Decision Diagrams

In [282], logical functions were represented by means of Reduced Ordered Multi-valued Decision Diagrams (ROMDDs, referred to as MDDs in the following). This representation, internally used in GINsim for efficiency purposes, facilitates the definition of algorithms for the analysis of logical models (e.g., see the stable state determination described in [282]). The use of MDDs also makes the definition of the P/T net representation of logical models easier and more concise than that proposed in for example, [65]. This representation, which is quite intuitive, is informally presented below.

Binary Decision Diagrams (BDDs) are a convenient data structure to represent Boolean functions [52]. A BDD is a rooted, directed, acyclic graph, encompassing *decision nodes* (labeled by a Boolean variable) and two leaves (also called terminal-nodes) labeled respectively, 0 (false) and 1 (true). Decision nodes have two successors (or children): the left (resp. right) outgoing edge represents an assignment of the variable to 0 (resp. to 1). Most BDDs are ordered and reduced, meaning that variables appear in the same order along all the paths from the root, and that all isomorphic (equal) subgraphs are merged and nodes having two isomorphic children are deleted (see Fig. 12.2 for an illustration). This structure has been naturally extended to handle discrete multi-valued functions, which are then represented by Multi-valued Decision Diagrams (MDDs), where the decision nodes have as many children as the number of their possible values and the leaves (or terminal nodes) are labeled by the values of the function [179]. The ordering and reduction rules defined for binary decision diagrams apply also to this multi-valued generalization (see [52] for further details). A path from the root to a leaf represents a (possibly partial) assignment of the decision variables for which the function takes the value carried by the leaf (see Fig. 12.2).

In the MDD representation of the function $\mathcal{K}_i$ of a regulatory component g_i, the decision variables are the variables associated to the regulators of g_i, and the leaves take their values in $[0, Max_i]$, which denotes the integer interval $\{0, \ldots, Max_i\}$. Note that the use of a MDD leads to a simplified expression of $\mathcal{K}_i$, but the resulting diagram and its complexity may vary depending on the ordering of the decision variables [179].

Finally, it is possible to consider a compact representation of the usual MDD by merging consecutive edges leading to the same child as explained hereafter.

Given $g_i \in \mathcal{G}$, for each decision variable x_j that appears in the diagram of $\mathcal{K}_i$, there are $Max_j + 1$ outgoing edges, implicitly labeled with the corresponding value

$$f_1(A,B,C) = A \vee (B \wedge C)$$

A	B	C	f_1
0	0	*	0
0	*	0	0
0	1	1	1
1	*	*	1

$$f_2(A,B) = \begin{cases} 0 \; if \; (A=0) \wedge (B=0) \\ 1 \; if \; ((A=0) \wedge (B=1)) \vee (A=1) \\ \quad \vee((A=2) \wedge (B=0)) \\ 2 \; if \; (A=2) \wedge (B=1) \end{cases}$$

A	B	f_2
0	0	0
0	1	1
1	*	1
2	0	1
2	1	2

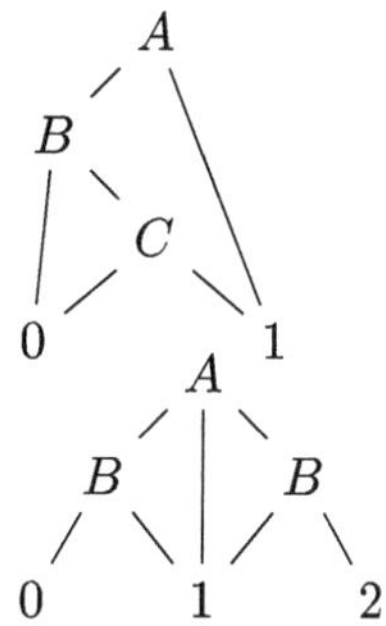

Fig. 12.2 Decision diagrams representing logical functions. The Boolean function f_1 (*top part*) of the Boolean variables A, B, C, evaluates 1 (i.e., true) if A or B and C ($\wedge$ stands for and, $\vee$ stands for or); f_1 can be equivalently defined by a truth table (where $*$ denotes all possible values, here 0 and 1) and by a binary decision diagram. For instance, A, B, C all set to 0 (*first line of the table*) corresponds to the path in the diagram following the leftmost edges going out A and B and leading to the leaf 0, indicating that, for these assignments of A, B, C, the function f_1 evaluates to 0. The multi-valued function f_2 (*bottom part*) evaluates to 0, 1 or 2, depending on the values of the discrete variables A and B. Similarly, f_2 can be defined by means of a table or a decision diagram. For instance, in this diagram, the variable assignments $A = 2$ and $B = 1$ (*last line of the table*) define the path comprising the rightmost edge going out A followed by the rightmost edge going out B, leading to the value 2

in $[0, Max_j]$. An alternative diagram can thus be considered by merging consecutive edges (i.e., labeled with consecutive values) towards the same child into a unique edge, which is then labeled with the integer interval of these consecutive values. Remaining edges are labeled by intervals containing a unique value for the decision variable. Then, in the resulting diagram, each decision path Φ (from the root node to a leaf labeled $v_\Phi \in [0, Max_i]$) corresponds to a set of assignments of the regulators $g_j \in Reg(i)$ for which the value of $\mathcal{K}_i$ is v_Φ (see Exercise 12.1):

- If path Φ encompasses an edge going out the decision variable x_j, the set of assignments of x_j is equal to the label $[\phi_j, \phi'_j] \subsetneq [0, Max_j]$ (called the Φ assignment interval for x_j) of the edge going out the decision variable x_j.
- If, along the path Φ, a decision variable x_j does not appear (due to the simplification of the MDD), it means that $\mathcal{K}_i(x) = v_\Phi$ does not depend on x_j.

Such diagrams were introduced in [377] as a generalization of MDDs and referred to as Interval Decision Diagrams (IDD). Here, we deal with a specific class of IDD, since our decision variables are discrete and bounded.

Remark that, because we have assumed that regulatory graphs do not encompass multi-arcs, in this IDD representation, each decision variable has exactly two children.

Figure 12.1 illustrates an LRG together with the functions $\mathcal{K}$s represented by means of IDDs. For a better readability, IDDs are often not fully reduced.

12.2.3 State Transition Graphs Associated to LRGs

The behavior of a logical regulatory graph is defined by the logical functions introduced in Definition 12.1. For any state of the system (i.e., a vector encompassing the levels of all the regulatory components), these functions indicate the target level of each component, that is the level to which it evolves. State Transition Graphs (STGs) constitute a classical and convenient way of representing the behavior of such systems. In these directed graphs, nodes represent states, and arcs represent transitions between states that amount to update one component level (increasing or decreasing it by one).

Definition 12.2 Given an LRG $\mathscr{R} = (\mathscr{G}, \mathscr{E}, \mathscr{K})$, its full *State Transition Graph (STG)* is a directed graph $(\mathscr{S}, \mathscr{T})$ such that:

- the set of nodes is the set of states $\mathscr{S}$ as defined above,
- $\mathscr{T} \subseteq \mathscr{S}^2$ defines the set of transitions (arcs) as follows. For all $(x, y) \in \mathscr{S}^2$:

$$(x, y) \in \mathscr{T} \iff \exists g_i \in \mathscr{G} \quad \text{s.t.} \quad \begin{cases} \mathscr{K}_i(x) \neq x_i, \\ y_i = \delta_i(x) =_{\text{def}} x_i + \text{sign}(\mathscr{K}_i(x) - x_i), \\ y_j = x_j \quad \forall j \neq i. \end{cases}$$

Given an initial state x^0, we further define the STG $(\mathscr{S}_{|x^0}, \mathscr{T}_{|x^0})$, which contains all states reachable from x^0:

- $x^0 \in \mathscr{S}_{|x^0}$,
- $\forall x \in \mathscr{S}_{|x^0}, \forall y \in \mathscr{S}, (x, y) \in \mathscr{T} \Rightarrow y \in \mathscr{S}_{|x^0}$ and $(x, y) \in \mathscr{T}_{|x^0}$.

The *updating function* δ_i of a regulatory component g_i, as defined above, specifies the update of x_i, the level of g_i; depending on the current state x of the network, the value of $\delta_i(x)$ is either x_i (no change, $x_i = \mathscr{K}_i(x)$), $x_i + 1$ (increase by one, $x_i < \mathscr{K}_i(x)$) or $x_i - 1$ (decrease by one, $x_i > \mathscr{K}_i(x)$).

In Fig. 12.3 the full STG of the LRG of Fig. 12.1 is displayed, together with two sub-graphs obtained for different initial states. Analyzing an STG, we can recover important properties of an LRG, among which:

- Single point attractors or stable states (e.g., corresponding to stable expression patterns) are nodes of the STG with no successor; in other words, they are states in which all component levels are equal to the target value indicated by the logical functions.
- Complex attractors (e.g., corresponding to oscillatory behaviors) are terminal strongly connected components encompassing more than one node, that is, sets such that all states are reachable from each other along directed paths, and there is no outgoing transition; in other words, the system is trapped in such a set once it has reached one of its state (see Exercise 12.2).
- Reachability of given attractors from initial conditions corresponds to the existence of path(s) in the STG.

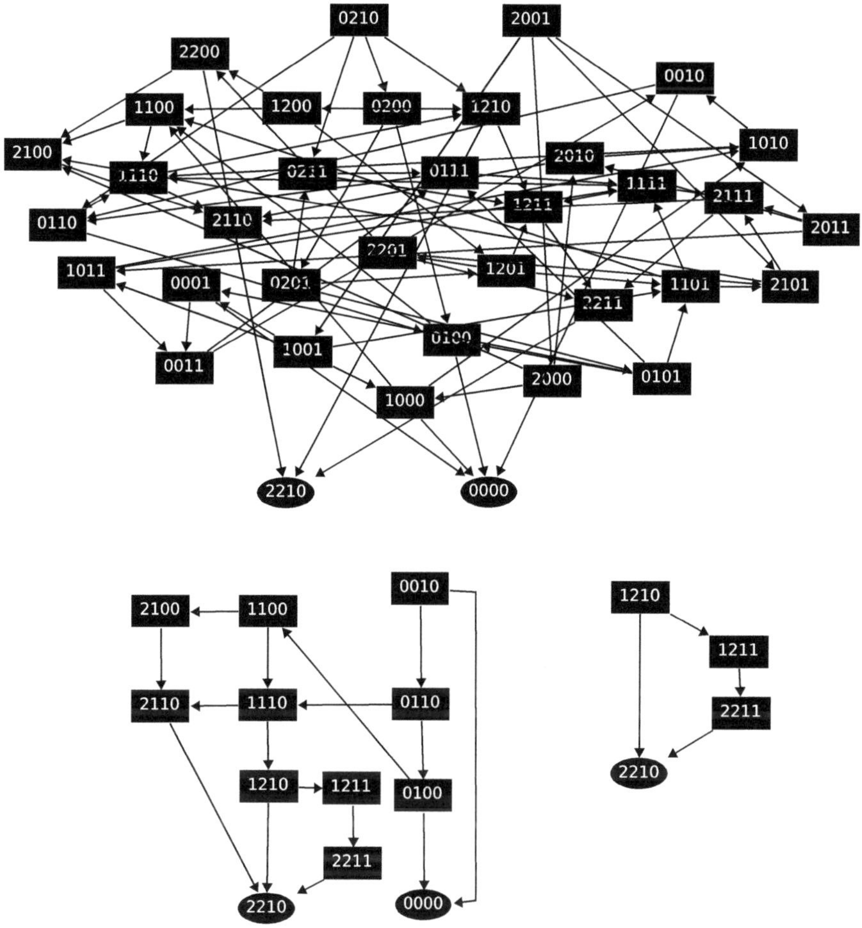

Fig. 12.3 State transition graphs corresponding to the LRG of Fig. 12.1. *On top*, the full STG, encompassing all 36 states. Note that there are two stable states indicated as *ellipse nodes*: $(r_{C0}, r_{C1}, r_{C2}, r_{C3}) = (2, 2, 1, 0)$ and $(0, 0, 0, 0)$. Other transient states are denoted as rectangular nodes. The two graphs in the *lower part* of the figure correspond to sub-graphs of the full STG for the initial states $(0, 0, 1, 0)$ (on the *left*) and $(1, 2, 1, 0)$ (on the *right*). Note that, for the initial state $(1, 2, 1, 0)$, one stable state is lost

12.3 P/T Petri Net Representation

The Definition 12.3 below explicitly defines a P/T net associated to an LRG, using the IDD representation of the functions $\mathscr{K}_i$s (as introduced in Sect. 12.2.1). Further details, basic properties and applications of this P/T net representation of LRGs are provided in [65] for the Boolean case, and in [66, 68] for the multi-valued one.

Definition 12.3 Given an LRG $\mathscr{R} = (\mathscr{G}, \mathscr{M}ax, \mathscr{E}, \mathscr{K})$, we define the corresponding *Multi-valued Regulatory Petri Net* (MRPN) as follows:

- For each $g_i \in \mathcal{G}$, two complementary places are defined, g_i and $\widetilde{g_i}$, satisfying, for all marking M:

$$M(g_i) + M(\widetilde{g_i}) = Max_i. \tag{12.1}$$

- For each $g_i \in \mathcal{G}$, for each path Φ from the root to a leaf of the IDD representing $\mathcal{K}_i$, at most two transitions are defined, one accounting for the increasing shift (denoted $t^+_{i,\Phi}$), the second accounting for the decreasing shift (denoted $t^-_{i,\Phi}$) (this simplifies when the leaf is associated with an extreme value, see below). Recall that Φ defines assignment intervals of the levels of g_j in $Reg(i)$: $x_j \in [\phi_j, \phi'_j]$, with $\phi_j, \phi'_j \in [0, Max_j]$ and $\phi_j \leq \phi'_j$.
- Transitions $t^+_{i,\Phi}$ and $t^-_{i,\Phi}$ are connected to:
  - place g_j, $j \in Reg(i)$, with a test arc weighted ϕ_j,
  - place $\widetilde{g_j}$, $j \in Reg(i)$, with a test arc weighted $Max_j - \phi'_j$.

  Transition $t^+_{i,\Phi}$ is further connected to:
  - place g_i, with an outgoing arc (increasing the level of g_i),
  - place $\widetilde{g_i}$, with an incoming arc weighted $Max_i - v_\Phi + 1$ (ensuring that the current level of g_i is less than the focal value v_Φ) and an outgoing arc weighted $Max_i - v_\Phi$ (accounting for the decreasing by one of the current marking of this complementary place).

  Symmetrically, transition $t^-_{i,\Phi}$ is further connected to:
  - place $\widetilde{g_i}$, with an outgoing arc (decreasing the level of g_i),
  - place g_i, with an incoming arc weighted $v_\Phi + 1$ (ensuring that the current level of g_i is greater than the focal value v_Φ) and an outgoing arc weighted v_Φ (accounting for the decreasing by one of the current marking).

From the definition above, it follows that, for all $g_i \in \mathcal{G}$ and Φ a path in the decision diagram associated to $\mathcal{K}_i$, when $v_\Phi = 0$ or $v_\Phi = Max_i$ (the value of the $\mathcal{K}_i$ for this assignment of the regulators is *extreme*), only one transition is relevant. Indeed, if $v_\Phi = 0$, transition $t^+_{i,\Phi}$ can be omitted as, by construction, there will never be $Max_i + 1$ tokens in place $\widetilde{g_i}$. Similarly, if $v_\Phi = Max_i$, transition $t^-_{i,\Phi}$ can be omitted as there will never be $Max_i + 1$ tokens in place g_i. Moreover, for $g_j \in Reg(i)$, when $\phi_j = \phi'_j$, from (12.1), it suffices to consider only one test arc (that towards place g_j for example).

Given an LRG $\mathcal{R}$ and an IDD representation of its logical functions, the Definition 12.3 uniquely specifies a P/T net. It can be shown that, given an initial state x^0, the STG $(\mathcal{S}_{|x^0}, \mathcal{T}_{|x^0})$ is isomorphic to the marking graph of the P/T net with the initial marking defined as $M_0(g_i) = x^0_i$ and $M_0(\widetilde{g_i}) = 1 - x^0_i$, for all components g_i (see proof in [68]). Hence, properties of an LRG can be derived from the analysis of the marking graph of its P/T net representation.

The IDD representation of the logical function leads to more compact Petri nets compared to those obtained by using decision trees or truth table representations (as in [66]). Different orderings of the variables in the IDD may generate different reductions. However, although the number of transitions may vary, it can be proved that the resulting dynamics (the marking graphs) are identical [68].

Figure 12.4 illustrates this P/T net representation for the LRG of the Fig. 12.1.

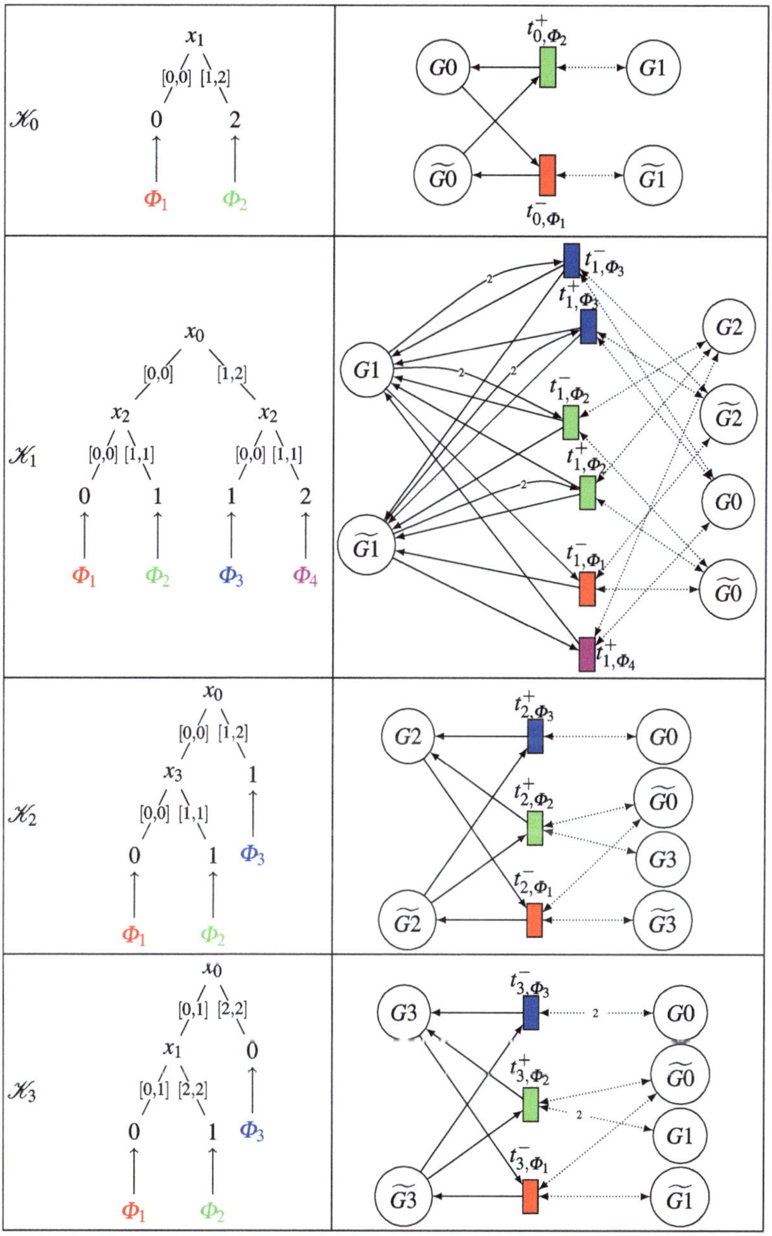

Fig. 12.4 P/T net representation for the LRG of the Fig. 12.1. In each row, the IDD giving the component logical function in displayed on the *left*, and the corresponding piece of P/T net is displayed on the *right*. Test arcs are depicted as *dotted lines*. Paths in the IDD are indicated with their respective transition(s), indicated by using the same color. For example, for $G0$, path Φ_1 (in *red*) corresponds to the situation where the absence of $G1$ leads to a decrease of $G0$'s value, represented in the P/T net by the transition in *red* t^-_{0,Φ_1}. For G_1, path Φ_3 gives rise to the two transitions t^-_{1,Φ_3} and t^+_{1,Φ_3} (in *blue*), since it leads to the intermediate value 1

12.3.1 Tools Support

GINsim is a software dedicated to the definition and analyze of regulatory networks logical models [132, 283]. With the representation presented in Sect. 12.3 we can also employ PN tools to analyze (larger) LRGs. GINsim has been equipped with export functionalities generating files in the INA [166] format, PNML (Petri net markup language) [306] and APNN (Abstract Petri Net Notation) [29].

In the case of the analysis of the segment-polarity system as defined in [335], the reachability analysis of the expected patterns was performed using INA. We have also assessed the high complexity of the STG of this model, using tools such as the model-checking tool DSSZ-MC [101] and Tina [393].

12.3.2 Related Works

Related works include the representation of logical regulatory networks by means of high-level Petri nets. In [66] a high-level PN representation of LRGs is defined, encompassing one place and one transition for each regulatory component. The even more compact representation defined in [81] enables the verification of the model coherence under various hypotheses (accounting for observed biological behaviors such as homeostasis, multistationarity, or even more specific temporal properties).

Based on similar principles as exposed in this section, Steggles et al. have defined a Petri net representation of Boolean regulatory graphs [374]. However, to account for Boolean networks as introduced by Kauffman [185] which are synchronously updated (meaning that all components are updated at once), a two-phase method is defined to ensure the synchronization of the updates. The first phase identifies all components that are called to change their values, the second phase performs this update synchronously.

Finally, the PN representation of regulatory networks as presented here, enables the delineation of integrated models of regulated metabolic pathways, considering a logical model of the regulatory level (a PN representation) linked to a classical PN model of the metabolic part. In [362], this approach is illustrated with a qualitative modeling of the biosynthesis of tryptophan (Trp) in *E. coli*.

12.4 Modules, Their Composition and High-Level Petri Net Representation

The framework presented in this section deals with collections of spatially distributed abstract modules (nonnested) which may be cells, compartments in cells or any subsystems that one wants to model separately. Each such module is defined by a regulatory network with identified inputs. The regulation functions take into account spatial configuration between modules.

In a first step, we introduce Logical Regulatory Modules (LRMs) as LRGs equipped with external input nodes and arcs. Then, we define how such modules can be spatially located and interconnected in order to form a Collection of interconnected Logical Regulatory Modules (CLRMs). Finally, given a collection of logical regulatory modules, we define its high-level Petri net representation.

12.4.1 Interconnecting Logical Regulatory Modules

In what follows, a *Logical Regulatory Module* is defined as a uniquely identified logical regulatory network associated with a set of *input* components that regulate internal ones. Notice that no output nodes are defined because any internal component may generate an external signal towards other modules; moreover, input components are not effective regulatory components in that they do not introduce any intermediary step in the signaling. These inputs are meant to combine (or integrate) external signals from other modules. Functions σ in Definition 12.4 perform such combinations by calculating the levels of the integrated signals, depending on the levels of the corresponding individual incoming signals and on their attributed weights. These weights, defined as real numbers on the interval [0; 1], encode neighboring relations, which in turn are defined when interconnecting the modules (see Definition 12.5). Hence, at this stage, in order to define a logical regulatory module, one has to specify the components that are likely to signal the module, and how input signals are combined through the functions σ. If needed, both constraints could be relaxed. In particular, postponing the specification of functions σ to the actual connection of modules would be a straightforward extension of the current framework.

Definition 12.4 Given Γ a domain of regulatory components, a *Logical Regulatory Module (LRM)* $\mathscr{M}$ is a tuple $(\mathscr{G}, \mathscr{E}, \gamma, \sigma, \mathscr{K})$, where

- $\mathscr{G} \subseteq \Gamma$ is the set of *internal components*;
- $\mathscr{E} \subseteq \mathscr{G} \times \mathscr{G}$ is the set of *internal interactions* between internal components;
- $\gamma \subseteq \Gamma$ is the set of *input components*;
- $\sigma = \{\sigma_{v,g} \mid (v, g) \in E \subseteq \gamma \times \mathscr{G}\}$ is a set of *integration functions*,

$$\sigma_{v,g} : \big(\{0, \dots, Max_v\} \times]0; 1]\big)^* \to \{0, \dots, Max_g\}$$

  computing the combined level for all inputs v regulating g. Their arguments are pairs (x_v, d_v) where each x_v is the level of v in one neighbor, weighted by $d_v > 0$;
- $\mathscr{K}$ is a set of logical functions defined on $\mathscr{G}$ giving, for each component $g \in \mathscr{G}$, its target level in $\{0, \dots, Max_g\}$, depending on the levels of its regulators. For an internal regulator r ($r \in \mathscr{G}$) the level used to evaluate $\mathscr{K}_g$ is x_r as usual, while for an input regulator v ($v \in \gamma$), the level used to evaluate $\mathscr{K}_g$ is the current value of $\sigma_{v,g}$. We denote by $args(\mathscr{K}_g)$ the set of arguments of $\mathscr{K}_g$.

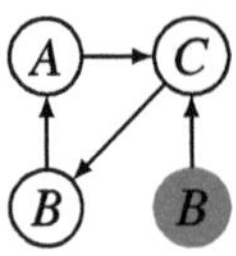

$$\mathcal{K}_A(x_B) =_{def} 1 \text{ if } (x_B = 1), \text{ else } 0$$
$$\mathcal{K}_B(x_C) =_{def} 1 \text{ if } (x_C = 1), \text{ else } 0$$
$$\mathcal{K}_C(x_A, \sigma_{B,C}) =_{def} 1 \text{ if } (x_A = 1 \wedge \sigma_{B,C} = 0), \text{ else } 0$$

Fig. 12.5 A "toy" LRM $\mathcal{M}$ with: $\mathcal{G} =_{\text{def}} \{A, B, C\}$, $\mathcal{E} =_{\text{def}} \{(A, C), (C, B), (B, A)\}$, $\gamma =_{\text{def}} \{B\}$, with, e.g., $\sigma_{B,C} =_{\text{def}} (y_j, d_j)_{j\geq 0} \mapsto \mathit{round}(\max(y_j \cdot d_j))$, and $\mathcal{K}$ being defined on the *right part* of the figure. The input node B is depicted as a *gray node*. We have $\mathit{args}(\mathcal{K}_C) = \{A, \sigma_{B,C}\}$

Figure 12.5 provides a simple example of an LRM. If $\sigma_{v,g} \in \mathit{args}(\mathcal{K}_g)$, then v is an input regulator of g; in the pictures, this is denoted by an arc toward the node g.

An LRM can be viewed as an *encapsulated* regulatory network with input nodes behaving as integrators of external signals of the corresponding components from other modules. For example, for the LRM depicted in Fig. 12.5, the level of external signal corresponding to B will be calculated as a weighted maximum over all the levels of B in neighboring modules (i.e., having a nonzero weight). Hence, this level can be evaluated only when the module is connected to other modules (see Definition 12.5). Nevertheless, at this stage, it is possible to recover a fully defined LRG, by setting the integration functions and specifying the behaviors of the input components (for example, as having constant levels).

Notice that, like in the Fig. 12.5, it is not required that $\mathcal{G} \cap \gamma = \emptyset$. For example, we may have $g \in \mathcal{G} \cap \gamma$ when the regulatory component g has both autocrine (acting on the same cell) and paracrine effects (acting on neighboring cells). Then $g \in \mathcal{G}$ accounts for the autocrine effect and $g \in \gamma$ accounts for the paracrine effect (as in the case of Wingless in the *Drosophila* segment polarity module depicted in Sect. 12.5).

We now proceed with collections of LRMs, which contain all the information needed to interconnect LRMs through their input nodes. It simply consists in defining a set of LRMs and a topological relation between these LRMs that establishes the actual connections between modules and allows the evaluation of the levels of input components.

Definition 12.5 A *Collection of interconnected Logical Regulatory Module* (or CLRM) is defined as a triplet $(\mathbb{I}, \mathbb{M}, \mathbb{T})$, where:

- $\mathbb{I} \subset \mathbb{N}$ is a finite set of integers (module identifiers);
- $\mathbb{M} = \{(m, \mathcal{M}) \mid m \in \mathbb{I}\}$ is a set of LRMs, each being associates to an identifier in $\mathbb{I}$;
- $\mathbb{T} : \mathbb{I} \times \mathbb{I} \setminus \{(m, m) \mid m \in \mathbb{I}\} \to [0; 1]$, is a topological (or neighboring) relation between modules in $\mathbb{M}$; the values of $\mathbb{T}$ can be interpreted as weights associated to external signals.

In a CLRM, one can define several copies of the same LRM (these copies are distinguished by their identifiers). This is the case in Fig. 12.6 that shows a CLRM encompassing three times the LRM of Fig. 12.5. Notice the dashed arcs that represent how each LRM signals its neighbors according to the topological relation.

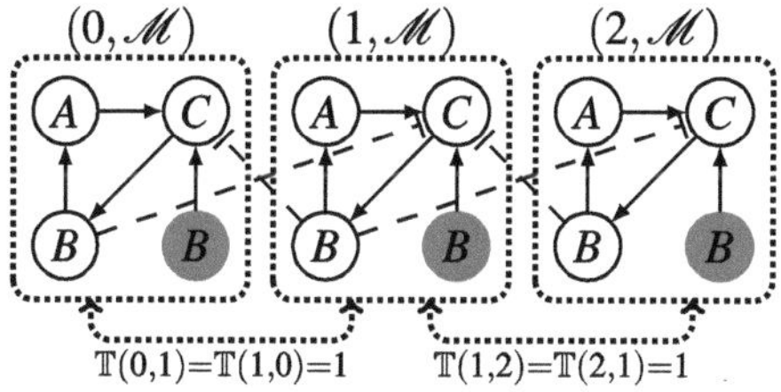

Fig. 12.6 A CLRM $(\mathbb{I}, \mathbb{M}, \mathbb{T})$ encompassing three copies of the LRM $\mathscr{M}$ of Fig. 12.5, where: $\mathbb{I} = \{0, 1, 2\}$, $\mathbb{M} =_{\text{def}} \{(0, \mathscr{M}), (1, \mathscr{M}), (2, \mathscr{M})\}$, and $\mathbb{T} =_{\text{def}} \{(0, 1) \mapsto 1; (1, 2) \mapsto 1; (2, 0) \mapsto 0\}$ completed to symmetry. Function $\sigma_{B,C}$ in module 0 depends on the level of B in module 1 (the sole connected module containing B) while $\sigma_{B,C}$ in module 1 depends on the levels of B in both modules 0 and 2 (which is depicted using *dashed arcs*)

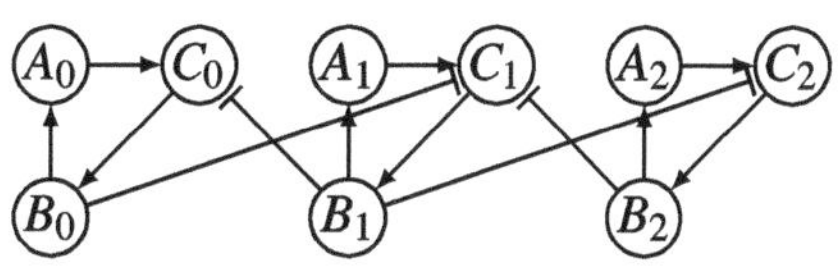

$$\sigma^0_{B,C} =_{def} \sigma_{B,C}((x_{B_1}, 1))$$
$$\sigma^1_{B,C} =_{def} \sigma_{B,C}((x_{B_0}, 1), (x_{B_2}, 1))$$
$$\sigma^2_{B,C} =_{def} \sigma_{B,C}((x_{B_1}, 1))$$

$$\mathscr{K}_{C_0}(x_{A_0}, x_{B_1}) =_{def} 1 \text{ if } (x_{A_0} = 1 \wedge x_{B_1} = 0), \text{ else } 0$$
$$\mathscr{K}_{C_1}(x_{A_1}, x_{B_0}, x_{B_2}) =_{def} 1 \text{ if } (x_{A_1} = 1 \wedge x_{B_0} = 0 \wedge x_{B_2} = 0), \text{ else } 0$$
$$\mathscr{K}'_{C_1}(x_{A_1}, x_{B_0}, x_{B_2}) =_{def} 1 \text{ if } (x_{A_1} = 1 \wedge (x_{B_0} = 0 \vee x_{B_2} = 0)), \text{ else } 0$$

Fig. 12.7 The LRG obtained from the CLRM of Fig. 12.6 with integration functions defined on the *right* and logical functions on the *bottom*. If we choose $\sigma^m_{B,C}$ as the maximal value of its argument levels ($\sigma^m_{B,C} = \max(x_{B_n} \cdot \mathbb{T}(n, m))$) as suggested in the caption of Fig. 12.5, then the logical function of C_0 is $\mathscr{K}_{C_0}$, and the logical function of C_1 is $\mathscr{K}_{C_1}$. Now, if $\sigma^m_{B,C} = \min(x_{B_n} \cdot \mathbb{T}(n, m))$, then, $\mathscr{K}_{C_0}$ remains the same, but $\mathscr{K}_{C_1}$ is replaced by $\mathscr{K}'_{C_1}$. Notice that $\mathscr{K}_{C_2}$ is similar to $\mathscr{K}_{C_0}$ (replace x_{A_0} with x_{A_2}), moreover, $\mathscr{K}_{A_i}$ and $\mathscr{K}_{B_i}$ for $0 \le i \le 2$ are exactly as in Fig. 12.5

Given a CLRM, the integration functions are expressible as logical terms. Indeed, the topological relations are fixed and the values of the σ functions depend on discrete bounded variables and take discrete bounded values. Hence, it is possible to recover an LRG (a logical model for the whole set of modules). For example, consider the CLRM in Fig. 12.6, its associated LRG is shown in Fig. 12.7.

12.4.2 High-Level Petri Net Representation

We first define a class of high-level Petri nets [202] especially crafted to our needs.

Intuitively, the main difference between such high-level Petri nets and the usual P/T Petri nets is that now places may carry tokens that have a value. In our particular case, these values will be pairs (i, v) where i is the identifier of a module and v the level of one of the components of this module.

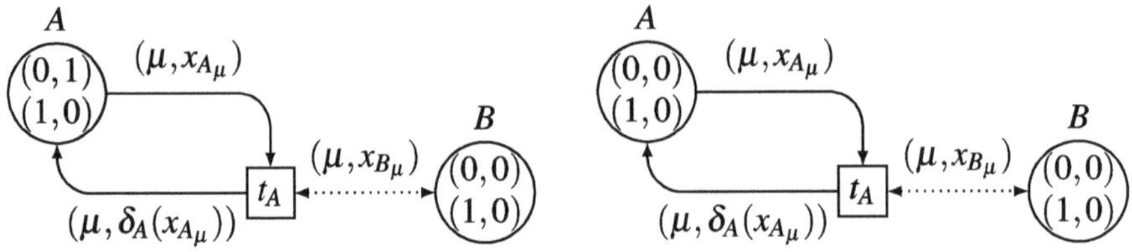

Fig. 12.8 The marking depicted on the *left* represents the fact that, for module 0, $x_A = 1$ and $x_B = 0$, and, for module 1, $x_A = 0 = x_B$. For this marking, transition t_A is enabled for the binding $\beta = (\mu \to 0, x_{A_\mu} \to 1, x_{B_\mu} \to 0)$. Suppose $\delta_A(1) = 0$ (δ_A being the updating function), the firing of t_A leads to the marking depicted on the *right* of the picture. This corresponds to a state change in the biological system

In order to consume and produce tokens when a transition fires, arcs are labeled with expressions involving variables; an empty label denotes the absence of arc. At fire time, it is necessary to *bind* (i.e., to map) each such variable to a concrete value so that the annotation on each arc can be evaluated to a collection of tokens. A *binding* β is a function that maps variables to values. We denote $\beta(expr)$ the evaluation of an expression *expr* under the binding β.

More precisely, a *high-level Petri net* is a tuple (S, T, ℓ), where:

- S is the set of places, each place is allowed to carry structured tokens from $\mathbb{N} \times \mathbb{N}$.
- T, disjoint from S, is the set of transitions.
- $\ell : (S \times T) \cup (T \times S)$ defines the labeling of the arcs by expressions.

In general, a *marking* M of a high-level Petri net is a mapping associating to each place $s \in S$ a multiset over $\mathbb{N} \times \mathbb{N}$ representing the tokens held by s at M. However, in our case, markings are always sets (in all evolutions, no token may be duplicated in the same place). So, we assume that arc labels always evaluate to sets of tokens.

A transition t may *fire* at a marking M with binding β if there are enough tokens in the input places of t: for all $s \in S$, $\beta(\ell(s,t)) \subseteq M(s)$. If so, t may fire and produce a new marking M', where M' is M in which consumed tokens are removed and produced tokens are added, that is, for all $s \in S$: $M'(s) =_{\mathrm{def}} (M(s) \setminus \beta(\ell(s,t))) \cup \beta(\ell(t,s))$, see Fig. 12.8.

Remark 12.4 For the sake of simplicity, Definition 12.6 below is valid only if modules do not overlap, that is, for $(m, \mathscr{M})$ and $(n, \mathscr{N})$ two LRMs in $\mathbb{M}$, $(\mathscr{M} \neq \mathscr{N}) \Rightarrow (\mathscr{G}^m \cap \mathscr{G}^n = \emptyset)$. This means that the set of all regulatory components is partitioned and we can focus on the evolution of each species in the context of the set of (identical) modules it belongs to.

Let $\mathscr{K}_g$ be the logical function of g. As previously, we define the updating function $\delta_g(x_{g_m}) =_{\mathrm{def}} x_{g_m} + sign(\mathscr{K}_g(\cdots) - x_{g_m})$ where $\mathscr{K}_g$ is computed with the appropriate arguments (including integration functions calls) as defined above.

Each internal regulatory component $g \in \mathscr{G}$ is modeled by a high-level Petri net place s_g that stores the numeric value of the corresponding level for each module m. In order to model several modules (each having a unique identifier), each place holds tokens of the form (m, x_g), where $m \in \mathbb{I}$ and x_g is the level of g in module m. The evolution of each regulatory component $g \in \mathscr{G}$ is implemented by a unique transition t_g that consumes the value of g in m (token (m, x_g) from place s_g), reads the values of the regulators of g, and produces the new level $\delta_g(x_{g_m})$ of g in m. Thus, transition t_g has all the necessary arcs to the places modeling g and its regulators.

Definition 12.6 Let $(\mathbb{I}, \mathbb{M}, \mathbb{T})$ be a CLRM. We denote $\mathscr{G} =_{\text{def}} \bigcup_{m\in\mathbb{I}} \mathscr{G}^m$ the set of components involved in the CLRM. The high-level Petri net associated to $(\mathbb{I}, \mathbb{M}, \mathbb{T})$ is (S, T, ℓ), defined as:

- $S =_{\text{def}} \{s_g \mid g \in \mathscr{G}\}$
- $T =_{\text{def}} \{t_g \mid g \in \mathscr{G}\}$
- For each $g \in \mathscr{G}$, let m denotes the LRM such that $g \in \mathscr{G}^m$, the arcs attached to each $t_g \in T$ are as follows:
  - First, a pair of arcs allows to read and update the current level of g for module $\mathscr{M}$, whose identifier m is captured by a net variable μ:
    - If $\sigma_{g,g} \notin args(\mathscr{K}_g)$, there is one arc (s_g, t_g) labeled by $\{(\mu, x_{g_\mu})\}$ and one arc (t_g, s_g) labeled by $\{(\mu, \delta_g(x_{g_\mu}))\}$.
    - Otherwise, there is one arc (s_g, t_g) labeled by $\{(\mu, x_{g_\mu})\} \cup \{(\eta, x_{g_\eta}) \mid \forall\eta : \mathbb{T}(\mu, \eta) > 0\}$ and one arc (t_g, s_g) labeled by $\{(\mu, \delta_g(x_{g_\mu}))\} \cup \{(\eta, x_{g_\eta}) \mid \forall\eta : \mathbb{T}(\mu, \eta) > 0\}$; it means that the levels of g in all modules η in the neighboring of m are read and only the level of g in m is updated.
  - Then, for each $g' \in \mathscr{G}^m \setminus \{g\}$, a test arc $(s_{g'}, t_g)$ is added to bind the parameters of the logical and integration functions, with net variable μ capturing as above the identity of module m:
    - If $(g', g) \in \mathscr{E}^m$ and $\sigma_{g',g} \notin args(\mathscr{K}_g)$ (g' is only an internal regulatory), this arc is labeled by $\{(\mu, x_{g'_\mu})\}$.
    - If $(g', g) \in \mathscr{E}^m$ and $\sigma_{g',g} \in args(\mathscr{K}_g)$ (g' is both an internal and external regulator), this arc is labeled by $\{(\mu, x_{g'_\mu})\} \cup \{(\eta, x_{g'_\eta}) \mid \forall\eta : \mathbb{T}(\mu, \eta) > 0\}$.
    - If $(g', g) \notin \mathscr{E}^m$ and $\sigma_{g',g} \in args(\mathscr{K}_g)$ (g' is only an external regulator), this arc is labeled by $\{(\eta, x_{g'_\eta}) \mid \forall\eta : \mathbb{T}(\mu, \eta) > 0\}$.

For the collection in Fig. 12.7, we obtain the Petri net depicted in Fig. 12.9.

12.4.3 Implementation

A prototype of the construction presented above has been implemented on the top of SNAKES toolkit [307]. SNAKES is a full featured Petri net library intended for quick prototyping; it uses Python programming language [313] to express the various Petri net annotations. Using this implementation:

- LRMs can be fully specified as Python classes.
- Arbitrary integration functions can be user-defined, some are predefined.
- CLRMs can be defined by composing LRM instances with arbitrary topologies.

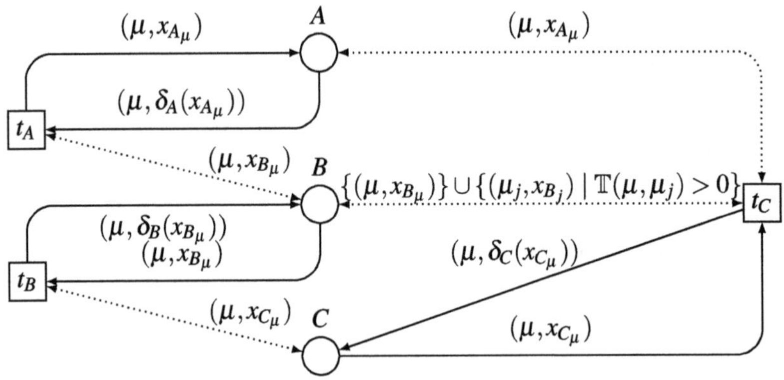

$$M_0(s_A) =_{def} \{(0,1),(1,0),(2,1)\}$$
$$M_0(s_B) =_{def} \{(0,1),(1,1),(2,0)\}$$
$$M_0(s_C) =_{def} \{(0,0),(1,1),(2,0)\}$$

Fig. 12.9 The high-level Petri net representation of the CLRM of Fig. 12.7. A possible initial marking may be M_0 as given below the Petri net

- CLRMs topology can be drawn with automatic layout.
- All possible stable states of a CLRM can be computed.
- Reachable states from an initial one can also be computed, with extraction of the reachable stable states.

12.5 Modeling Interconnected LRMs Using High-Level Petri Nets

In this section, we illustrate the modeling of interconnected logical regulatory modules by means of high-level Petri nets. In a first step, we analyze nine cases of collections of several copies of our toy LRM (as defined in Fig. 12.5). We show how variations of the topological or the integration functions can affect the behaviors of the whole model. Then we discuss the application of our framework to the modeling of the segment-polarity module involved in the segmentation of the *Drosophila* embryo.

In both cases, the considered collection of modules is composed of copies of the same LRM. This, indeed, will mostly be the case when modeling patches of identical cells, but one should be aware that the proposed framework does not impose such a restriction.

12.5.1 Interconnecting Occurrences of the Toy LRM

Let consider the toy LRM $\mathscr{M}$ as defined in Fig. 12.5. In this section, we analyze the behaviors of a series of collections encompassing a number of occurrences of $\mathscr{M}$.

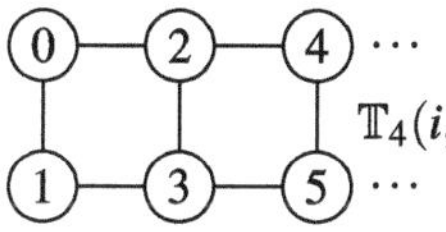

$$\mathbb{T}_4(i,j) =_{def} \mathbb{T}_4(j,i) =_{def} \begin{cases} 1 & \text{if } j=i+2 \vee (i \bmod 2 = 0 \wedge j=i+1) \\ 0 & \text{otherwise} \end{cases}$$

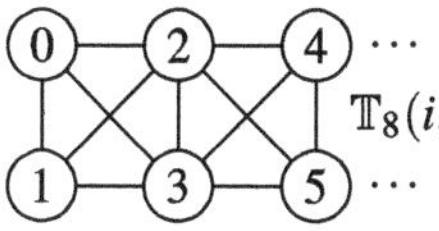

$$\mathbb{T}_8(i,j) =_{def} \mathbb{T}_8(j,i) =_{def} \begin{cases} 1 & \text{if } j=i+1 \vee j=i+2 \\ & \qquad \vee (i \bmod 2 = 0 \wedge j=i+3) \\ 0 & \text{otherwise} \end{cases}$$

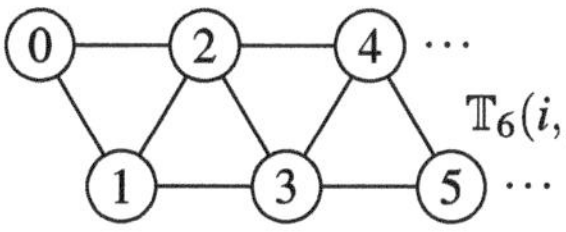

$$\mathbb{T}_6(i,j) =_{def} \mathbb{T}_6(j,i) =_{def} \begin{cases} 1 & \text{if } j=i+1 \vee j=i+2 \\ 0 & \text{otherwise} \end{cases}$$

Fig. 12.10 The various topologies considered for 2-width tapes. On the *left*, the graphical view showing the organization of the connections. Definitions on the *right* side rely on the assumption that modules on the *top line* of the tape are numbered with consecutive even numbers, while modules on the *bottom line* are numbered (following the same direction) with consecutive odd numbers. All topologies considered here are symmetrical, the subscript k indicates the maximal number of neighbors in the topology $\mathbb{T}_k$ in an extended grid (e.g., with the topological relation $\mathbb{T}_4$, any module would have at most 4 neighbors)

All the considered collections are based on a tape of modules of width 2, with various lengths and topologies as depicted in Fig. 12.10. Each topological relation $\mathbb{T}_k$ is called in such a way because modules may have neighbors in k directions. Moreover, we denote *Toy*(n,k) the CLRM containing $2 \cdot n$ copies of $\mathcal{M}$ arranged using $\mathbb{T}_k$ on a tape, as illustrated in Fig. 12.10.

All stable states of systems *Toy*(n,k) are such that the levels of the components in each module are equal (components A, B and C all equal to 0 or all to 1). So, to depict such a module stable state, we shall print a black and white grid, where black (resp. white) positions correspond to modules whose components are all 1 (resp. all 0), these positions respecting the topological relations of the modules. To simplify the presentation, we have also considered initial states that can be represented in this way.

Figure 12.11 shows the stable states that can be reached from chosen initial states. We can observe variations in the number of reachable states (which correspond to the size of the marking graph) and of reachable stable states; moreover, the stable states themselves are different.

All these examples have been obtained using the integration function $\sigma_{B,C}$ defined as $(y_i, d_i)_{1 \le i \le k} \rightarrow round(\max(y_i \cdot d_i \mid 1 \le i \le k))$ (i.e., $\sigma_{B,C}$ evaluates to the maximal value of B in neighboring modules; that is to say $\sigma_{B,C}$ is 1 provided that at least one neighboring module encompasses a component B which level is 1). We may consider instead a function that returns 1 if at least two components B in the neighboring modules are equal to 1, and return 0 otherwise. In this case, the behaviors of the collections might be rather different; in particular, the initial states considered for *Toy*$(3,4)$ and *Toy*$(4,4)$ (Fig. 12.11, fourth and seventh rows) turn to be stable.

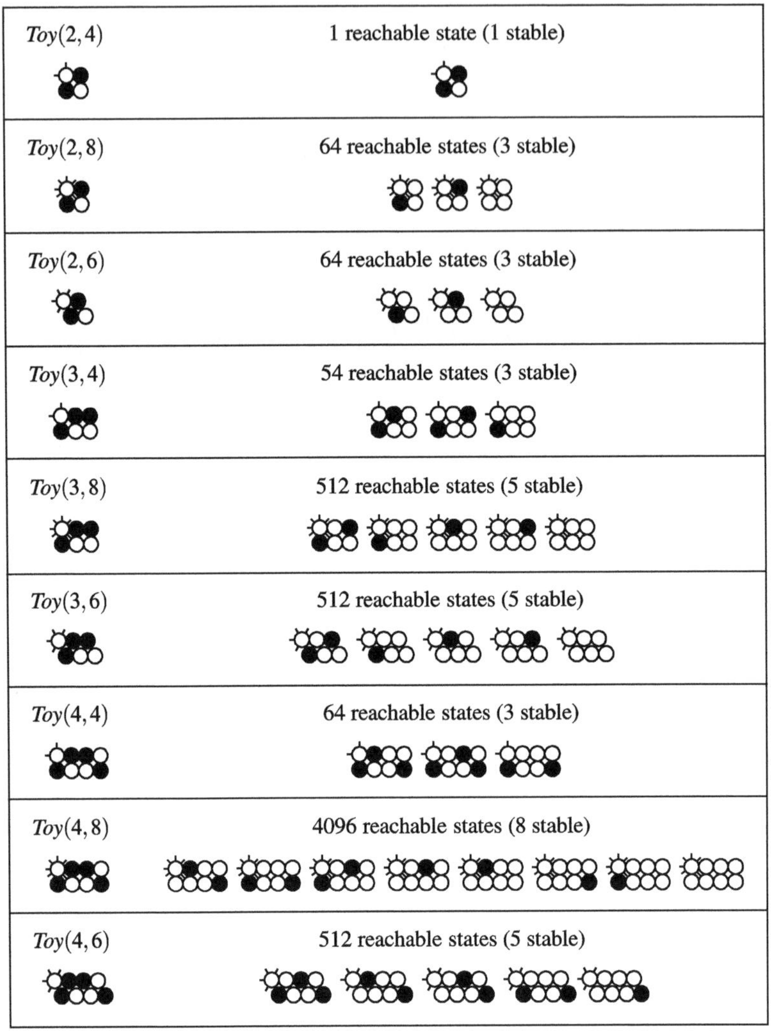

Fig. 12.11 Impact of the topological relation on the behaviors of the collections *Toy*(n, k), defined on the LRM of Fig. 12.5. In each row, a different combination of values of n and k (i.e., different number of modules and topological relation) is considered: on the *left*, the initial collection state is depicted, that is, the initial state of each of its modules (*black* if internal A, B, C are all 1, *white* if they are all 0), and on the *right*, the total number of reachable states is given and the reachable stable states are depicted. The *top-left* module of each state is decorated with links showing the directions of its potential neighbors, for example, in $\mathbb{T}(3, 8)$, a module has possibly 8 neighbors (see Fig. 12.10). This series of experiments shows that, for the same initial condition, a collection of interconnected modules can behave differently, depending on the signaling capacities (expressed here in terms of topological relations). For example, in the first collection $\mathbb{T}(2, 4)$, the modules are not able to signal along the diagonal, contrary to the collection $\mathbb{T}(2, 8)$. In the first, the initial pattern is stable, whereas it is not in $\mathbb{T}(2, 8)$, which instead will stabilize in one of three other patterns. The integration function $\sigma_{B,C}$ considered here takes the maximal value of B over the neighboring modules

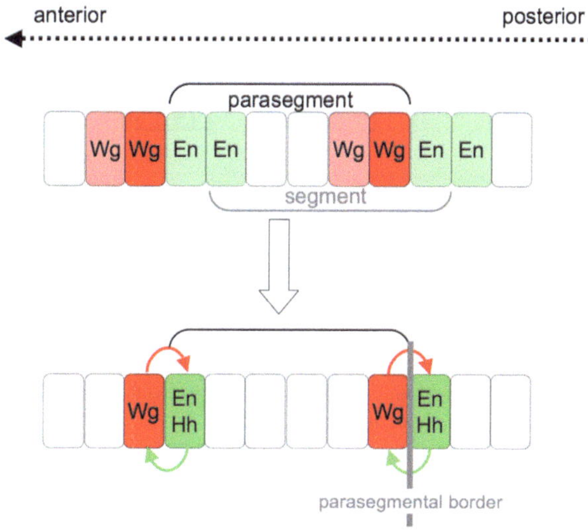

Fig. 12.12 Schematic illustration of the patterns of expression along the antero-posterior axis, in the last step of the genetic hierarchy controlling the *Drosophila* segmentation: (*top*) the antero-posterior axis; (*middle*) initial activation of En (Engrailed) and Wg (Wingless) by the pair-rule signals in stripes of cells (here, we considered that six cells are sufficient to represent the different regions of a parasegment, which includes the posterior part of a segment and the anterior part of the next segment); (*bottom*) the consolidation and refinement of the parasegmental border by the action of the polarity genes, which requires cell-cell communication

12.5.2 The Drosophila Segment-Polarity Module

Early development of the fruit fly embryo is an ideal, very well-studied system for developmental biologists (see [128, 417] for good introductions to this topic). The embryo is organized into a series of segments along its antero-posterior axis (from the head to the tail). These segments will give rise to adult structures (legs, halteres, wings, etc.). This organization is initiated by maternal morphogens, which control few dozens of genes involved in the initial segmentation. These genes have been split into several classes. The first classes, called *gap*, *pair-rule* and *segment-polarity* modules, constitute a temporal hierarchical genetic system. Segment-polarity genes are under the control of the pair-rule genes. Their patterns of expression define the anterior and posterior parts of the segments and they are responsible for the stabilization of the borders between embryonic segments (see Fig. 12.12). The segment-polarity module involves about twenty genes, cross-regulations and intercellular signalings. It has been the subject of a wealth of theoretical modeling studies, in both continuous (e.g., [87, 168]) and logical (e.g., [5, 69, 335]) frameworks.

In [335], a logical model is defined and analyzed, based on an intracellular interaction network of a dozen of components, submitted to two external inputs (the Wingless (Wg) and Hedgehog (Hh) signalings). Six copies of this module have been interconnected in a stripe to allow the representation of the different gene expression domains flanking the parasegmental borders (*parasegments* correspond to portions

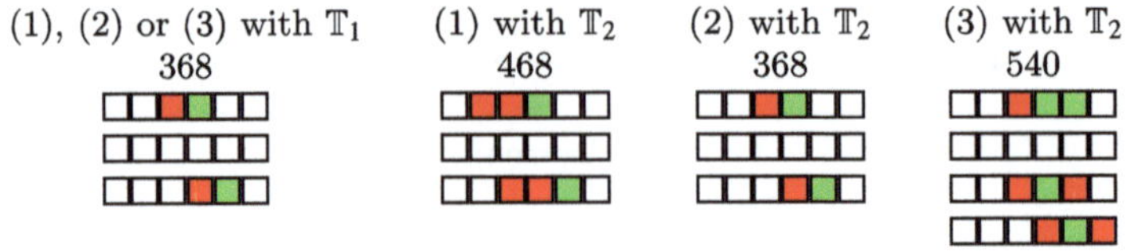

Fig. 12.13 Reachable states and reachable stable states for various configurations: (1) $\sigma_{\mathrm{Hh,Wg}} = \sigma_{\mathrm{Wg,Wg}} = \sigma_{\mathrm{Wg,En}} = \sigma_{\max}$; (2) $\sigma_{\mathrm{Hh,Wg}} = \sigma_{\mathrm{Wg,Wg}} = \sigma_{\mathrm{Wg,En}} = \sigma_{\max\geq 1}$; (3) $\sigma_{\mathrm{Hh,Wg}} = \sigma_{\max\geq 1}$ and $\sigma_{\mathrm{Wg,Wg}} = \sigma_{\mathrm{Wg,En}} = \sigma_{\max\geq 1/2}$. For each configuration, we have indicated the total number of reachable state and drawn the reachable stable states. *White squares* correspond to cells where all the component levels are zero; *red squares* correspond to cells where $x_{\mathrm{Wg}} = 2$ and $x_{\mathrm{Hh}} = x_{\mathrm{En}} = 0$; and *green squares* correspond to cells where $x_{\mathrm{Wg}} = 0$ and $x_{\mathrm{Hh}} = x_{\mathrm{En}} = 1$. Note that the expected wild-type pattern as shown in the lower part of Fig. 12.12, with a Wg expressing cell (red) and an En expressing cell (green) flanking the border, is recovered with topology $\mathbb{T}_1$ and with $\mathbb{T}_2$ if choosing $\sigma_{\max\geq 1}$

of two adjacent segments, see Fig. 12.12). The cells are numbered from 1 to 6, with cell 3 denoting the cell just anterior to the border, cell 4 the cell just posterior to the border (see Fig. 12.15). Connections between these cells are set up through Hh and Wg signals, and in the wild-type case, these intercellular interactions are restricted to neighboring cells. This fully defines the topological and integration functions of our collection of six modules.

In [335], the whole logical regulatory graph encompassing the six interconnected modules has been translated into its standard Petri net representation. This allowed us, by using INA [166], to verify the reachability of expected stable states from a given initial pattern set up by the pair-rule module (see Fig. 12.15). The proper construction of the model, by correctly interconnecting the six modules was rather cumbersome, justifying the development of a framework to support modular modeling.

In the context of this chapter, we will rely on a very much reduced version of the intracellular network. This reduced network has been obtained by the application of a systematic reduction method defined in [284], which preserves the essential dynamical properties (in particular the stable states). Our aim is not to give details on the properties of the segment-polarity module but rather to illustrate the potential of our framework. The LRM is depicted in Fig. 12.14.

We study the reachable stable states for a collection of six cells arranged on a line as depicted in Fig. 12.15. The initial state (defined by the action of the previous pair-rule module, as illustrated in Fig. 12.12) is specified in Fig. 12.15. We consider two topological relations (or signaling distances) and three integration functions, defined below.

- $\mathbb{T}_1$ defined such that $\mathbb{T}_1(i,j) =_{\mathrm{def}} 1$ if $j = i+1$, $\mathbb{T}_1(i,j) =_{\mathrm{def}} 0$ otherwise, and $\mathbb{T}(j,i) =_{\mathrm{def}} \mathbb{T}(i,j)$ meaning that cells can only signal their immediate neighbors (plain lines in Fig. 12.15).
- $\mathbb{T}_2$ defined such that $\mathbb{T}_2(i,j) =_{\mathrm{def}} 1$ if $j = i+1$, $\mathbb{T}_1(i,j) =_{\mathrm{def}} 1/2$ if $j = i+2$, $\mathbb{T}_j(i,j) =_{\mathrm{def}} 0$ otherwise, and $\mathbb{T}(j,i) =_{\mathrm{def}} \mathbb{T}(i,j)$, meaning that cells can signal

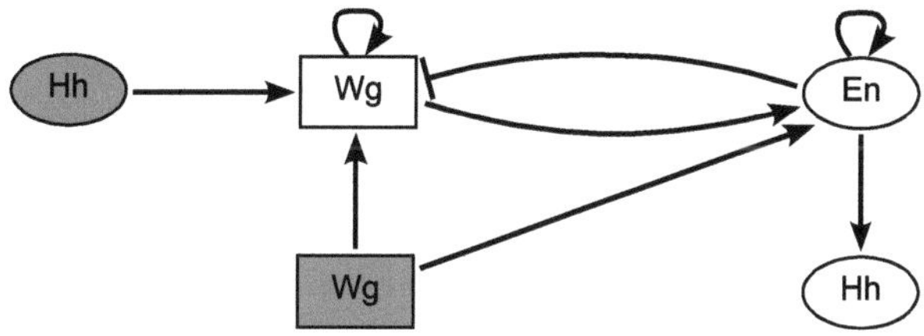

Node	Target level	Logical rule
Wg	2	$((En = 0 \wedge Wg = 0 \wedge \sigma_{Wg,Wg} = 2) \vee (En = 0 \wedge Wg > 1)) \wedge \sigma_{Hh,Wg} = 1$
	0	otherwise
En	1	$((Wg = 0 \wedge \sigma_{Wg,En} = 2) \vee Wg > 1) \wedge En = 1$
	0	otherwise
Hh	1	$En = 1$
	0	otherwise

Fig. 12.14 A reduced version of the segment-polarity logical regulatory module (obtained from the model of 12 components of Sánchez [335], applying the reduction method presented in [284]). In this network, we kept the components exerting external signals (Wg and Hh) and the read-out genes of the action of this regulatory network (En and Wg). The external signals (coming from neighboring cells) are denoted by *gray nodes*. *Rectangular nodes* denote multi-valued components (Wg) and *ellipse nodes* denote Boolean components. The logical functions of the three internal components are given as logical rules

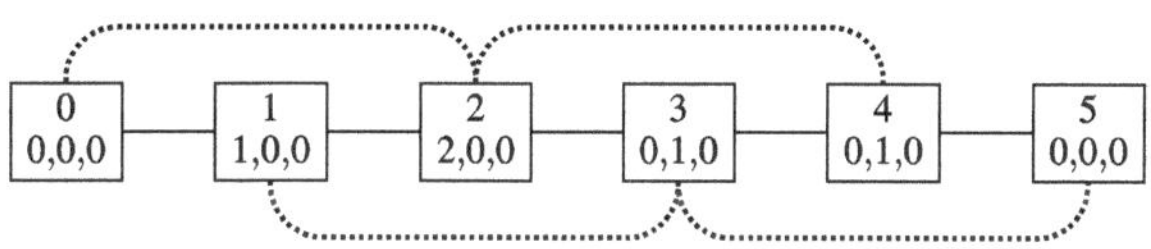

Fig. 12.15 A stripe of six cells with three cells accounting for the posterior region of a parasegment (cells 1, 2, and 3) and three cells accounting for the anterior region of the next parasegment (cells 4, 5, and 6), the parasegmental border being established between cells 3 and 4. In each cell, the initial levels of the components are indicated; each cellular state is given as a triple $x_{\mathrm{Wg}}, x_{\mathrm{En}}, x_{\mathrm{Hh}}$. *Plain lines* denote where $\mathbb{T}_1$ and $\mathbb{T}_2$ are equal to 1 and *dotted lines* denote where $\mathbb{T}_2$ is equal to $1/2$ (in other words, these lines denote the signaling capacities between the cells)

their immediate neighbors (plain lines), but also their second neighbors, with a lower weight (dotted lines in Fig. 12.15).

The integration functions we consider are:

- Maximum weighted level:

$$\sigma_{\max} =_{\mathrm{def}} (v_i, d_i)_{1 \le i \le k} \rightarrow \mathit{round}\big(\max(v_i \cdot d_i \mid 1 \le i \le k)\big).$$

- Maximum level of direct neighbors:

$$\sigma_{\max \ge 1} =_{\mathrm{def}} (v_i, d_i)_{1 \le i \le k} \rightarrow \max(v_i \mid 1 \le i \le k, d_i = 1).$$

- Maximum level for neighbors with weight at least 1/2:

$$\sigma_{\max\geq 1/2} =_{\text{def}} (v_i, d_i)_{1\leq i\leq k} \rightarrow \max(v_i \mid 1 \leq i \leq k, d_i \geq 1/2).$$

For example, let consider the initial state depicted in Fig. 12.15, topology $\mathbb{T}_2$ and $\sigma_{\mathrm{Wg,Wg}}$, which integrates the Wg signal acting on Wg. If $\sigma_{\mathrm{Wg,Wg}} = \sigma_{\max}$, the value of the integrated signal is the maximal weighted level of Wg in neighboring modules (including second neighbors), hence for the module 4, this value is 1: Wg is 0 in both direct neighbors and in the second neighbor (module 2), Wg is 2 with a weight equal to 1/2. If $\sigma_{\mathrm{Wg,Wg}} = \sigma_{\max\geq 1}$, the value of the integrated signal is the maximal level of Wg in direct neighbors (weight greater than or equal to 1), hence is the case of module 4 it evaluates to 0. Finally, if $\sigma_{\mathrm{Wg,Wg}} = \sigma_{\max\geq 1/2}$, the value of the integrated signal is the maximal level of Wg in neighboring cells (direct or not, that is, weight greater than or equal to 1/2), hence in the case of module 4, it evaluates to 2.

Figure 12.13 shows the results for various combinations of these parameters. When $\mathbb{T}_1$ is used, all neighbor signals have a weight 1, so we have $\sigma_{\max} = \sigma_{\max\geq 1} = \sigma_{\max\geq 2}$.

In [335], the model considered for the six cell stripe was based on the assumption of short range Wg and Hh signals. Hence, to analyze the case of nkd loss-of-function (naked cuticle is one of the segment-polarity genes), it was necessary to consider a rewired network, accounting for the increased diffusion of Wg. In the framework presented here, such a change would only consist in modifying the topological function.

12.6 Conclusions

In this chapter, we have presented a modeling framework combining the logical formalism with Petri nets (PNs). The standard Petri net representation of logical regulatory graphs, although not very legible, enables the use of existing PN analysis tools such as INA. In this respect, GINsim provides export facilities to generate files in the format expected by PN tools (e.g., INA [166]). Hence, a first advantage of the PN representation of logical regulatory graphs is the possibility to analyze them using existing PN tools.

A challenging problem arises when considering regulated metabolic networks. PNs open the way to a qualitative integrated modeling of regulated metabolic pathways as proposed in [362], properly connecting PN models of the biochemical pathway and the regulatory control (this being modeled as a logical regulatory graph expressed in terms of a PN).

Developmental processes relate to cell differentiation and pattern formation. In particular, in this chapter, we have delineated a framework that allows the modeling of patches of communicating cells. To illustrate the potential of this framework, we have considered the segment-polarity module involved in the segmentation of the *Drosophila* embryo. In such processes, one has to consider connections of several (intra)cellular regulatory networks. We propose to define these individual regulatory networks as logical regulatory modules (LRMs), identifying input signals they can receive from the outside. Then, a collection of such modules (a CLRM) can be defined, by setting up the number and type of LRMs, as well as the rules governing their interconnections. From such a CLRM, a large logical regulatory network can be recovered. This procedure could be easily implemented in GINsim.

More importantly, we have defined a compact and legible high-level Petri net (HLPN) representation of a CLRM. Implementation of this construction has been provided. HLPNs provide a flexible framework, which allows to easily model different configurations of a patch of cells as well as topological relations (e.g., the range of a signal).

It is worth noting here, that the proposed framework allows the consideration of other situations than communicating cells within a patch. For example, in large regulatory networks, modularity arises from physical delimitation or cellular compartments (e.g., nucleus, cytoplasm) or from functional role. Recently, a modular logical modeling of the budding yeast cell cycle has been delineated in [116]. In this case, modules corresponding to regulatory networks with distinct functional roles in the cell cycle control, have common components. Hence, while composing these modules, one has to properly set up the logical rules of these common components. Based on the methodology proposed in [116], the framework presented in this chapter might be extended to handle overlapping modules.

Although modularity is now recognized as an important feature for large biological networks, little formal work and tool development support combination or composition of regulatory networks (see [357] and references therein). In [357], the authors address this question and distinguish between fusion, composition, aggregation and flattening as distinct processes for building larger models from smaller ones. The HLPN framework presented in Sect. 12.4 can be viewed as a model aggregation, in that it is a reversible process (the modules are conserved). In terms of modeling tools, it is worth noting that ProMoT, a tool that eases the definition and edition of modular models, supports the logical formalism [265]. It is based on principles that are quite similar to those presented in Sect. 12.4.

The main motivation to develop the framework proposed here is to study how existing cellular processes are controlled. However, it could be useful in the field of synthetic biology that aims at designing novel artificial biological systems (see [100] and references therein). Indeed, synthetic biology relies on the concept of modularity, by conceiving building blocks and combining them [2]. So far, synthetic biology mainly designed intracellular gene networks, but synthetic multicellular systems involving cell-cell communication emerged in recent years (see, e.g., [28]).

The complexity of regulatory networks dealt by modelers calls for the development of original and efficient computational means. Here, we have defined a framework that greatly facilitates the definition of models encompassing interconnected regulatory modules. However, we still need to make progresses to analyze such large models. Combining the logical and PN formalisms, as well as taking advantage of the modular structure of the models should allow the development of more efficient tools.

12.7 Problems

12.1 Give the MDD representation of the function f_2 given in Fig. 12.2 considering the variables A and B in the reverse order (i.e., first B, then A). Transform the obtained MDD into an IDD (labeling the edges with integer intervals, and merging them when possible as explain in Sect. 12.2.2).

12.2 Consider the genetic regulatory network called *repressilator* as defined by Elowitz and Leibler [109], which consists of three genes (denoted here A, B and C), connected in a negative circuit (each component represses its successor in the circuit, and is repressed by its predecessor).

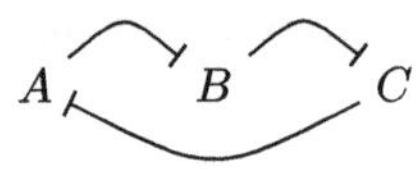

Assuming Boolean levels for A, B and C, all interaction thresholds are equal to 1, and the components share the same logical rule: $\mathscr{K}(0) = 1$ and $\mathscr{K}(1) = 0$, that is, if the repressor is at level 0 (absent), the regulated component target level is 1 (present), and the other way around. Draw the full STG of this model and verify that it encompasses a unique complex attractor, reachable from any initial state.

12.3 Consider an LRG encompassing a component C regulated by A and B, both positively auto-regulated, with $Max_A = Max_B = 1$ and $Max_C = 2$. The logical functions $\mathscr{K}_A$, $\mathscr{K}_B$ and $\mathscr{K}_C$ are given by their MDD representations below (note that $\mathscr{K}_A$ and $\mathscr{K}_B$ are the identity function). Give the logical expressions defining these three functions (using the connectors $\wedge$ and $\vee$ as in Fig. 12.2). Give the P/T representation of this LRG (verify that the autoregulations define transitions that are useless since they are never enabled, see [68]).

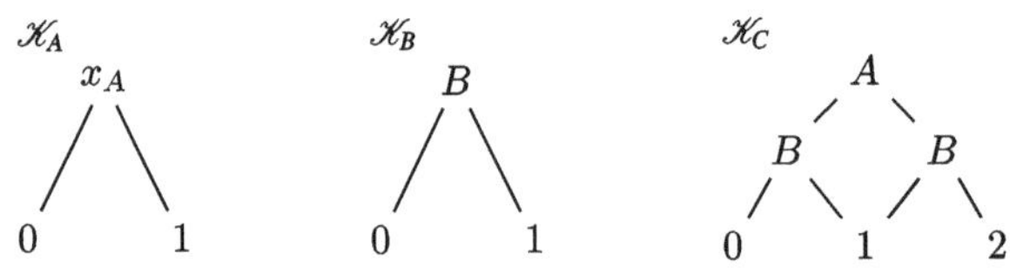

12.4 Find another enabling binding for the high-level Petri net depicted in Fig. 12.8. Which module is involved by this binding and what are the corresponding levels x_A and x_B? Assume that $\delta_A(x_A) = 1$ and fire the transition accordingly.

12.5 Construct the high-level Petri net representation of the CLRM depicted in Fig. 12.15, including the initial marking. Indications:

1. Start with determining the arguments of each regulatory function, separating internal and external regulators.
2. Draw each Petri net transition separately, together with the places for the needed regulatory components.

3. Merge these Petri net parts by collapsing the places that implement the same regulators.
4. Add the initial marking, corresponding to the expression levels indicated in Fig. 12.15.

Acknowledgements We thank G. Batt, A. Naldi, E. Remy, S. Soliman, D. Thieffry for fruitful discussions. This work was supported by the French Research Agency (project ANR-08-SYSC-003).

Chapter 13
A Case Study of HFPN Simulation: Finding Essential Roles of Ror Gene in the Interaction of Feedback Loops in Mammalian Circadian Clock

Natsumi Mitou, Hiroshi Matsuno, Satoru Miyano, and Shin-Ichi T. Inouye

Abstract Mammalian circadian clock is composed of two feedback loops, a *Per-Cry* and *Clock-Bmal* loops. The role of *Rev-Erb* gene, which interconnects these two feedback loops by the inhibition of *Bmal* from PER/CRY complex, has been investigated through biological experiments as well as computational simulations. However, for the role of *Ror* gene, which exerts contrary effect on the same target gene *Bmal* as the *Rev-Erb*, enough consideration has not been paid so far. This paper first improves the previous hybrid functional Petri net (HFPN) model of the circadian clock so that both of the *Per-Cry* and *Clock-Bmal* loops can participate in the maintenance of the circadian oscillations. This improvement is incomplete, however, because a fixed level of PER/CRY eliminates all the circadian oscillations. Although this problem can be resolved by the introduction of *Ror* into the HFPN model, another inconsistency remains, *Bmal* oscillation is not abolished by the knock-out of the *Cry*. Then we further incorporate a hypothetical path into the

N. Mitou
Kyuden Business Solutions Co. Inc., Fukuoka 810-0004, Japan
e-mail: natsumi.mitou@qdenbs.com

H. Matsuno (✉)
Graduate School of Science and Engineering, Yamaguchi University, Yamaguchi 753-8512, Japan
e-mail: matsuno@sci.yamaguchi-u.ac.jp

S. Miyano
Human Genome Center, Institute of Medical Science, University of Tokyo, Tokyo 108-8639, Japan
e-mail: miyano@ims.u-tokyo.ac.jp

S.-I.T. Inouye
Institute for Time Studies, Yamaguchi University, Yamaguchi 753-8511, Japan
e-mail: inouye@c-able.ne.jp

I. Koch et al. (eds.), *Modeling in Systems Biology*, Computational Biology 16,
DOI 10.1007/978-1-84996-474-6_13,

HFPN model, succeeding in eliminating this inconsistency while keeping complementary actions of two feedback loop.

Abbreviation of Biological Terms

Bmal:	Bmal1 gene
BMAL:	Bmal1 protein
Clock:	Clock gene
CLOCK:	Clock protein
Cry1:	Cryptochrome 1 gene
Cry2:	Cryptochrome 2 gene
Cry:	combination of *Cry1* and *Cry 2* genes
CRY:	*Cry* protein
Cyc:	Cycle gene
CYC:	Cycle protein
dClk:	*Drosophila* Clock gene
dCLK:	*Drosophila* Clock protein
Per1:	Period 1 gene
Per2:	Period 2 gene
Per3:	Period 3 gene
Per:	combination of *Per1*, *Per2*, and *Per3* genes
PER:	*Per* protein
Tim:	Timeless gene
TIM:	Timeless protein
Rev-erb:	Rev-erbα gene
REV-ERB:	Rev-erbα protein
Rorα:	retinoic acid receptor-related orphan nuclear receptor α gene
RORα:	retinoic acid receptor-related orphan nuclear receptor α protein
SCN:	suprachiasmatic nuclei

13.1 Introduction

Circadian rhythms, which regulate physiology and behavior of living organisms, are endogenous oscillations with a period close to 24 h. In mammalian case, they are driven by a central circadian clock located in the suprachiasmatic nuclei (SCN) of the hypothalamus [326, 341]. Recent molecular biological studies have disclosed that the circadian rhythm of the SCN is generated at the level of the gene expression and protein synthesis [206]. Molecular bases of circadian rhythms including the mammalian one is well summarized in the paper [103] that describes common elements in the design of circadian oscillators over various organisms from bacteria to mammal. Readers who are not familiar with the biology of circadian rhythms are recommended to read this paper.

The central logic of the circadian clock is an autoregulatory transcriptional and translational feedback loop that consists of negative and positive elements. The negative elements constitute an oscillation process whose product feeds back to slow down the rate of the process itself, and the positive elements keep the oscillator from winding down through the activation of the product of the negative elements. In mammals, positive elements in circadian clocks are the proteins CLOCK and BMAL (BMAL1), and negative elements PER (period) and CRY (cryptochrome) [103]. The PER/CRY complex exert this repressive effect by binding to a complex of CLOCK/BMAL that activates *Per* and *Cry* gene expression. This autoregulatory feedback loop is common over a diverse range of living organisms from bacteria to plants, insects, and mammals. Besides, it has been revealed that some eukaryotes have more sophisticated circadian clock, namely "interlocked feedback loops" [151].

The interlocked feedback loop was firstly illustrated in the *Drosophila* circadian clock [135]. *Drosophila* circadian feedback loop is composed of two interlocked negative feedback loops: a *Period* (*Per*)-*Timeless* (*Tim*) loop, which is activated by a complex of CYCLE (CYC) and *Drosophila* CLOCK (dCLK) proteins and repressed by a complex of PER and TIM proteins, and a *dClk* loop, which is repressed by a CYC/dCLK complex and derepressed by a PER/TIM complex. The mammalian circadian clock has a similar structure of *Per-Cry* and *Clock-Bmal* loops which correspond to the *Per-Tim* and the *Cyc-dClk* loops in *Drosophila*, respectively, except that *Rev-Erb* gene and its product are participate in bridging these two loops; PER/CRY represses the transcription of *Rev-Erb* whose product REV-ERB represses the *Bmal* transcription [309].

There have been many theoretical models of circadian clocks proposed for *Synechococcus* [384, 385], *Neurospora* [225, 365, 368], *Drosophila* [20, 137, 226, 400], and mammals [30, 31, 117, 167, 227]. All of these models employed differential equations as a formalization method. Although a model with differential equations allows us to use the existing computational tools such as Mathematica and MATLAB, it is usually hard for biologists without mathematical background to use such computational tools. In contrast, due to the graphical nature of its representation method, Petri net is an acceptable formalization for such ordinary biologists, enabling them to understand computational models of biochemical networks smoothly. Especially, hybrid Petri net-based method (Chap. 6 of this book and [248]) is a promising formalization method, because this method not only inherits this graphical property of Petri nets, but also makes it possible to use continuous elements as well as discrete elements.

Retinoic Acid-related Orphan Nuclear receptor α (*Ror*) is a gene that was identified as an essential gene for the circadian clock of mammals, whose regulation is similar to *Rev-Erb* except that the gene product ROR activates *Bmal* (REV-ERB represses *Bmal*) [3, 337]. The *Rev-Erb* and *Ror* exert contrary effects on the *Bmal* expression; REV-ERB inhibits *Bmal*, but ROR activates it. Hence, the *Ror* plays an important role in the interlocked feedback loops in mammals.

A computational model of the interlock feedback loop of mouse should include the effects of *Ror* as well as *Rev-Erb*. However, in the models in [30, 228], only the effect of *Rev-Erb* was described in an indirect manner, that is, just as an interaction from CLOCK/BMAL to *Bmal*. In addition, the effect of *Ror* was not considered in these models. In the models in [31, 227, 255], the *Rev-Erb* interaction was explicitly described, but no description about the *Ror* interaction. Although the model presented in [117] includes both of the *Rev-Erb* and *Ror* interactions, the difference between *Rev-Erb* and *Ror* in the effect on circadian oscillations was not investigated. Hence, the purpose of this paper is to investigate the roles of *Rev-Erb* and *Ror* interactions in the oscillatory behaviors of the mammalian circadian clock by means of computer simulations using hybrid functional Petri net (HFPN) [251]. The HFPN can be treated as a special case of hybrid functional Petri net with extension (HFPNe) (cf. Chap. 6) which has high potential for modeling various kinds of biological processes such as localization of molecules, molecular modifications, and alternating splicing.

This paper starts with the construction of an HFPN model which consists of basic 5 genes; *Per*, *Cry*, *Rev-Erbα*, *Clock*, and *Bmal*. Section 13.2 demonstrates this construction in a step-by-step manner with presenting 3 simple exercises. Readers can study the construction process of HFPN model through these exercises.

In Sect. 13.3, we first point out the problem in the previous HFPN model of [255] in which the *Clock-Bmal* loop did not contribute to the oscillation maintenance of circadian rhythms. A modification on this HFPN model through appropriate parameter choices enabled the *Clock-Bmal* loop to participate in the circadian oscillation. However, a fixed level of PER/CRY in a simulation eliminated the circadian oscillations, indicating the incompleteness of this modification for the robust oscillations with the interlocked model.

This incompleteness was resolved, in Sect. 13.4, by the introduction of *Ror* into the HFPN model. However, a discrepancy in gene-knock-out still remains in this *Ror*-introduced model; *Bmal* oscillation was not abolished by the disruption of *Cry* gene. Then, we further introduced a hypothetical interaction "ROR excites the *Bmal* expression when PER/CRY level are above a certain threshold" into the HFPN model. This hypothetical path eliminated the inconsistency on the *Cry* disruption, while keeping complementary actions of two interlocked loops; forced fixation of PER/CRY or CLOCK/BMAL level did not abolish the circadian oscillation.

We further examined the availability of this HFPN model by simulating two biologically confirmed facts in *Rev-Erb* deficient mouse; *Bmal* oscillation level staying at higher level and a shorter period of oscillations.

13.2 Modeling Molecular Circadian Oscillator in Mouse with Hybrid Functional Petri Net

Although hundreds of genes in mRNA levels are found to be oscillated in mammalian cells [90], the minimum set of genes that are responsible for the central circadian clock regulation are now considered to be 5 genes, *Period*, *Cryptochrome*, *Rev-Erbα*, *Clock*, and *Bmal1*.

This section first describes the regulatory interaction of these 5 genes along with their biological functions. Afterwards, this regulatory interaction is modeled with the HFPN with presenting some exercises that help readers to understand the construction process of the HFPN model.

13.2.1 Circadian Clock Oscillator with Five Genes

The genes that are involved in this intracellular system are called clock genes, which include *Per (Per1, Per2, Per3), Cry (Cry1, Cry2), Rev-Erb (Rev-Erbα), Clock* and *Bmal (Bmal1)* [103]. To make a model simple, *Per1, Per2 and Per3* are combined into single *Per* in the following discussion. Similarly, *Cry1 and Cry2* are treated together as *Cry*. Each mRNA produces a corresponding protein, PER, CRY, REV-ERB, CLOCK or BMAL.

In the following, brief introductions of these 5 genes are given with the necessary descriptions for the following discussion.

Per gene: *Per* (*Period*) is a gene, expressing high transcription level in daytime, whose activity is initiated by the binding of CLOCK/BMAL heterodimer at the E-box regulatory element that is located at the upstream of the *Per*.

Cry gene: *Cry* (*Cryptochrome*) gene has the same E-box regulatory element as *Per*, so it is regulated by CLOCK/BMAL. CRY interferes the action of CLOCK/BMAL after forming a complex with PER followed by the migration into the nucleus, thus, resulting in the repression of *Per*, *Cry*, and *Rev-Erb*. In the *Cry*-deficient mouse, constant low-level expression of *Bmal* was observed [360].

Rev-Erbα gene: *Rev-Erb* (*Rev-Erbα*) gene, which is peaking in daytime, is also controlled by CLOCK/BMAL due to the existence of the E-box regulatory element at its upstream position. REV-ERB inhibits the *Bmal* transcription. It was reported that disruption of *Rev-Erb* caused the low level oscillation of *Bmal* relative to the normal mouse and the shortening of free running period in the activity of mice [309].

Bmal and Clock genes: *Bmal* gene, whose transcription level peaks in the night, expresses antiphase oscillation to *Per*. *Clock* keeps almost constant expression through a day. BMAL and CLOCK form a heterodimer that induces the activation of *Per*, *Cry*, and *Rev-Erb* by binding to the E-box regulatory elements of them.

Although we have conducted simulations on the HFPN model [255] that was constructed based on these biological facts of the 5 genes, the antiphase relationship between *Per* and *Bmal* could not be realized with this HFPN model. This inconsistency has been resolved by the introduction of the hypothetical interaction "PER/CRY activates *Bmal* transcription," where PER/CRY represents a heterodimer of PER and CRY [255]. Figure 13.1 is the resulting gene regulatory map of these 5 genes and these products together with this hypothetical path.

13.2.2 Construction of an HFPN Model

Figure 13.2 shows an HFPN that models the part of *Per-Cry* interactions in the genetic interaction map of Fig. 13.1. Construction of this HFPN model was made based on the rules listed below. The firing speed of a continuous transition is given

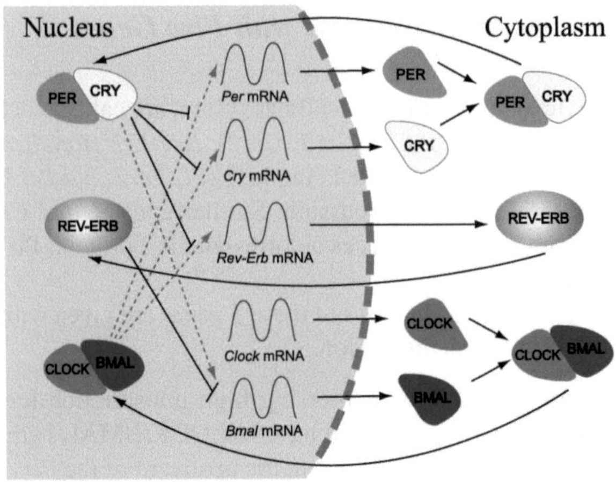

Fig. 13.1 Genetic interactions of 5 clock genes based on the biological facts and the hypothesis that resolves the inconsistency of antiphase relationship between *Per* and *Bmal* [255]. *Normal arc* represents molecular reactions such as transcription, migration, and complex formation. *Dotted arc* represents the activation of a gene by a positive transcription factor: for example, PER/CRY complex acts as a positive transcription factor which activates *Bmal* gene. Arc with short bar at its top represents the repression of a gene by a negative transcription factor: for example, REV-ERB protein acts as a negative transcription factor which represses *Bmal* gene

Fig. 13.2 An HFPN that models *Per-Cry* interaction process in Fig. 13.1

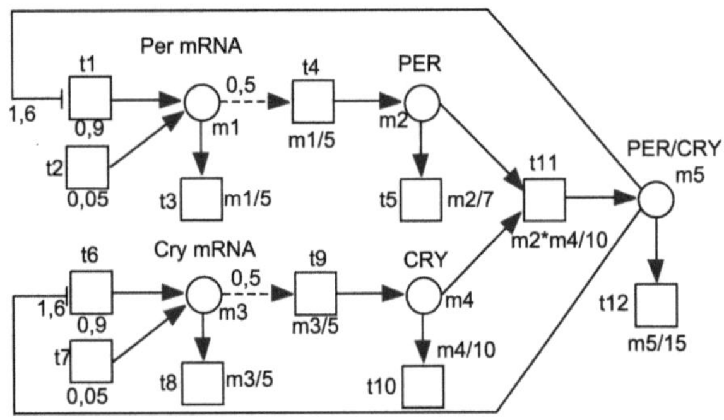

according to a simple arithmetic formula of either mX/a or $(mX \times mY)/a$, where mX (mY) is a variable representing the content of the input place of the transition, and a is a constant by which the firing speed is controlled. For example, $m2$ in Fig. 13.2, the content of continuous place PER, is increased at the rates of $m1/5$ and decreased at the rate of $m2/7$, where $m1$ is the content of continuous place *Per* mRNA. The formula $(mX \times mY)/a$ represents the rate of a complex formation, the formula "$(m2 \times m4)/10$" found at continuous transition t_{11} in Fig. 13.2 is an example of this.

- Each substance such as mRNA and protein is corresponded to a continuous place.
- At each transition, we assign a function of a form as $mX/10$ to define the speed of the corresponding reaction. For example, the translation speed of PER has been determined by the formula $m1/5$, where $m1$ represents the concentration of *Per* mRNA.
- Complex forming rate is generated using the formula of a form, $mX * mY/10$. For example, the formula $m2 * m4/10$ has been assigned at continuous transition t_{11} as a complex forming rate of PER ($m2$) and CRY ($m4$).
- Continuous transition without outgoing arc is used to represent the natural degradation rate of a substance such as mRNA, protein, and protein complex. For example, as the degradation rates of *Cry* and its products, the formulas $m3/5$, $m4/10$, and $m5/15$ have been assigned to transitions t_8, t_{10}, and t_{12}, respectively.
- A test arc is used to represent a translation process because no consumption of mRNA occurs in the translation process. In Fig. 13.2, there are two test arcs for the translation processes of *Per* and *Cry*. A value that accompanies to a test arc (0.5 in each of test arcs) represents the threshold of the mRNA; the translation process continues during the period while the content of a continuous place representing mRNA exceeds this threshold.
- Negative feedback loop can be modeled with an inhibitory arc, whose action is controlled by the threshold value attached to the inhibitory arc. In Fig. 13.2, two inhibitory arcs can be found, which are used to model feedback loops from PER/CRY in the cytoplasm to the transcriptions of *Per* and *Cry*. During the period while the content of PER/CRY place exceeds 1.6, both of these two inhibitory arcs stop the firing of transitions t1 and t6 that represent the transcriptions of *Per* and *Cry*, respectively.

Transitions t2 and t3 are used to model the provision of basal amount of *Per* and *Cry* that are continuously produced in a cell.

The *Per-Cry* regulatory mechanism constitutes the negative feedback loop; the product PER/CRY of *Per* and *Cry* inhibits the transcriptions of the *Per* and the *Cry* themselves. However, more precisely, PER/CRY does not directly inhibit these *Per* and *Cry* transcriptions, it interferes the action of CLOCK/BMAL that activates the *Per* and the *Cry*. We extend the HFPN model of *Per-Cry* negative feedback by adding *Clock-Bmal* regulatory interactions.

Up to now, two complex forming process, the PER/CRY process and the CLOCK/ BMAL process, have been constructed with HFPN. One important gene, *Rev-Erb*, remains to be modeled, which bridges these two complex forming processes. That is, PER/CRY suppresses the transcription of the *Rev-Erb* when it is activated by CLOCK/BMAL, and REV-ERB suppresses the transcription of *Bmal*.

All transcription-translation components of the 5 genes are ready to model in the HFPN forms. A whole circadian HFPN model is complete to construct after accomplishing the exercise below.

13.3 Effect of the Two Interlocked Loops in Circadian Gene Regulations in Mouse

13.3.1 A Problem of the Previous HFPN Model in an Oscillation Maintenance

In a previous paper [255], we have analyzed oscillatory behaviors of the clock genes based on the model of the mouse circadian clock shown in Fig. 13.3. As was shown there, the HFPN model, when a hypothetical facilitatory interaction from PER/CRY on *Bmal* transcription was assumed, successfully yielded stable rhythmic changes in clock mRNA and protein levels. Most of the results were consistent with reported results of biochemical experiments including proper phase relations among these (Fig. 13.4). Abundance of mRNA and proteins of the clock genes shows stable oscillations. For example, the peaks of *Bmal* mRNA were found at the midpoint of successive *Per* or *Cry* peaks. In addition, this model reproduced no rhythms when *Cry* gene was completely eliminated, that is in *Cry*-deficient animals (Fig. 13.5).

However, as well as the reproduction of biochemical results, the previous simulation also raised questions. Simulation results in Fig. 13.6 illustrate that, although fixation of CLOCK/BMAL at a certain level did not influence on the rhythmicity of *Per* and *Cry*, the CLOCK/BMAL and REV-ERB loop could not rescue the rhythms when PER/CRY level was fixed at a constant. That analysis showed that the oscillation could be brought about only by PER/CRY loop, leaving the other loop CLOCK/BMAL and REV-ERB nonfunctional in terms of the maintenance of oscillation.

13.3.2 Modification in Parameters for Constituting Interlocked Loops

Close examination of system dynamics indicates a possible source of the remaining problems. Although CLOCK/BMAL could switch on and off to allow the excitatory effect on *Per*, *Cry* or *Rev-Erb*, its level was found to always be above the threshold (Fig. 13.7). Accordingly, the starting model did not make use of the capability of CLOCK/BMAL to turn the pathway on and off, since the CLOCK/BMAL level stayed higher than the threshold and continuously linked to the rest of the system at all time. This gave us a hint to improve the stability of the system: Since biological system is made in such an efficient and economical way, could any interaction have no role in the biological system? Why the molecular circadian system is composed of the overlapping two loops, instead of one loop, if the one feedback loop is sufficient for the system to oscillate? Although a biochemical study suggested the BMAL level always above the threshold to be the case [223], we explored, by the tools of Cell Illustrator [60, 274], the consequence of the presence of the two complementary overlapping loops on the molecular circadian system of the mouse:

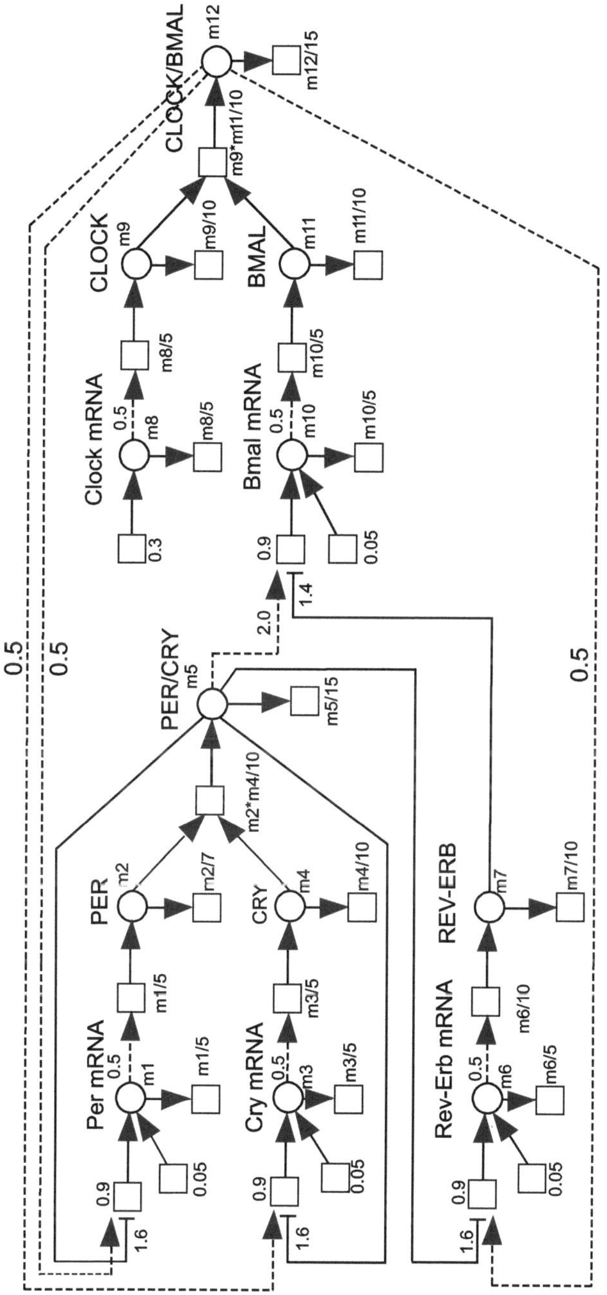

Fig. 13.3 The starting HFPN model originally published in Matsuno et al. [255]

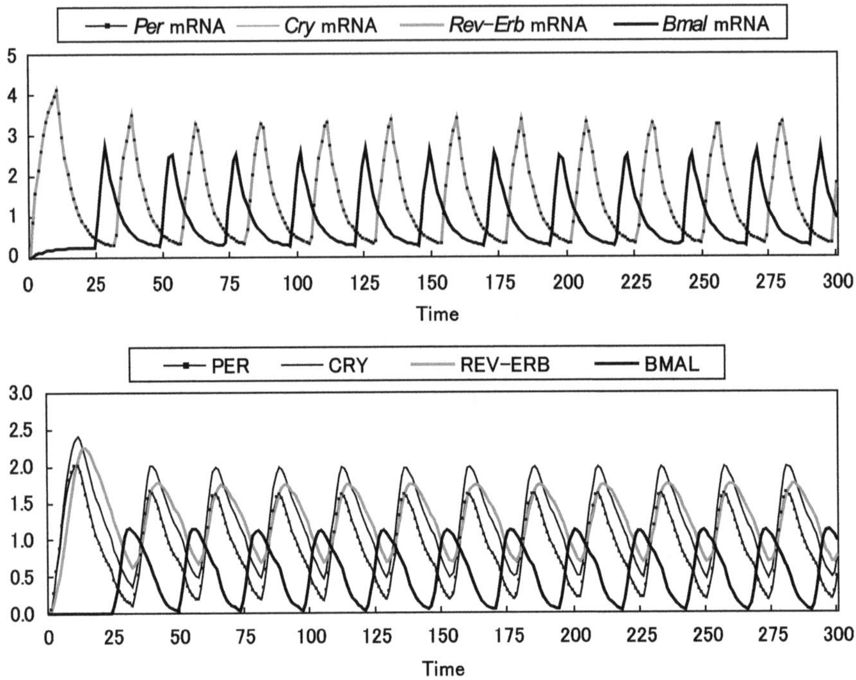

Fig. 13.4 Oscillations of five clock gene mRNA and proteins in the starting model of Fig. 13.3

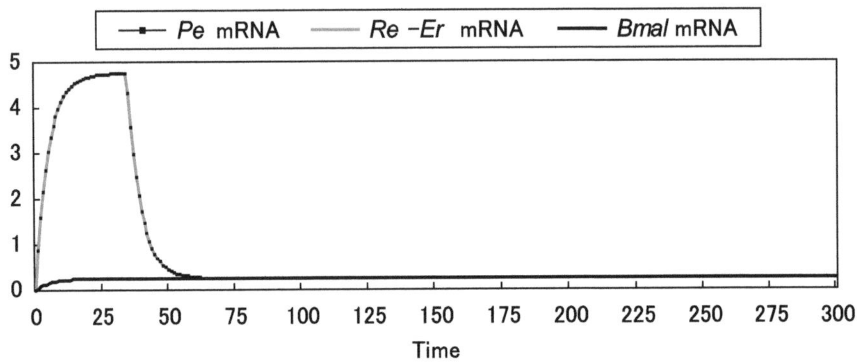

Fig. 13.5 Removal of *Cry* gene (*Cry* knock out) wiped out circadian rhythms in the starting model of Fig. 13.3

One loop consists of the feedback pathway from CLOCK/BMAL to *Per* and *Cry*, the other from CLOCK/BMAL to *Rev-Erb*.

In order to explore the possibility that the second loop through *Rev-Erb* functions cooperatively to make the system more robust, together with the first loop, we have adjusted parameters of CLOCK/BMAL threshold as shown Fig. 13.8, so as to

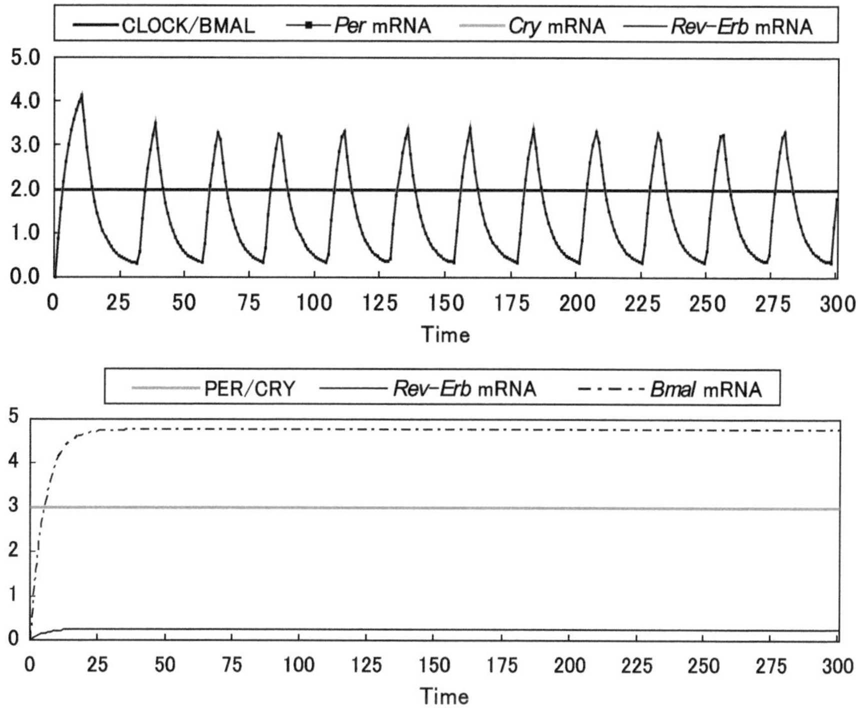

Fig. 13.6 Although fixation of CLOCK/BMAL at a level still maintains circadian rhythmicity (*upper panel*), constant level of PER/CRY could no longer sustain a rhythm (*lower panel*)

make the level of CLOCK/BMAL above the threshold during a part of the cycle and below during the rest of the time. With the careful choice of the parameters, CLOCK/BMAL loop became to be switched on and off within a cycle (Fig. 13.9). Simulation with the help of Cell Illustrator under these condition showed that this modified interlocked system is also able to generate and maintain the stable rhythms of the system, as is shown in Fig. 13.10. These results demonstrated that the addition of the functional second loop is compatible with the rhythmicity. Furthermore, this model successfully predicted total loss of the rhythms in *Cry* knock-out animals.

However, the simulation results also showed that, even the two loops work together, a fixed level of PER/CRY eliminated the rhythmicity (Fig. 13.11), indicating the loop via *Rev-Erb* alone could not sustain circadian rhythms. This might be due to the simplification made on the excitatory effect of PER/CRY on *Bmal*. In the starting model, whereas a constant high level of PER/CRY facilitated *Bmal*, it inhibits *Per*, *Cry*, *Rev-Erb* transcriptions, suppressing the rhythms. On the other hand, a constant low level of PER/CRY is unable to stimulate *Bmal* transcription and *Bmal* can not accumulate under these conditions enough to initiate rhythmic oscillations of the clock gene components.

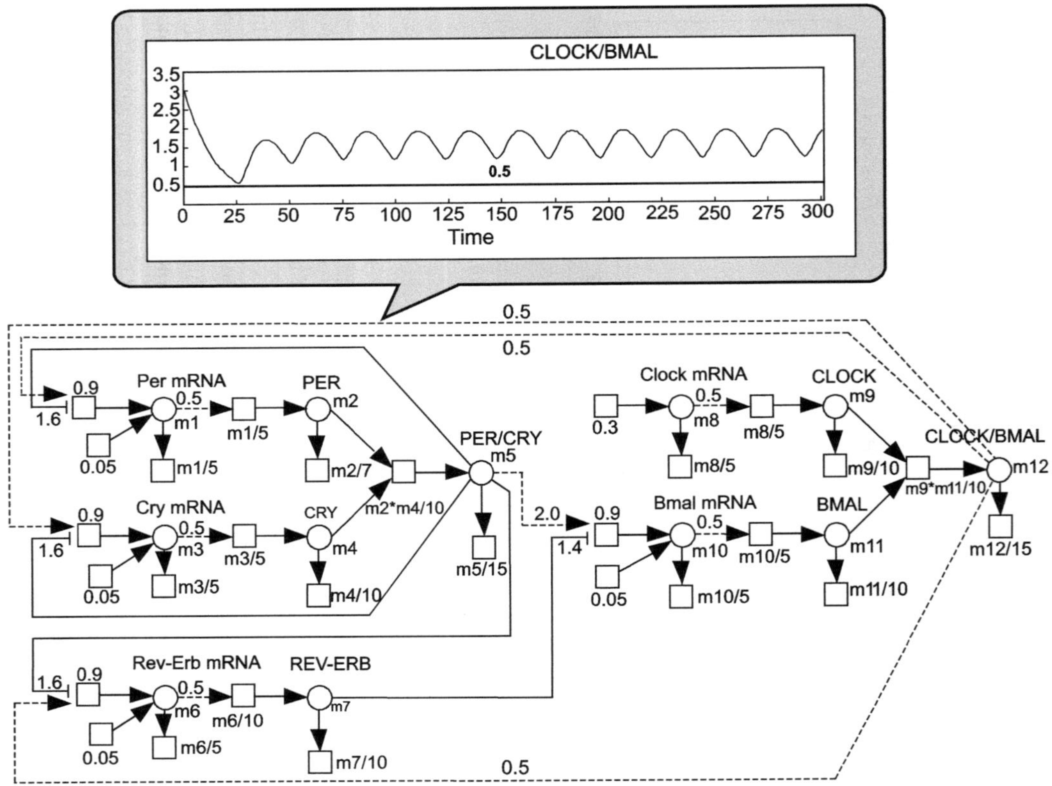

Fig. 13.7 In the starting model, the CLOCK/BMAL level stays above the threshold at all time

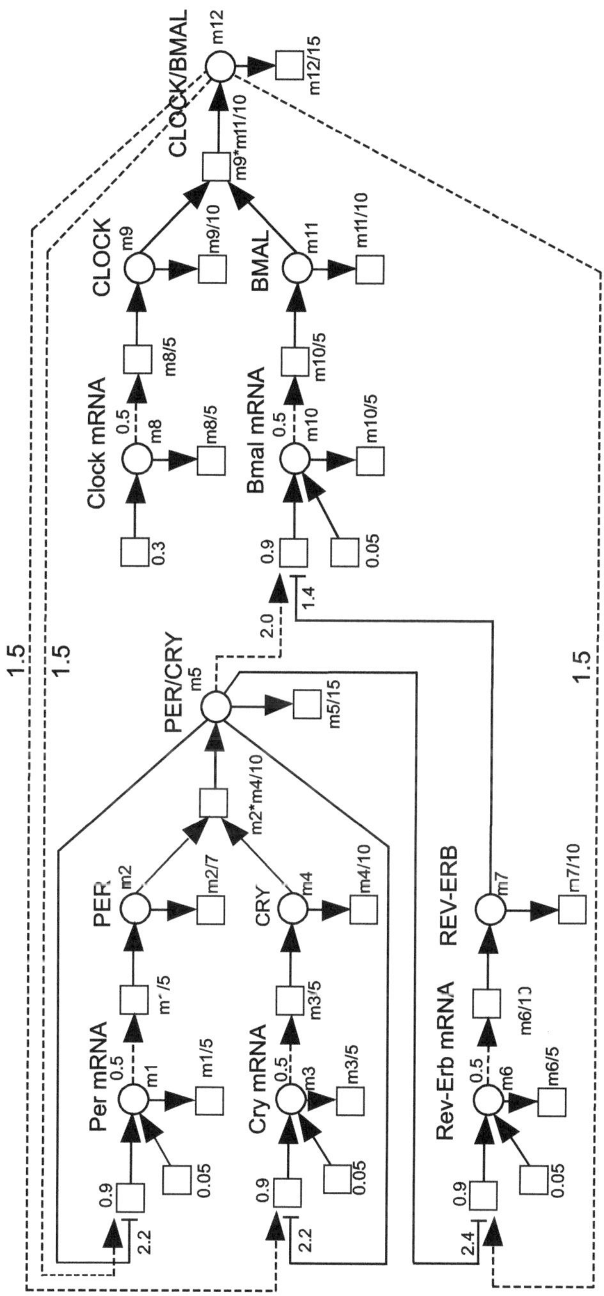

Fig. 13.8 Interlocked model of the circadian clock of mice

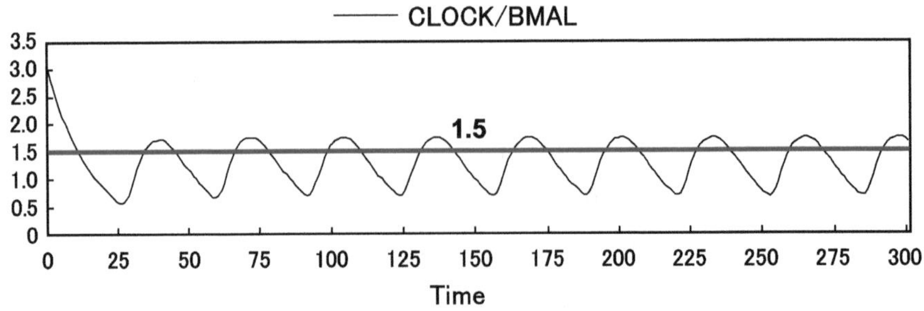

Fig. 13.9 The level of CLOCK/BMAL crosses the threshold within a cycle in the interlocked model

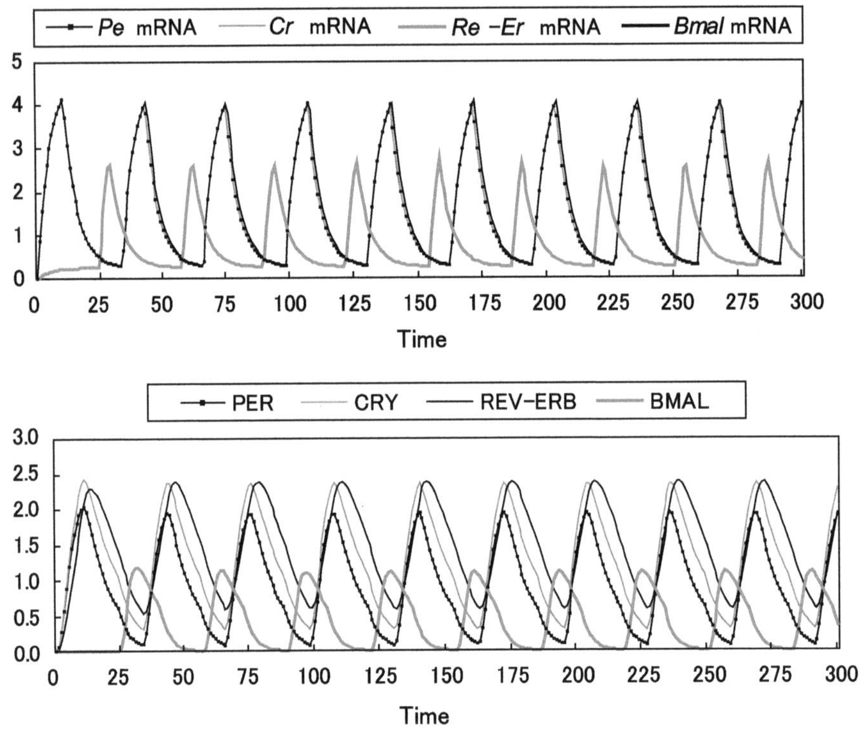

Fig. 13.10 The interlocked model also provides stable oscillations of both mRNA and proteins of the clock genes, consistent with the experimental results

13.4 Function of *Ror* in the Interlocked Model

13.4.1 Introduction of Ror

As described in the previous section, a careful choice of parameters that make the positive loop switching on and off was unable to make the loop via *Rev-Erb* com-

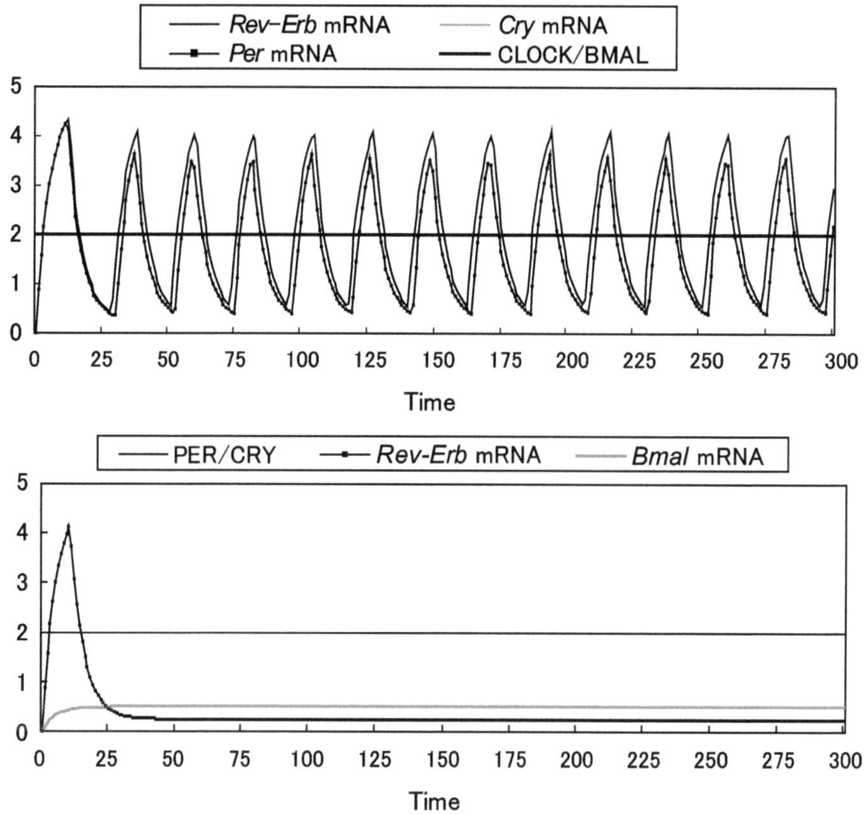

Fig. 13.11 The CLOCK/BMAL through REV-ERB loop could not sustain the rhythmicity when the other loop, the loop through PER/CRY, was blocked in this interlocked model (*lower*), while rhythmicity survives the blocking the link at CLOCK/BMAL (*upper panel*)

plementary to the negative loop, as far as the oscillations were concerned. This was due to two different roles assigned to one protein complex: PER/CRY suppressed *Per*, *Cry*, *Rev-Erb* transcriptions and at the same time enhanced the transcription of *Bmal*. Hence, we further introduced an element of the biological clock, namely *Ror* [3, 337], shown in Fig. 13.12 to separate the distinct effects of PER/CRY and examined if this assignment of *Ror* to the new indirect interaction improved the stability of the system. *Ror* has been found in the SCN, but the functional roles of *Ror* remains to be elucidated. *Rev-Erb* has attracted more interests than *Ror*, probably because inhibitory interaction on *Bmal* transcription seemed to be necessary to obtain the observed inverted phase relation of *Bmal* in the clock system. The *Bmal* level is low when the PER/CRY is high and high when PER/CRY low. So it is reasonable to imagine that important interaction should be inhibitory exerted from day-peak mRNA like REV-ERB. On the other hand, ROR tend to excite transcription of *Bmal*, so it is hard to imagine that excitatory element induced inverted phase of *Bmal* oscillation.

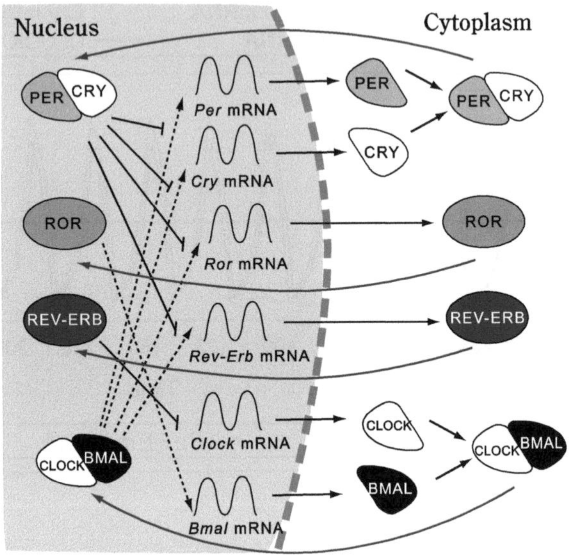

Fig. 13.12 A second interlocked model that includes the pathway by which *Ror* mediates enhancement of *Bmal*

Figure 13.13 shows the HFPN model of Fig. 13.12. In this model, PER/CRY complex exerts inhibitory influence on the transcription of *Per*, *Cry*, *Rev-Erb* and *Ror*, and in turn, ROR facilitatory on *Bmal* independently of PER/CRY. In other words, the second feedback loop now consists not only of *Rev-Erb* but also of *Ror*. The delay component for the translation to ROR shown in the small white box labeled ROR-delay of Fig. 13.13 was introduced on the following reason.

When facilitatory and inhibitory effects act on at the same time, inhibitory influence surpasses facilitatory one and blocks the link completely. Hence, the same phase of the oscillation and the same thresholds of ROR and REV-ERB did not induce a rhythm in the *Bmal* level. Different thresholds, as in the case of Fig. 13.14, where the threshold of ROR was set 2.4, whereas the threshold of REV-ERB 1.4, did not solve the problem. On the other hands, addition of the delay component shown in the small white box successfully reproduced consistent behavior of the rhythms, which are shown in Fig. 13.15.

This system exhibited stable rhythms with appropriate phase relationship as shown in Fig 13.16. For example, *Bmal* peaks were found at the midpoints between two peaks of *Per* or *Cry*.

Moreover, this interlocked model with *Ror* added successfully produced complementary salvation of the rhythms. The system keeps on oscillating when either the CLOCK/BMAL or PER/CRY level was fixed at a constant level, as shown in Fig. 13.17. Whereas negative feedback loop via PER/CRY maintain the rhythms without the feedback through *Rev-Erb* and *Ror*, the second loop through CLOCK/BMAL and REV-ERB and ROR, in turn, provide effective interaction causing oscillations of the clock genes even when the PER/CRY level was kept constant.

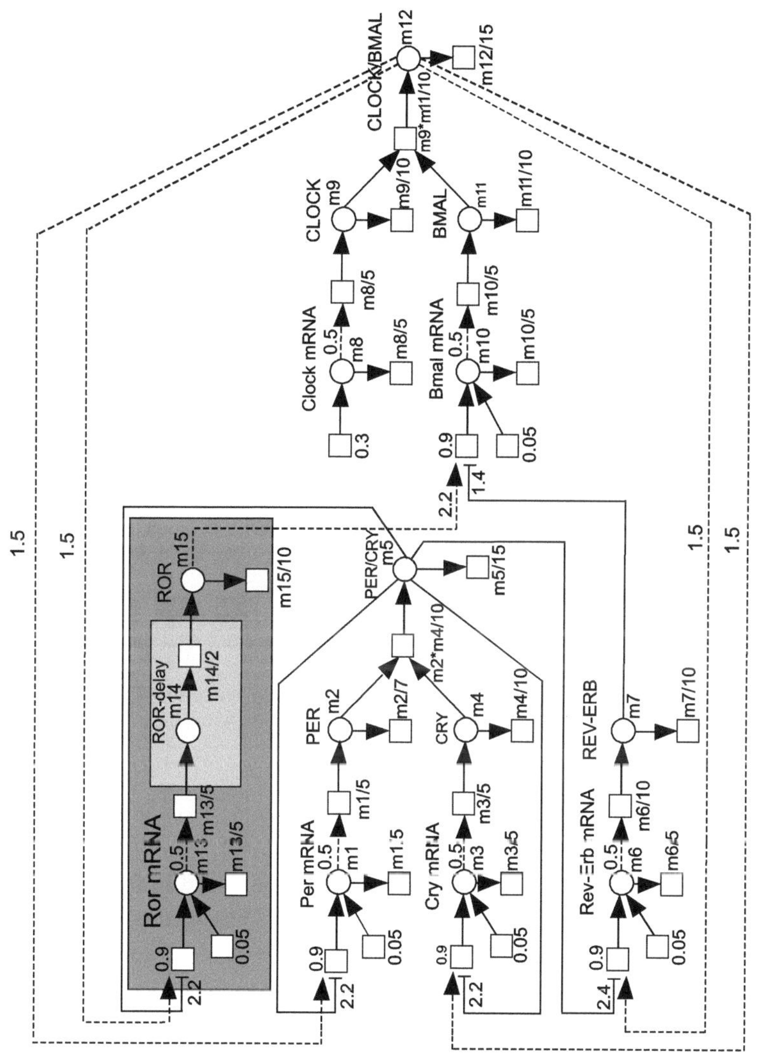

Fig. 13.13 HFPN realization of the interlocked model with Ror shown in Fig. 13.12. Time delay between *Ror* mRNA and ROR protein are also included

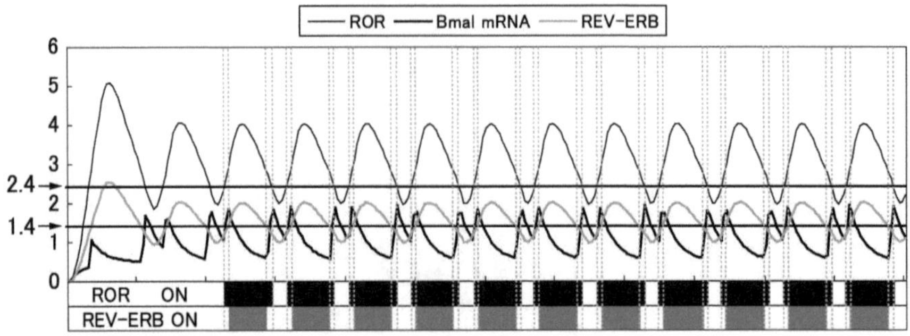

Fig. 13.14 Simulation results in the case of the different thresholds for ROR activation (2.4) and REV-ERB (1.4) repression. Normal oscillation of *Bmal* is not obtained. The *black portions* in the *upper black-white bar* represent the period when ROR activates *Bmal*, and the ones in the *lower bar* the period when REV-ERB represses *Bmal*

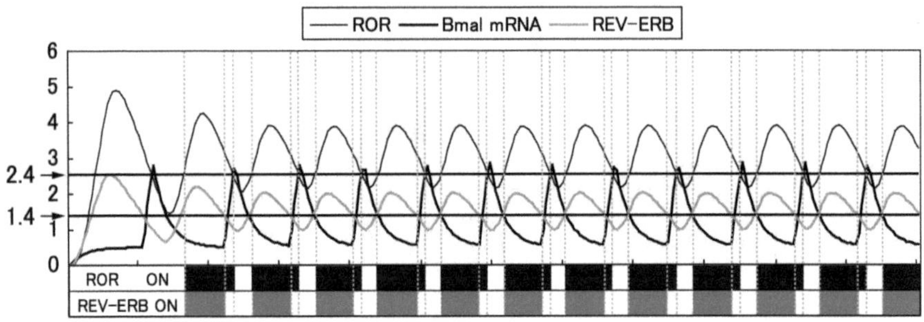

Fig. 13.15 Simulation results in the case of delayed ROR compared with REV-ERB in phase. Normal *Bmal* oscillation is obtained. Two *black-white bars* have the same meaning as the ones in Fig. 13.14

Although most of the biochemical results were reproduced in this model, one important biochemical characteristic still remained to be worked out. In this *Ror*-included interlocked model of Fig. 13.12, deletion of *Cry* gene did no longer abolish the rhythmicity, as shown in Fig. 13.18.

To resolve the discrepancy, a clue came from the report of genetic experiments showing that in *Cry*-deficient mice, *Bmal* level was lower than in control mice [360]. This indicates that, *Cry*, either directly or indirectly, must be involved in the enhancement of *Bmal* transcription, which is missing in the interlocked and *Ror* included model of Fig. 13.12. Implementation of this interaction would solve the persistent discrepancies remaining in the model up to the present point.

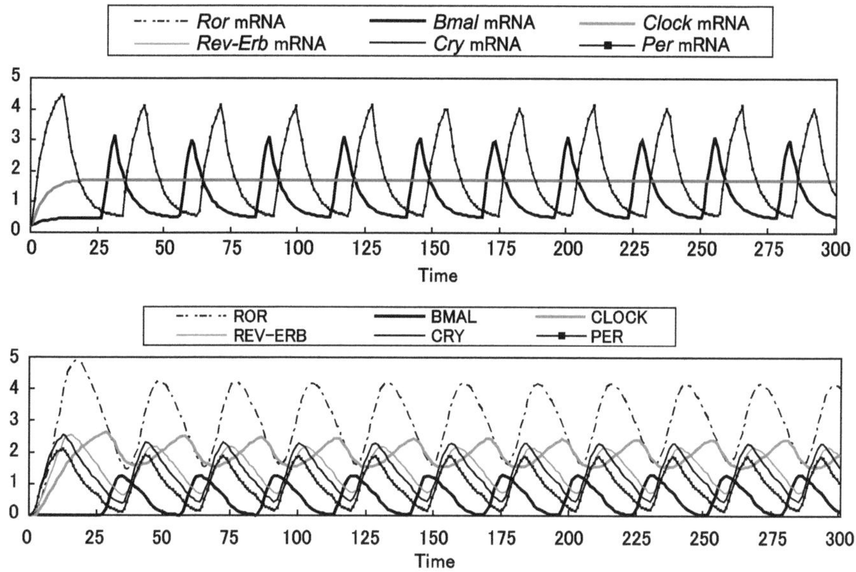

Fig. 13.16 Simulation results of clock gene mRNAs and proteins by the simulation based on the interlocked model with *Ror* of Fig. 13.12. They show stable rhythms, consistent with the biochemical observation

13.4.2 Inclusion of a Hypothetical Path: ROR Excites Bmal Only When PER/CRY Level Are Above the Threshold

In order to achieve elimination of the rhythm in *Cry*-deficient mice in the HFPN model, we introduced a hypothetical path indicated by a test arc in bold in Fig. 13.19, that represents the facilitatory effects of ROR exerting on *Bmal* transcription, which become effective only when the PER/CRY level is beyond the threshold (in this case 0.5). Although biochemical substrate of the interaction has not been identified, this modification resolved most of the remaining discrepancies in the circadian system.

Computer simulation showed that this model reproduced rhythms of the clock genes with proper phase relationship to each other (Fig. 13.20). In addition, forced fixation of PER/CRY or CLOCK/BMAL level did not abolish the rhythms, making the positive and negative loops complementary to each other of the circadian clock system (Fig. 13.21). Furthermore, the model successfully reproduced arrhythmicity of the *Cry*-deficient system, with the *Bmal* level sticking at a low value (Fig. 13.22). All these characteristics are consistent with the biochemical experiments using *Cry*-deficient mice.

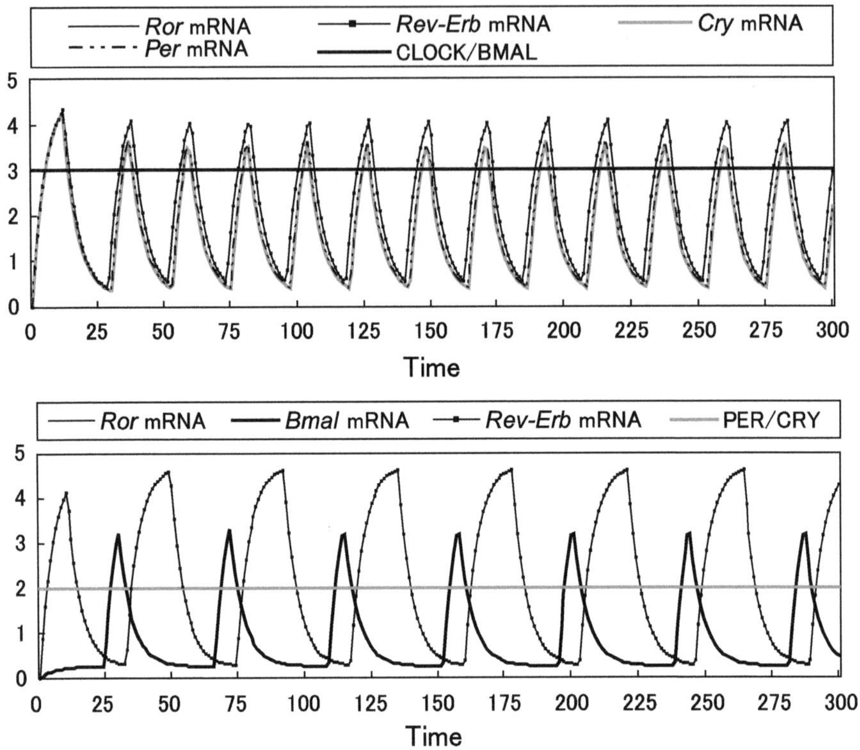

Fig. 13.17 Complementarity is maintained in the interlocked model with *Ror* of Fig. 13.12. Blocking of neither CLOCK/BMAL nor PER/CRY loop eliminates oscillation of the system

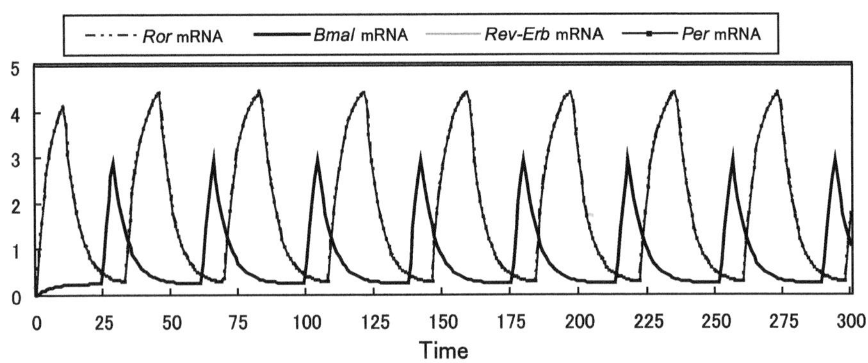

Fig. 13.18 In the interlocked model with *Ror* of Fig. 13.12, *Cry* deficiency do not abolish the rhythms of *Per* and *Bmal*

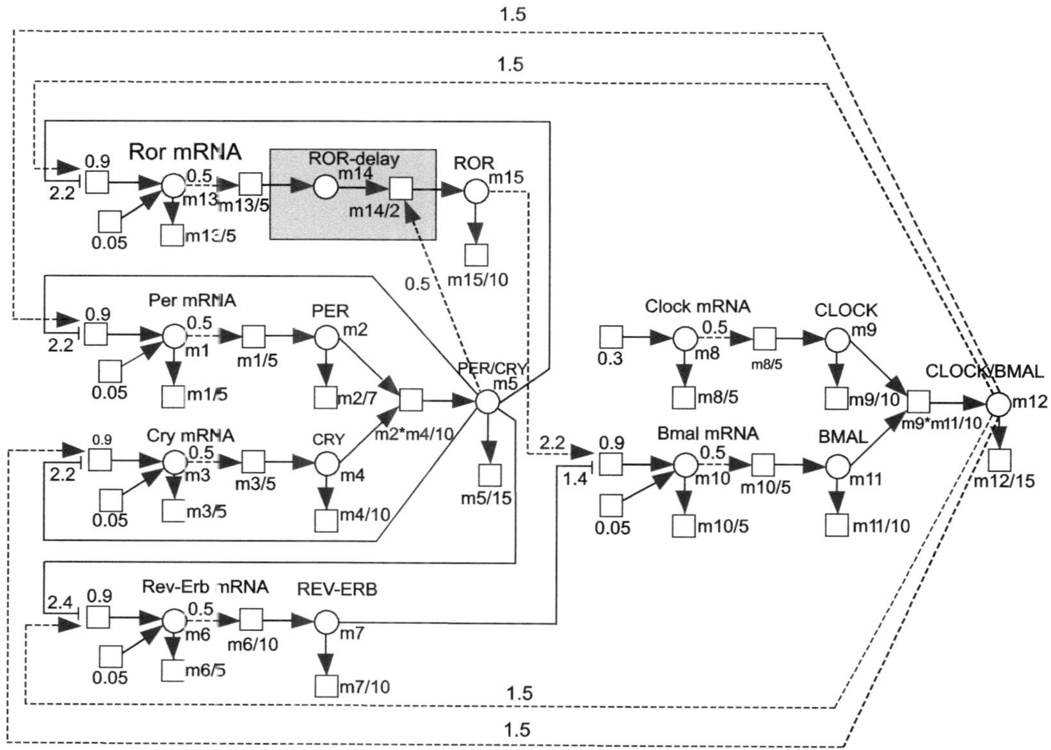

Fig. 13.19 Modified interlocked model with *Ror*. Now *Ror* excitation on *Bmal* is effective only if the PER/CRY level is higher than the prescribed value. This final modification yields system characteristics consistent with the biochemical experiments, as is shown in Fig. 13.20

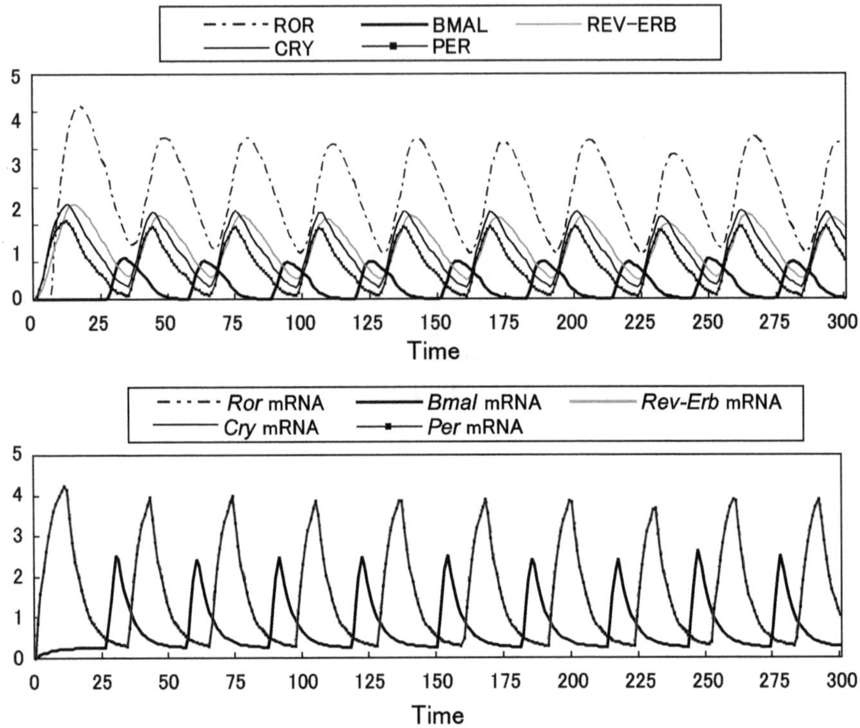

Fig. 13.20 Simulation results of clock gene mRNA and proteins of the modified interlocked model with *Ror* of Fig. 13.19

13.4.3 Simulation for Rev-Erb Knockout Mice

It is noteworthy that the final modified interlocked model with *Ror* described above (Fig. 13.8) successfully reproduced clock gene behavior not only in *Cry* knockout mice but also in *Rev-Erb* knockout mice. The results of simulation when *Rev-Erb* was deficient were shown in the lower panel of Fig. 13.23, which displayed a *Bmal* oscillation whose level stayed higher and never became below the peaks in control. Amplitude was also found smaller than that in control. This is observed fact in the biochemical results [309]. Moreover, our model also explained a shorter period of oscillation when *Rev-Erb* was deleted, as shown in Fig. 13.24 [309]. The modified interlocking model with *Ror* exhibits a shorter period of oscillations, as compared with that in the starting model. Black line indicate the *Per* in the starting model and gray line in the modified interlocking model with *Ror*. This tendency was also reported in a biochemical experiment. In other words, the present model indicates that *Rev-Erb* contribute to the elongation of the period of the system oscillation.

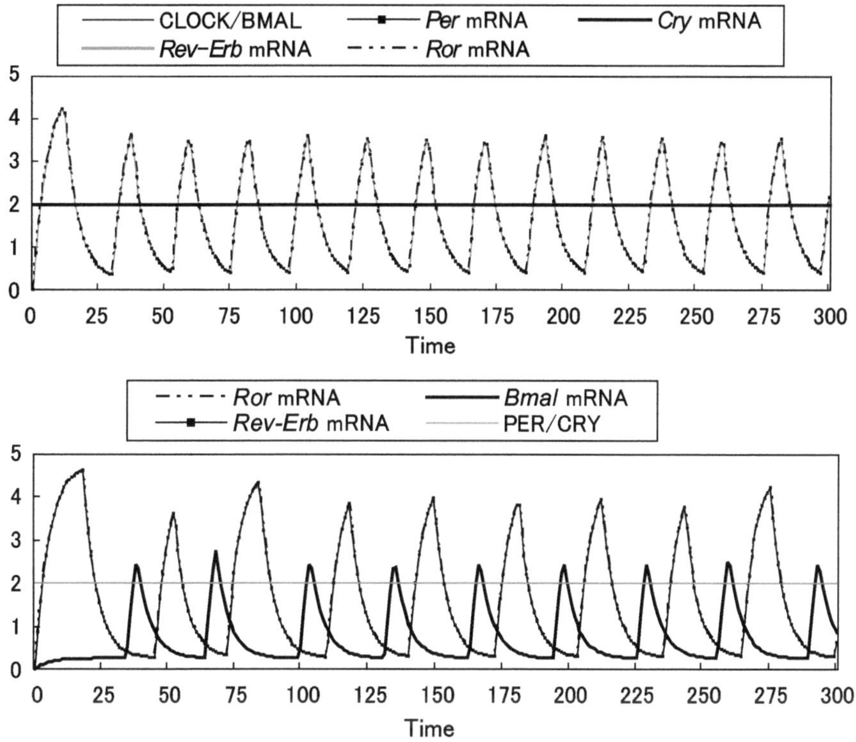

Fig. 13.21 Complementarity is maintained in the modified interlocked model with *Ror* of Fig. 13.19

13.5 Conclusions

We have described an in-depth search for the critical missing component or interaction of the circadian clock of mice. Starting from the simple model (Fig. 13.3), we employed two characteristic behaviors as guiding criteria to perform the exploratory analysis: complementarity of the positive and negative loop and total loss of the rhythms in *Cry* deficiency. Since any components in the circadian system are complicatedly intermingled, it is meaningless to single out one component as a critical element for the system behavior. However, our present analysis suggested that often-neglected components are essential for the system to maintain its appropriate behaviors.

To bring the circadian system into being, we found *Ror* plays a critical role. Unless *Ror* is incorporated in the HFPN, the presence of two, positive and negative, loop did not contribute to the robustness of the system. However, only with *Ror* in the model, blockade of either loop does not abolish the oscillation.

This design of the system also predicts such observed properties as the inverse phase relation of *Bmal* to *Per* or *Cry* level and higher level of *Bmal* in *Rev-Erb* knockout mice. Moreover, efforts to comply the two criteria produced the model

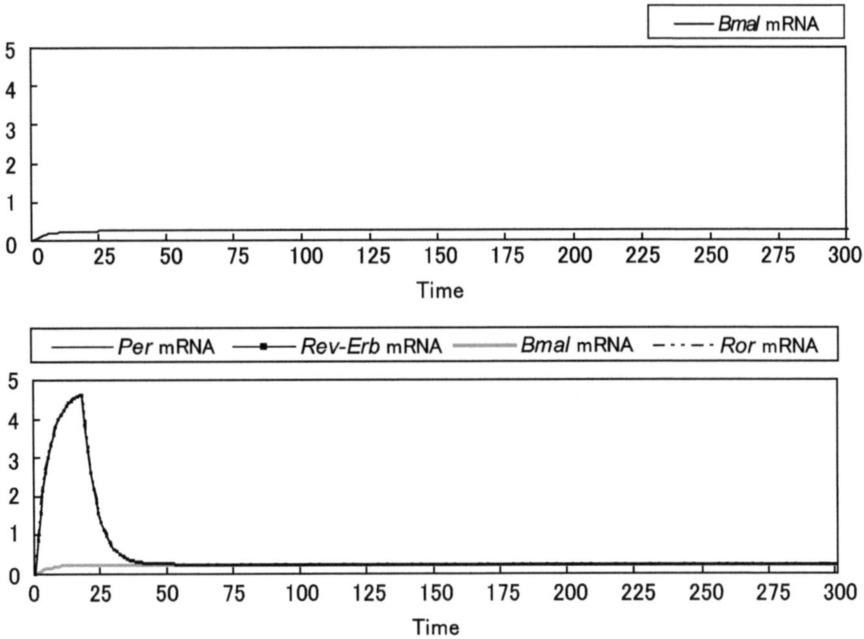

Fig. 13.22 The modified interlocked model with *Ror* reproduces abolishment of the rhythms in *Cry*-deficient mice, as is demonstrated in biochemical experiments

also explains other characteristics behaviors of the circadian system, namely, higher *Bmal* level in *Rev-Erb* deficient mice. The model also provides a suggestion that *Rev-Erb* contributes to the longer period of oscillation. The present exploration of system properties of circadian clock demonstrated the power and possibility of the HFPN-based computational simulation into the understanding of the complex biological systems.

13.6 Problems

13.1 In the same way as the constructing process of *Per-Cry* HFPN model, add the transcription, translation, complex formation, and degradation processes of *Clock* and *Bmal* into the HFPN model of Fig. 13.2. Set the transition speeds and threshold values of the HFPN according to the following comments for reactions.

- For the transcription and translation processes of *Clock*,
  - the transcription speed is 0.3 and no basal production,
  - the degradation speed of *Clock* mRNA is 1/5th of its concentration,
  - the translation speed of CLOCK is 1/5th of its concentration,
  - the degradation speed of CLOCK is 1/10th of its concentration, and
  - the threshold value to activate the translation is 0.5.

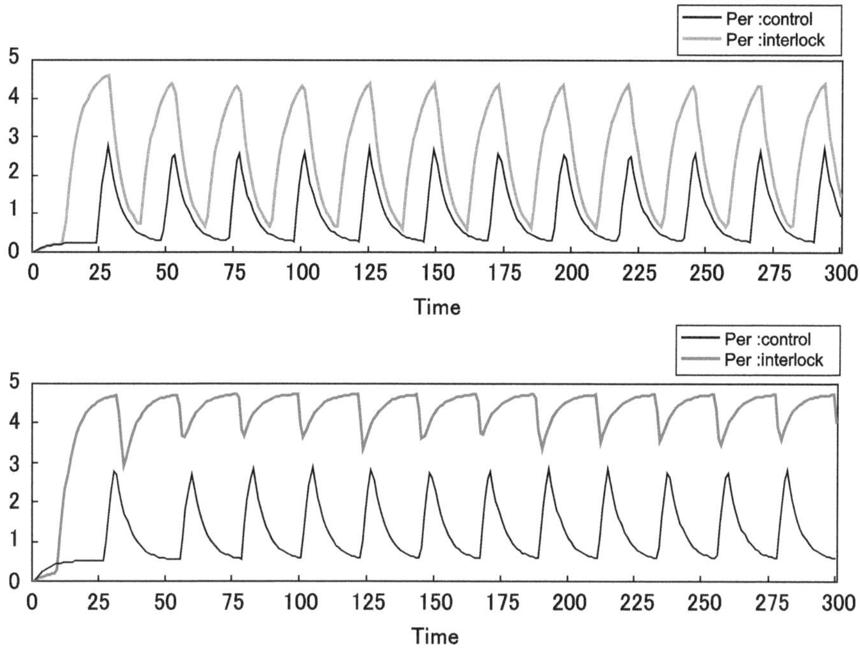

Fig. 13.23 Rhythms of *Bmal* in the mouse with *Rev-Erb* (*black*) and those in the *Rev-Erb* knock-out mouse (*gray*). *Upper illustrations* shows the levels in the starting model of Fig. 13.3. *Lower illustration* in the modified interlocking model with *Ror* depicted in Fig. 13.19. *Lower panel* shows a *Bmal* rhythm staying higher than that in the control mouse

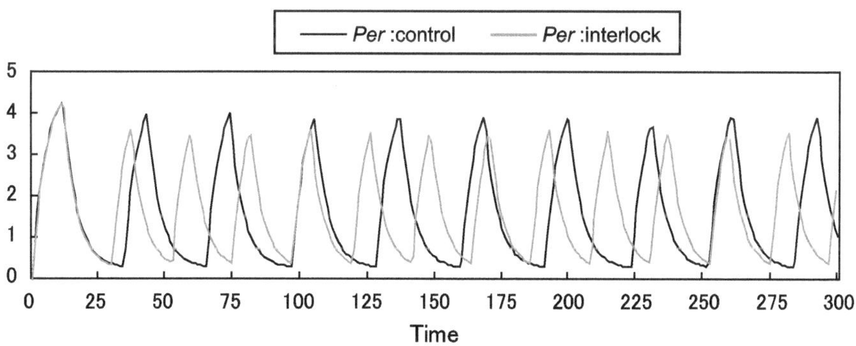

Fig. 13.24 Rhythms of *Per* in the mouse of the starting model in Fig. 13.3 (*black*) and those in the modified model with the hypothetical interaction of ROR (*gray*). It is of note that the period of *Per* in the modified model is found much longer than that in the starting model in Fig. 13.3

- For the transcription and translation processes of *Bmal*,
  - the transcription speed is 0.9 and the basal production speed is 0.05,
  - the degradation speed of *Bmal* mRNA is 1/5th of its concentration,
  - the translation speed of BMAL is 1/5th of its concentration,

  - the degradation speed of BMAL is 1/10th of its concentration, and
  - the threshold value to activate the translation is 0.5.
- For the complex formation process of CLOCK and BMAL,
  - the speed of the complex formation of CLOCK and BMAL is 1/10th of the product of these concentrations and
  - the degradation speed of CLOCK/BMAL is 1/15th of its concentration.

13.2 Add an HFPN model of the *Rev-Erb* transcription and translation process into Fig. 13.2, and thereafter, set transcription speeds and a threshold value according to the following comments.

- For the transcription and translation processes of *Rev-Erb*,
  - the transcription speed is 0.9 and the basal production is 0.05,
  - the degradation speed of mRNA is 1/5th of the concentration of *Rev-Erb* mRNA,
  - the translation speed is 1/10th of the concentration of *Rev-Erb* mRNA,
  - the degradation speed is 1/10th of the concentration of REV-ERB, and
  - the threshold value to activate the translation is 0.5.

13.3 Describe the regulatory interactions of gene activation/inactivation by using test and inhibitory arcs with thresholds values according to the following comments. A value in parentheses of each sentence is a threshold value to be assigned to a test or inhibitory arc.

- PER/CRY activates *Bmal* (2.0),
- PER/CRY represses *Rev-Erb* (1.6),
- REV-ERB represses *Bmal* (1.4),
- CLOCK/BMAL activates *Per* (0.5),
- CLOCK/BMAL activates *Cry* (0.5), and
- CLOCK/BMAL activates *Rev-Erb* (0.5).

Acknowledgements This work was partially supported by Grant-in-Aid for Scientific Research on Priority Areas "Systems Genomics" (17017008) and Grant-in-Aid for Scientific Research (B) (19300103) from the Ministry of Education, Culture, Sports, Science and Technology of Japan.

Chapter 14
Prediction of Network Structure

Annegret Wagler

Abstract For many aspects of health and disease, it is important to understand different phenomena in biology and medicine. To gain the required insight, experiments are performed and the resulting experimental data have to be interpreted. This leads to the Network Reconstruction Problem, the challenging task to generate all models that explain the observed phenomena. As in systems biology, the framework of Petri nets is often used to describe models for the regulatory mechanisms of biological systems, our aim is to predict all the possible network structures being conformal with the given experimental data. We discuss a combinatorial approach proposed by Marwan et al. (Math. Methods Oper. Res. 67:117–132, 2008) and refined by Durzinsky et al. (Proc. of CMSB 2008, LNBI, vol. 5307, pp. 328–346, Springer, Berlin, 2008) to solve this problem. In addition, we also present an algorithm by Durzinsky et al. (J. Theor. Comput. Sci., 2009) that, based on these results, generates a complete list of all potential networks reflecting the experimentally observed behavior.

14.1 Introduction

Predictive models of biological systems and phenomena are of high scientific interest and practical relevance, but not always easy to obtain due to their inherent complexity. One fundamental question in this context is to detect the local mechanisms of interaction starting from the experimentally observed global behavior of a biological system, that is, to predict all the possible network structures being conformal with the given experimental data. This is the topic of this chapter.

Structure and function of the studied system can be probed by stimulating one or several of its elements and by measuring a set of parameters as a function of time

A. Wagler (✉)
Magdeburg Center of Systems Biology (MaCS) and Institute for Mathematical Optimization, Otto-von-Guericke Universität Magdeburg, Universitätsplatz 2, 39106 Magdeburg, Germany
e-mail: wagler@imo.math.uni-magdeburg.de

I. Koch et al. (eds.), *Modeling in Systems Biology*, Computational Biology 16, DOI 10.1007/978-1-84996-474-6_14, in order to see how this stimulation propagates throughout the system. The task is to predict the network structure from such experimental time series data, that is, to determine in which way the measured components interact with each other.

In practice, this is typically done heuristically, based on subjectively plausible interpretations of the data. Mathematical approaches help here to rigorously interpret the data with the help of a mathematical model and to state provable assertions on properties.

Many different mathematical models and methods for their reconstruction have been considered. For instance, Kholodenko et al. [188] use differential equation systems to describe the dynamics of gene regulatory networks based on a linear model which is unraveled by measuring the effects of perturbations to the system. As a discrete version of differential equations, Laubenbacher and Stigler [218] propose an algebraic approach that models the observed phenomena in terms of a finite-state polynomial model, which allows the use of algebraic tools to find the set of all solutions. The outcome, however, depends on the chosen order of the parameters and might claim an interaction between parameters which are in fact not coupled. Krishna and Guo [213] describe a statistical method which estimates the likelihood of one component effecting another one by Granger causality and generates a dependency graph, which provides only the information which components are coupled, but not how they interact.

All these approaches yield potentially incomplete information as neither network elements are considered that are causally involved but have not been measured, nor a certificate can be given that no other model can explain the phenomena.

The aim of this chapter is to present a combinatorial approach proposed by Marwan et al. [245] and refined by Durzinsky et al. [105] for reconstructing models of biological systems from experimental data, that overcomes these difficulties. In particular, this approach also provides a certificate to detect whether the given data do not suffice to find a coherent model of the studied biological system or phenomenon. In this situation, additional elements are considered in the reconstruction process that are causally involved but have not been measured. Based on these results, Durzinsky et al. [106] developed an algorithm that generates a complete list of all potential models reflecting the experimentally observed behavior, using the minimal number of additional elements.

To model the studied biological systems and phenomena, we use the framework of Petri nets to benefit from the following advantages. First, Petri nets are a coherent model to describe all different motifs that occur in biological systems (see Part I Petri nets and the references in the portal mentioned therein [70, 71, 106, 208, 303]). Second, the mathematical description of both the interactions in terms of network structure and the dynamics in terms of state changes allows easy handling and assembly of reaction chains. Finally, the graphical representation of Petri nets is often used to model biological systems and processes, as it shows the structure and connectivity of the parts of the systems directly.

We now describe the problem setting in a formally correct way. Recall that a Petri net N is given by

$$N = (P, T, F, W, m_0)$$

with the sets P of places and T of transitions, the flow relation $F \subseteq (P \times T) \cup (T \times P)$ providing arcs of the form (p,t) or (t,p), a mapping $W : F \to \mathbb{N}$ assigning weights to the arcs from F, and the initial marking $m_0 : P \to \mathbb{N}$ assigning each place p its initial token load $m_0(p)$.

The places $p \in P$ represent the set of studied components (as proteins or their conformational states, enzymes etc.), the transitions $t \in T$ their interactions (as chemical reactions, activations, causal dependencies etc.).

Recall further that a marking m enables a transition t if each arc (p,t) from a pre-place of t is weighted with $W(p,t) \le m(p)$ tokens. The occurrence of t changes the marking m to a marking m' by consuming $W(p,t)$ tokens from each pre-place p of t and producing $W(t,p')$ tokens in each post-place p' of t.

Markings stand for system states, and the dynamic behavior of a Petri net is described step-wise, by converting states into new states upon the occurrence of transitions, starting from the state corresponding to the initial marking m^0.

For our purpose, it is convenient to associate with each marking m a *state vector* $x \in \mathbb{N}^{|P|}$ with $x_p = m(p)$ for all places $p \in P$, and to express the occurrence of a transition t with the help of its update vector. In addition, we consider, without loss of generality, simple Petri nets only where no pair of place and transition is linked in both directions (i.e., if there is an arc $(p,t) \in F$, then $(t,p) \notin F$ holds, and vice versa). Therefore, we redefine the update vector of a transition t by $r^t \in \mathbb{Z}^{|P|}$ having

$$r^t_p = \begin{cases} -W(p,t) & \text{if } (p,t) \in F, \\ W(t,p) & \text{if } (t,p) \in F, \\ 0 & \text{else} \end{cases}$$

as its entries. Then the change from a state x to a state x' by the occurrence of transition t can be expressed as $x + r^t = x'$. Moreover, we can rewrite each Petri net $N = (P, T, F, W, m_0)$ by

$$N = \left(G, x^0\right) \quad \text{with } G = (P \cup T, F, W)$$

where G is the associated *network* and x^0 the state vector in $\mathbb{N}^{|P|}$ encoding the initial marking. In mathematical terms, $G = (P \cup T, F, W)$ is a weighted directed bipartite graph. A directed graph has a set of nodes (here the set $P \cup T$ of places and transitions) and directed arcs in F linking its nodes. G is bipartite as its node set $P \cup T$ is partitioned into two subsets, the set P of places and the set T of transitions, and the arcs in F link a place $p \in P$ exclusively with a transition $t \in T$ (or vice versa). Recall that we do not have any pair of place and transition linked in both directions. The integral arc weights reflect the stoichiometric coefficients of the reactions.

Our aim is to predict the possible network(s) $G = (P \cup T, F, W)$ for the studied system, where the set $P = \{p_1, \dots, p_n\}$ of these components is chosen which is expected to be crucial for the observed phenomenon.

An experiment is performed by triggering the system in some state x^0 (by external stimuli like the change of nutrient concentrations or the exposition to some pathogens), thereby generating an initial state x^1 of the system.[1] Then one observes how the system reacts by changing its states according to the stimulus, thereby measuring the values of the components in P at certain time points. Some measured data are of a discrete nature, for example, a gene can be expressed or not, an enzyme can be present or not, a protein occurs in one of its conformational states. If some measured data are continuous, for example, concentrations of certain metabolites, then it is necessary to appropriately discretize the continuous data into finitely many discrete states. For that, the chosen level of discretization has to be fine enough to preserve all dynamic features of the time course (in particular, all local maxima or minima of the values), but rough enough to be robust to noise in the experimental data.

This yields a sequence $x^1, \dots, x^k$ of (discrete or discretized) states reflecting the time-dependent responses of the system to the stimulation in x^1 denoted by

$$\mathscr{X}(x^1) = (x^0; x^1, \dots, x^k).$$

Note that we also provide the state x^0 as the starting point for the stimulation, which will be needed later on.

As several experiments, starting from different initial states in a set $\mathrm{Ini} \subseteq \mathbb{Z}_+^{|P|}$, may be necessary to describe the whole phenomenon, we use *experimental time series data* of the form

$$\mathscr{X}' = \{\mathscr{X}(x^1) : x^1 \in \mathrm{Ini}\}.$$

We assume $\mathscr{X}'$ to be reproducible, thus we require $\mathscr{X}(x^{i_1}) = \mathscr{X}(x^{j_1})$ for $x^{i_1} = x^{j_1}$.

Example 14.1 As running example of this chapter, we will consider the *light-induced sporulation of Physarum polycephalum*. The reconstruction process for different experimental settings exploring this light-induced sporulation is explained in Durzinsky et al. [104].

The developmental decision of starving *P. polycephalum* plasmodia to exit the vegetative plasmodial stage and to enter the sporulation pathway is controlled by environmental factors like visible light [372]. One of the photoreceptors involved in the control of sporulation *Sp* is a phytochrome-like photoreversible photoreceptor protein which occurs in two stages P_{fr} and P_r. If the dark-adapted form P_{fr} absorbs far-red light *FR*, the receptor is converted into its red-absorbing form P_r, which causes sporulation [216]. If P_r is exposed to red light *R*, it is photoconverted back to the initial stage P_{fr}, which prevents sporulation. Note that the changes between the stages P_{fr} and P_r can be experimentally observed due to a change of color.

This experimental setting uses the set $P = \{FR, R, P_{fr}, P_r, Sp\}$ of studied components. As we can deal for all components in P only with their availability, we represent all (observed) states by vectors of the form

$$x = (x_{FR}, x_R, x_{P_{fr}}, x_{P_r}, x_{Sp})^T$$

---

[1] We use super-indices to reflect the order of the states x^j and to allow sub-indices for the entries x^j_p in a particular component p.

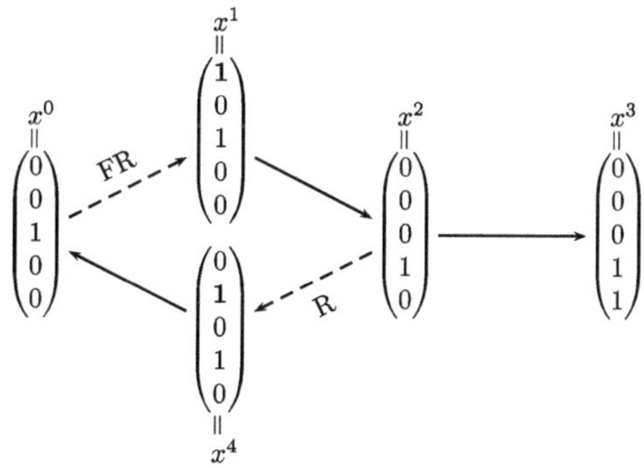

Fig. 14.1 A scheme illustrating the stimulations and the experimental observations (*dashed arrows* stand for stimulations, *solid arrows* for the observed responses of the system)

having 0/1-entries only. The above described behavior of the system can, therefore, be represented by a scheme as depicted in Fig. 14.1. This scheme can be interpreted as experimental time series data $\mathscr{X}'$ with initial states in $\text{Ini} = \{x^0, x^1, x^4\}$ and the following sequences of measured states:

- $\mathscr{X}(x^0) = (x^0)$ as no sporulation occurs without stimulation (dark control),
- $\mathscr{X}(x^1) = (x^0; x^1, x^2, x^3)$ as response to irradiation with far-red light at state x^0,
- $\mathscr{X}(x^4) = (x^2; x^4, x^0)$ showing that sporulation can be suppressed by irradiation at state x^2 with red light.

Thus, the experimental setting is given by $(P, \mathscr{X}')$. The task is to find all networks $G = (P \cup T, F, W)$ with P as set of places which are appropriate to explain the experimentally observed behavior reported in $\mathscr{X}'$.

In the best case, two consecutively measured states $x^j, x^{j+1} \in \mathscr{X}'$ are also consecutive system states, that is, x^{j+1} can be obtained from x^j by a single transition. This is, however, in general not the case (and depends on the chosen time points to measure the states in $\mathscr{X}'$). Typically, x^{j+1} is obtained from x^j by a sequence

$$x^j = y^1, y^2, \ldots, y^{k+1} = x^{j+1}$$

of some length $k \geq 1$ using the transitions from a subset $T' \subseteq T$ to change from an intermediate state y^l to y^{l+1}, where the intermediate states are not reported in $\mathscr{X}'$.

In a network G fitting the experimental data, this can be interpreted as follows. With G, an incidence matrix $C \in \mathbb{Z}^{|P|\times|T|}$ is associated, where each row corresponds to a place $p \in P$ of the network, and each column to a transition $t \in T$. In particular, we associate a transition $t \in T$ with the corresponding column C_{*t} of C having

$$c_{pt} = \begin{cases} -W(p,t) & \text{if } (p,t) \in F, \\ W(t,p) & \text{if } (t,p) \in F, \\ 0 & \text{else} \end{cases}$$

as entries.[2] Note that C_{*t} corresponds exactly to the update vector of transition t. That x^{j+1} is reached from x^j by a sequence using the transitions from a subset $T' \subseteq T$ is, therefore, equivalent to obtain the state vector x^{j+1} from x^j by adding the corresponding columns C_{*t} of C for all $t \in T'$:

$$x^j + \sum_{t \in T'} C_{*t} = x^{j+1}.$$

In terms of the incidence matrix C, the above equation can be rewritten as

$$x^{j+1} - x^j = C\lambda$$

where the vector $\lambda \in \mathbb{Z}_+^{|T|}$ indicates which transitions are involved to obtain the studied sequence. Hence, to fit the experimental data, a network G has to satisfy the following definition.

Definition 14.1 A network $G = (P \cup T, F, W)$ with incidence matrix $C \in \mathbb{Z}^{|P| \times |T|}$ is *conformal* with $\mathscr{X}'$ if, for each $x^1 \in \text{Ini}$ and any two consecutive states $x^j, x^{j+1} \in \mathscr{X}(x^1)$, the linear equation system $x^{j+1} - x^j = C\lambda$ has an integral solution $\lambda \in \mathbb{Z}_+^{|T|}$.

In other words, a network $G = (P \cup T, F, W)$ is conformal with $\mathscr{X}'$ if T contains appropriate transitions such that, for any pair x^j, x^{j+1} of consecutively measured states in $\mathscr{X}'$, state x^{j+1} can be reached from x^j by an intermediate sequence using transitions from T. This leads to the following problem.

Problem 14.1 (Network Reconstruction Problem) Given $(P, \mathscr{X}')$, generate all the networks $G = (P \cup T, F, W)$ being conformal with $\mathscr{X}'$.

In Sect. 14.2, we discuss two approaches for solving the Network Reconstruction Problem, depending on the kind and quality of the given data, which both construct the complete set of networks being conformal with the given data. We first outline the principle ideas of a combinatorial reconstruction approach proposed by Marwan et al. [245] and discuss the resulting difficulties. Then we present a refinement of this approach for the case of *monotone data* which was developed by Durzinsky et al. [105, 106] to overcome these difficulties.

In Sect. 14.3, we present the resulting algorithm for reconstructing conformal networks from monotone experimental time series data from Durzinsky et al. [106], including a formal pseudo code description.

Besides predicting the structure of the studied biological system in terms of conformal networks, it is also interesting to consider the dynamics of the studied system. In Sect. 14.4, we discuss a stronger version of conformality also taking some

[2] In usual mathematical terms, the column of a matrix C indexed by t is denoted by $C_{\cdot t}$. As this notion interferes with the notion $\bullet t$ of preplaces of a transition t, we here use C_{*t} instead of $C_{\cdot t}$.

dynamic aspects into account, as suggested in Marwan et al. [245] and refined by Torres et al. [395]. That way, we can rule out networks as appropriate models which are conformal but not strongly conformal with the given data.

We finally summarize all these methods for solving the Network Reconstruction Problem in Sect. 14.5, and discuss the practical impact of the presented results.

14.2 Methods to Predict Network Structures

We first outline the principle ideas of a combinatorial reconstruction approach proposed by Marwan et al. [245] (see Sect. 14.2.1). Then we discuss more detailed a refinement of this approach suggested by Durzinsky et al. [105, 106] for the case of monotone data (see Sect. 14.2.2). The latter approach involves some ideas in order to keep both the running time of the algorithm and the number of produced solutions as small as possible.

14.2.1 A General Combinatorial Reconstruction Approach

We now outline the major steps of a reconstruction algorithm proposed by Marwan et al. [245] for turning experimental data into a complete list of networks being conformal with the phenomena observed.

Recall that the input $(P, \mathscr{X}')$ of the algorithm consists of the set $P = \{p_1, \dots, p_n\}$ of studied components and experimental time series data $\mathscr{X}' = \{\mathscr{X}(x^1) : x^1 \in \text{Ini}\}$.

Typically, we have components of different types; we call $p \in P$ an input (resp. output) component if tokens in p can not be produced (resp. consumed) by the system. All components subject to external stimulation are input components, whereas output components reflect irreversible modifications of the system. All other components $p \in P$ are free and are required as intermediates to describe the internal response of the studied system to the external stimulus. Accordingly, we partition the set P into

$$P = P_I \cup P_O \cup P_F$$

where P_I (resp. P_O, resp. P_F) is the subset of all input (resp. output, resp. free) components. Recall further that we encode stimuli within the initial states $x^1 \in$ Ini (by setting the values of the components in P_I accordingly), instead of using interface reactions $r \in \mathbb{Z}^n_+$ with $r_p > 0$ for $p \in P_I$.

An initial step of the algorithm is to extract the observed internal responses from the experimental data. For that, we define the set

$$\mathscr{D} := \{d^j = x^{j+1} - x^j : j > 0, x^j, x^{j+1} \in \mathscr{X}(x^1), x^1 \in \text{Ini}\}$$

(recall that the change from x^0 to x^1 within the sequence $\mathscr{X}(x^1) = (x^0; x^1, \dots, x^k)$ corresponds to a stimulation, but not to a response of the system).

In order to find the complete list of all networks $G = (P \cup T, F, W)$ being conformal with $\mathscr{X}'$, we shall find the corresponding matrices $C \in \mathbb{Z}^{|P|\times|T|}$ such that each $d \in \mathscr{D}$ has a representation of the form

$$C\lambda = d$$

where the Parikh vector $\lambda \in \mathbb{Z}_+^{|T|}$ indicates how often the columns of C are used in order to represent d. The difficulty is that both, the matrix C and the vector λ, are unknown.

The first step is to describe the set of potential reaction vectors (update vectors of the potential transitions) which might constitute the columns of C. For that, capacity bounds u for the number of tokens in each place of the network are required. In biological systems, internal elements can be considered to be bounded, as the value x_p of any state refers to the measured concentration of the studied element, which can only increase up to a certain maximum u_p. Indeed, many of the elements are even binary ($u_p = 1$) to distinguish between the presence ($x_p = 1$) and absence ($x_p = 0$) of the element $p \in P$ only. For external elements $p \in P_I$, the value x_p^1 in an initial state x^1 provides a natural upper bound u_p for the number of tokens in place p for any system state (or the maximum of x_p^1 taken over all initial states x^1, if several time series data are considered simultaneously).

This shows that every column $r^t = C_{*t}$ of such a matrix C has to satisfy $-u \leq r^t \leq u$, as all potential system states have to belong to the box

$$\mathscr{X} = \{x \in \mathbb{Z}^n : 0 \leq x_p \leq u_p \, \forall p \in P\}.$$

This can be refined by considering the different types of components, taking into account that an input (resp. output) component can not be produced (resp. consumed) by the system. Moreover, every nonnegative column r^t would correspond to a transition which produces substances without consuming something. As we encode stimuli within the initial state(s), we can exclude all vectors in $\mathbb{Z}_+^n$ from the set of all potential reactions in order to consider for the reconstruction process only vectors corresponding to internal reactions of the system.

Accordingly, we define the *reaction space* as

$$\mathscr{R} = \left\{ r \in \mathbb{Z}^{|P|} : \begin{array}{c} -u_p \leq r_p \leq 0 \, \forall p \in P_I \\ 0 \leq r_p \leq u_p \, \forall p \in P_O \\ -u_p \leq r_p \leq u_p \, \forall p \in P_F \end{array} \right\} \setminus \mathbb{Z}_+^{|P|}$$

where, depending on the biological system, other constraints may be present and restrict this set further. For instance, if $P' \subseteq P$ is a P-invariant of the system, then $\sum_{p \in P'} r_p = 0$ follows for all vectors $r \in \mathscr{R}$.

Example 14.2 In order to apply the reconstruction approach to the experimental time series data $\mathscr{X}'$ described in Example 14.1, we illustrate here how to built the according reaction space. The experimental setting uses again the set $P = \{FR, R, P_{fr}, P_r, Sp\}$ of studied components and we represent all (observed) states by vectors of the form $x = (x_{FR}, x_R, x_{P_{fr}}, x_{P_r}, x_{Sp})^T$. As we can deal for all components in P only with their availability, we have $u_p = 1$ for all $p \in P$. Due to the experimental setting, we have a partition of P into $P_I = \{FR, R\}$, $P_F = \{P_{fr}, P_r\}$, $P_O = \{Sp\}$. As P_{fr} and P_r are two different stages of one photoreceptor protein, they form a p-invariant $P' = \{P_{fr}, P_r\}$. Thus, we have

$$\mathscr{R} = \left\{ r \in \mathbb{Z}^5 : \begin{array}{r} -1 \le r_p \le 0 \; \forall p \in P_I \\ -1 \le r_p \le 1 \; \forall p \in P_F \\ 0 \le r_p \le 1 \; \forall p \in P_O \\ r_{P_{fr}} + r_{P_r} = 0 \end{array} \right\} \setminus \mathbb{Z}^5_+$$

and, thus, the following vectors form the set $\mathscr{R}$:

	1	2	3	4	5	6	7	8	9	10	11	12	13	14	15	16	17	18	19	20	21	22
FR	−1	−1	−1	−1	−1	−1	−1	−1	−1	−1	−1	−1	0	0	0	0	0	0	0	0	0	0
R	−1	−1	−1	−1	−1	−1	0	0	0	0	0	0	−1	−1	−1	−1	−1	−1	0	0	0	0
P_{fr}	−1	−1	0	0	1	1	−1	−1	0	0	1	1	−1	−1	0	0	1	1	−1	−1	1	1
P_r	1	1	0	0	−1	−1	1	1	0	0	−1	−1	1	1	0	0	−1	−1	1	1	−1	−1
Sp	0	1	0	1	0	1	0	1	0	1	0	1	0	1	0	1	0	1	0	1	0	1

Recall that we exclude the two reaction vectors from $\mathbb{Z}^5_+$ equal to $(0,0,0,0,0)^T$ and $(0,0,0,0,1)^T$ corresponding to the "empty" reaction and a spontaneous sporulation without stimulation, respectively.

As next step, the idea is to represent the vector d by building all conic integer combinations

$$d = \sum_{r^t \in \mathscr{R}} \lambda_t r^t, \quad \lambda_t \in \mathbb{Z}_+ \tag{14.1}$$

of reaction vectors r^t in $\mathscr{R}$; we denote by $\Lambda(d)$ the set of all integral solutions of this linear equation system. For each vector $\lambda \in \Lambda(d)$, let

$$\mathscr{R}_{d,\lambda} = \left\{ r^t \in \mathscr{R} : \lambda_t \neq 0 \right\}$$

be the (multi-)set of reaction vectors used for this solution λ.

The difficulty is that, in general, the system (14.1) has infinitely many solutions, as we can extend any representation of d by an integral solution $\eta \neq \mathbf{0}$ of the homogeneous system

$$\mathbf{0} = \sum_{r^t \in \mathscr{R}_H} \eta_t r^t, \quad \eta_t \in \mathbb{Z}_+ \tag{14.2}$$

where $\mathscr{R}_H$ is the subset of all vectors $r \in \mathscr{R}$ with $r_p = 0$ for all $p \in P_I \cup P_O$. We denote the set of all homogeneous solutions $\eta \neq \mathbf{0}$ by $\Lambda(\mathbf{0})$. As $d = d + \mathbf{0}$ holds, every solution $\lambda \in \Lambda(d)$ can be clearly extended by any homogeneous solution $\eta \in \Lambda(\mathbf{0})$ to a new solution $\lambda + \eta \in \Lambda(d)$. However, not all conic combinations of solutions $\lambda \in \Lambda(d)$ and $\eta \in \Lambda(\mathbf{0})$ are of interest for our purpose, as for instance

$\mathscr{R}_{d,\lambda} \subseteq \mathscr{R}_{d,\lambda+\eta}$ and $\mathscr{R}_{d,\lambda+\eta} = \mathscr{R}_{d,\lambda+k\eta}$ holds for all $k \geq 1$. In particular, there are only finitely many different sets $\mathscr{R}_{d,\lambda}$, $\lambda \in \Lambda(d)$ (as the finite set $\mathscr{R}$ has only a finite number of subsets). Nevertheless, the occurrence of homogeneous solutions substantially increases the number of different representations for each $d \in \mathscr{D}$.

Example 14.3 We apply the above steps to the experimental time series data $\mathscr{X}'$ described in Example 14.1, again using vectors of the form $x = (x_{FR}, x_R, x_{P_{fr}}, x_{P_r}, x_{Sp})^T$ to represent system states.

Extracting the difference vectors from the sequences $\mathscr{X}(x^1) = (x^0; x^1, x^2, x^3)$ and $\mathscr{X}(x^4) = (x^2; x^4, x^0)$ of measured states yields:

$$d^1 = x^2 - x^1 = \begin{pmatrix} -1 \\ 0 \\ -1 \\ 1 \\ 0 \end{pmatrix}, \qquad d^2 = x^3 - x^2 = \begin{pmatrix} 0 \\ 0 \\ 0 \\ 0 \\ 1 \end{pmatrix}, \qquad d^3 = x^0 - x^4 = \begin{pmatrix} 0 \\ -1 \\ 1 \\ -1 \\ 0 \end{pmatrix}$$

(recall that the stimulations $x^0 \xrightarrow{FR} x^1$ and $x^2 \xrightarrow{R} x^4$ are not subject to reconstruction).

In order to represent the difference vectors d^j as conic combinations of vectors in the reaction space $\mathscr{R}$ constructed in Example 14.2, we have to find all integral solutions λ of the linear equation system (14.1) for d_j and $j \in \{1, 2, 3\}$.

We start with d^1 and observe that for its representation only vectors $r \in \mathscr{R}$ can be used with $r_R = 0$ and $r_{Sp} = 0$. This yields the following minimal representations

$$d^1 = \begin{pmatrix} -1 \\ 0 \\ -1 \\ 1 \\ 0 \end{pmatrix} = \begin{pmatrix} -1 \\ 0 \\ 0 \\ 0 \\ 0 \end{pmatrix} + \begin{pmatrix} 0 \\ 0 \\ -1 \\ 1 \\ 0 \end{pmatrix} = \begin{pmatrix} -1 \\ 0 \\ 1 \\ -1 \\ 0 \end{pmatrix} + 2 \begin{pmatrix} 0 \\ 0 \\ -1 \\ 1 \\ 0 \end{pmatrix}$$

with $\mathscr{R}_{d^1,\lambda^1} = \{r^7\}$, $\mathscr{R}_{d^1,\lambda^2} = \{r^9, r^{19}\}$ and $\mathscr{R}_{d^1,\lambda^3} = \{r^9, r^{11}\}$ using the numbering of the reaction vectors from Example 14.2.

To represent d^2, we can use vectors $r \in \mathscr{R}$ with $r_R = 0$ and $r_{FR} = 0$, which yields the minimal representations

$$d^2 = \begin{pmatrix} 0 \\ 0 \\ -1 \\ 1 \\ 0 \end{pmatrix} + \begin{pmatrix} 0 \\ 0 \\ 1 \\ -1 \\ 1 \end{pmatrix} = \begin{pmatrix} 0 \\ 0 \\ -1 \\ 1 \\ 1 \end{pmatrix} + \begin{pmatrix} 0 \\ 0 \\ 1 \\ -1 \\ 0 \end{pmatrix}$$

with $\mathscr{R}_{d^2,\lambda^1} = \{r^9, r^{22}\}$ and $\mathscr{R}_{d^2,\lambda^2} = \{r^{20}, r^{21}\}$.

For d^3, vectors $r \in \mathscr{R}$ with $r_{FR} = 0$ and $r_{FR} = 0$ are appropriate and yield the minimal representations

$$d^3 = \begin{pmatrix} 0 \\ -1 \\ 1 \\ -1 \\ 0 \end{pmatrix} = \begin{pmatrix} 0 \\ -1 \\ 0 \\ 0 \\ 0 \end{pmatrix} + \begin{pmatrix} 0 \\ 0 \\ 1 \\ -1 \\ 0 \end{pmatrix} = \begin{pmatrix} 0 \\ -1 \\ -1 \\ 1 \\ 0 \end{pmatrix} + 2 \begin{pmatrix} 0 \\ 0 \\ 1 \\ -1 \\ 0 \end{pmatrix}$$

with $\mathscr{R}_{d^3,\lambda^1} = \{r^{17}\}$, $\mathscr{R}_{d^3,\lambda^2} = \{r^{15}, r^{21}\}$ and $\mathscr{R}_{d^3,\lambda^3} = \{r^{13}, r^{21}\}$.

Every minimal representation can be further extended by any solution of the homogeneous system (14.2). As all vectors r in $\mathscr{R}_H$ have to satisfy $r_p = 0$ for all $p \in P_I \cup P_O$, we obtain:

$$\mathscr{R}_H = \left\{ \begin{pmatrix} 0 \\ 0 \\ -1 \\ 1 \\ 0 \end{pmatrix}, \begin{pmatrix} 0 \\ 0 \\ 1 \\ -1 \\ 0 \end{pmatrix} \right\}.$$

Thus, there is exactly one homogeneous solution η using both vectors r^{19} and r^{21} from $\mathscr{R}_H$. Taking all combinations with different sets of used reactions into account yields 6 different representations for d^1, 4 different representations for d^2, and 6 different representations for d^3.

Note that, except for some pathological cases with $|P| \leq 2$, there always exists at least one solution of the linear equation system (14.1) for each $d \in \mathscr{D}$ (see [105]). The next step of the approach consists in combining the reactions used for the representations of the single difference vectors to conformal networks.

By construction, selecting one solution $\lambda \in \Lambda(d)$ for each $d \in \mathscr{D}$ and taking the union of the corresponding sets $\mathscr{R}_{d,\lambda}$ yields the columns of an incidence matrix of a conformal network. In order to encode which solution $\lambda_i \in \Lambda(d) = \{\lambda_1, \ldots, \lambda_{|\Lambda(d)|}\}$ has been selected for each $d \in \mathscr{D}$, define a vector $\kappa \in \mathbb{Z}_+^{|\mathscr{D}|}$ with $1 \leq \kappa_d \leq |\Lambda(d)|$. Then

$$C(\kappa) = \bigcup_{d \in \mathscr{D}} \mathscr{R}_{d,\lambda_{\kappa_d}}$$

defines the incidence matrix of the network corresponding to the selected solutions $\lambda_{\kappa_d} \in \Lambda(d)$ for each $d \in \mathscr{D}$. The complete list of all conformal networks is given in terms of their incidence matrices by

$$\mathscr{C} = \left\{ C(\kappa) : \kappa \in \left[1, \left|\Lambda(d)\right|\right]^{|\mathscr{D}|} \right\}.$$

Remark 14.1 Note that each $C(\kappa)$ indeed consists of the union of the selected sets $\mathscr{R}_{d,\lambda_{\kappa_d}}$ in the set-theoretic sense, hence $C(\kappa)$ does not contain any equal columns (and, thus, the resulting network $G(\kappa)$ no two transitions having identical update vectors). That way, it may happen that $C(\kappa) = C(\kappa')$ holds even for different selections $\kappa \neq \kappa'$. Hence, $\mathscr{C}$ may contain less different solution alternatives than the $\prod_{d \in \mathscr{D}} |\Lambda(d)|$ possible selections.

Example 14.4 We apply the last step of the reconstruction approach to the data given in Example 14.1. For that, recall from Example 14.3 that there are three difference vectors d^1, d^2, d^3 with the following different representations:

d^j	λ^i	$\mathscr{R}_{d^j,\lambda^i}$
d^1	λ^1	$\{r^7\}$
	λ^2	$\{r^9, r^{19}\}$
	λ^3	$\{r^9, r^{11}\}$
	$\lambda^4 = \lambda^1 + \eta$	$\{r^7, r^{19}, r^{21}\}$
	$\lambda^5 = \lambda^2 + \eta$	$\{r^9, r^{19}, r^{21}\}$
	$\lambda^6 = \lambda^3 + \eta$	$\{r^9, r^{11}, r^{19}, r^{21}\}$
d^2	λ^1	$\{r^9, r^{22}\}$
	λ^2	$\{r^{20}, r^{21}\}$
	$\lambda^3 = \lambda^1 + \eta$	$\{r^9, r^{19}, r^{21}, r^{22}\}$
	$\lambda^4 = \lambda^2 + \eta$	$\{r^{19}, r^{20}, r^{21}\}$
d^3	λ^1	$\{r^{17}\}$
	λ^2	$\{r^{15}, r^{21}\}$
	λ^3	$\{r^{13}, r^{21}\}$
	$\lambda^4 = \lambda^1 + \eta$	$\{r^{17}, r^{19}, r^{21}\}$
	$\lambda^5 = \lambda^2 + \eta$	$\{r^{15}, r^{19}, r^{21}\}$
	$\lambda^6 = \lambda^3 + \eta$	$\{r^{13}, r^{19}, r^{21}\}$

Thus, taking also the homogeneous solution η using the two vectors r^{19} and r^{21} into account yields 6 different representations for d^1, 4 different representations for d^2, and 6 different representations for d^3.

This leads to $6 \cdot 4 \cdot 6 = 144$ different selection vectors. For instance, the following matrices and networks are associated with the combinations corresponding to $\kappa = (1,1,1)^T$ and $\kappa' = (2,2,3)^T$:

$$C(\kappa) = \begin{pmatrix} -1 & 0 & 0 & 0 \\ 0 & 0 & 0 & -1 \\ -1 & 1 & -1 & 1 \\ 1 & -1 & 1 & -1 \end{pmatrix}, \qquad C(\kappa') = \begin{pmatrix} -1 & 0 & 0 & 0 & 0 \\ 0 & 0 & 0 & 0 & -1 \\ 0 & -1 & 1 & -1 & -1 \\ 0 & 1 & -1 & 1 & 1 \\ 0 & 1 & 0 & 0 & 0 \end{pmatrix}$$

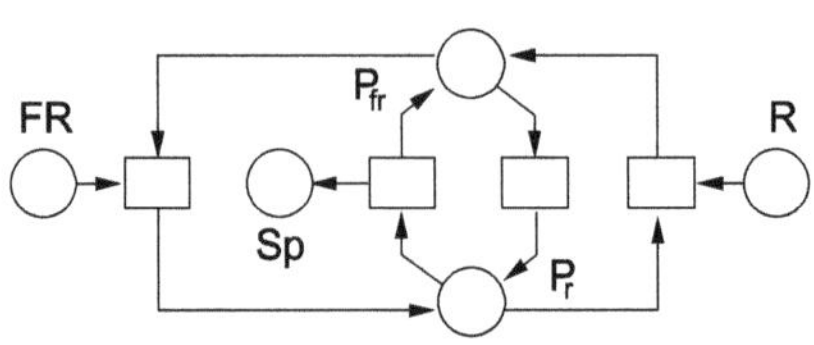

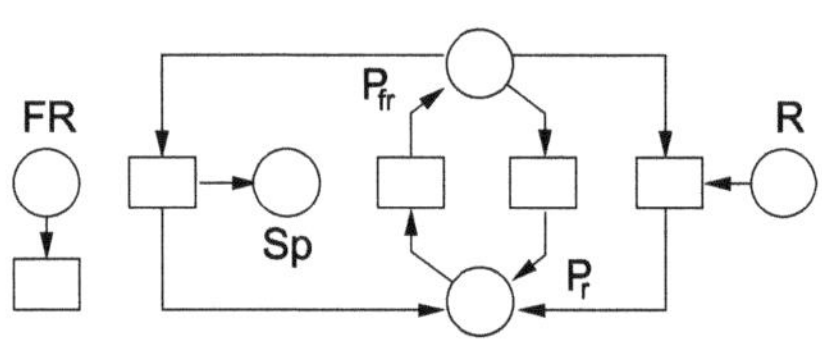

Note that not all possible selections yield different networks as for some choices, the reaction sets coincide (for instance, the matrix $C(\kappa')$ from above already contains both vectors from $\mathscr{R}_H$ such that any combination of the selected inhomogeneous solutions with η results in the same network).

For the sake of minimality, it was proposed in [245] to take into acount only these solutions in $\Lambda(d)$ having an inclusion-wise minimal set $\mathscr{R}_{d,\lambda}$ of used reactions. This leads to conformal networks also being minimal in the sense of involved reactions, and helps to not artificially increase the total number of conformal networks.

Using in the above example only these minimal inhomogeneous representations, there are 3 different representations for d^1, 2 different representations for d^2, and 3 different representations for d^3 left which can be combined to 18 (minimal) conformal networks (instead of 144 networks).

Remark 14.2 The homogeneous solution in the above example can be excluded by another argument, too. It can be experimentally observed that no other change between the two stages P_{fr} and P_r of the photoreceptor has been occurred than reported in the time series data. Hence, the values of P_{fr} and P_r must not oscillate in potential intermediate states between two consecutively measured states x^j and x^{j+1}. This excludes the homogeneous solution from the consideration. For the same reason, both (inhomogeneous) representations λ^1 and λ^2 of d^2 can be excluded. As any reconstructed network must use one of the representations of d^2, none of them is biologically meaningful in the sense that it truly reflects the observed phenomenon.

This shows that this general reconstruction approach is not appropriate to produce biologically meaningful networks as soon as the intermediate sequences linking two consecutively measured states of the experimental data are restricted.

14.2.2 An Approach for the Case of Monotone Data

We now turn to the case of experimental data where the network elements have been measured so accurately that an oscillation of their values between two measured states can be excluded a priori.

Definition 14.2 We say that the experimental data $\mathscr{X}'$ are *monotone* if for each $x^1 \in$ Ini and any two consecutively measured states $x^j, x^{j+1} \in \mathscr{X}(x^1)$ the intermediate states y^l of the sequence $x^j = y^1, y^2, \dots, y^k, y^{k+1} = x^{j+1}$ satisfy

- $y_p^1 \le y_p^2 \le \cdots \le y_p^k \le y_p^{k+1}$ for all $p \in P$ with $x_p^j \le x_p^{j+1}$ and
- $y_p^1 \ge y_p^2 \ge \cdots \ge y_p^k \ge y_p^{k+1}$ for all $p \in P$ with $x_p^j \ge x_p^{j+1}$.

In particular, the intermediate states of this sequence satisfy

$$y_p^1 = y_p^2 = \cdots = y_p^k = y_p^{k+1}$$

for all $p \in P$ with $x_p^j = x_p^{j+1}$ and belong to the box

$$\mathscr{X}(d) = \{y \in \mathscr{X} : y_p \in [x_p^j, x_p^{j+1}] \text{ for all } p \in P\}.$$

Remark 14.3 Note that in the general case, we have monotonicity for the places in $P_I \cup P_O$ as their values do certainly not oscillate (recall that the system can neither produce a token in a place in P_I nor consume a token from a place in P_O). Here, monotonicity is also assumed for all places in P_F. This is necessary if the experimental observations show that the values of the places in P_F do not oscillate in potential intermediate states between two consecutively measured states. This situation occurs if the measurements keep track of all local maxima and minima of the time course for every single component.

For the case of monotone data, Durzinsky et al. [105] showed that the general problem to represent a vector $d \in \mathscr{D}$ can be refined as follows [105]. As the values of the components can not oscillate in the intermediate states between x^j and x^{j+1}, it suffices to represent the vector $d = x^{j+1} - x^j$ using vectors $r \in \mathscr{R}$ only which are sign-compatible with d, that is, which belong to the following set

$$\text{Box}(d) = \left\{ r \in \mathbb{Z}^n : \begin{array}{r} 0 \le r_p \le d_p \text{ if } d_p > 0 \\ d_p \le r_p \le 0 \text{ if } d_p < 0 \\ r_p = 0 \text{ if } d_p = 0 \\ \sum_{p \in P'} r_p = 0 \ \ \forall P' \in \mathscr{P} \end{array} \right\}$$

where $\mathscr{P}$ is the family of all P-invariants $P' \subseteq P$ of the system. Note that in this setting, upper bounds u_p for the values $|r_p|$ are not required as we clearly have $|r_p| \le |d_p| \le |u_p|$ for all $p \in P$ due to monotonicity.

In fact, Durzinsky et al. [105] showed that no homogeneous solutions have to be considered in the monotone case and that the set $\Lambda(d)$ consists of all integral solutions λ of the system

$$d = \sum_{r^t \in \mathscr{R}_0(d)} \lambda_t r^t, \quad \lambda_t \in \mathbb{Z}_+ \tag{14.3}$$

using only vectors from

$$\mathscr{R}_0(d) = \text{Box}(d) \setminus \{r, d - r \in \text{Box}(d) : r \in \mathbb{Z}_+^{|P|}\}.$$

The reason is that none of the vectors $d - r \in \text{Box}(d)$ with $r \in \mathbb{Z}_+^{|P|}$ can be used in any representation (as the vectors in $r \in \mathbb{Z}_+^{|P|}$ are excluded from consideration a priori as they correspond to interface reactions).

Example 14.5 The experimental time series data $\mathscr{X}'$ described in Example 14.1 satisfy the monotonicity property, as it can be experimentally observed that no other change between the two stages P_{fr} and P_r of the photoreceptor occurs than reported in the time series data.

We now represent the difference vectors d^1, d^2, d^3 reflecting the responses of the studied system according to the above rules.

Recall that $x^2 - x^1 = d^1 = (-1,0,-1,1,0)^T$ holds. We construct the sets $\mathrm{Box}(d^1)$ and $\mathscr{R}_0(d^1)$ of reaction vectors and obtain:

	$\mathrm{Box}(d^1)$				$\mathrm{Box}(d^1)\cap\mathbb{Z}^5_+$	$\mathscr{R}_0(d^1)$		
FR	−1	−1	0	0	0	−1	−1	0
R	0	0	0	0	0	0	0	0
P_{fr}	−1	0	−1	0	0	−1	0	−1
P_r	1	0	1	0	0	1	0	1
Sp	0	0	0	0	0	0	0	0

Using the reactions from $\mathscr{R}_0(d^1)$ only yields the two representations

$$d^1 = \begin{pmatrix} -1\\ 0\\ -1\\ 1\\ 0 \end{pmatrix} = \begin{pmatrix} -1\\ 0\\ 0\\ 0\\ 0 \end{pmatrix} + \begin{pmatrix} 0\\ 0\\ -1\\ 1\\ 0 \end{pmatrix}$$

with $\mathscr{R}_{d^1,\lambda^1} = \{r^7\}$ and $\mathscr{R}_{d^1,\lambda^2} = \{r^9, r^{19}\}$ using the numbering of the reaction vectors from Example 14.2. Note that the third representation from Example 14.4 is not valid anymore as the involved reaction r^{11} does not belong to $\mathscr{R}_0(d^1)$.

For $x^3 - x^2 = d^2 = (0,0,0,0,1)^T$, the sets $\mathrm{Box}(d^2)$ and $\mathscr{R}_0(d^2)$ are as follows:

	$\mathrm{Box}(d^2)$		$\mathrm{Box}(d^2)\cap\mathbb{Z}^5_+$		$\mathscr{R}_0(d^2)$
FR	0	0	0	0	
R	0	0	0	0	
P_{fr}	0	0	0	0	
P_r	0	0	0	0	
Sp	0	1	0	1	

We have $\mathscr{R}_0(d^2) = \mathrm{Box}(d^2) \setminus \{r, d - r \in \mathrm{Box}(d^2) : r \in \mathbb{Z}^5_+\}$ by construction, which implies $\mathscr{R}_0(d^2) = \emptyset$ (as the reactions used for the representations in Example 14.4 are not present here). Hence, no representation of d^2 is possible anymore.

For $x^0 - x^4 = d^3 = (0,-1,1,-1,0)^T$, the sets $\mathrm{Box}(d^3)$ and $\mathscr{R}_0(d^3)$ are as follows:

	$\mathrm{Box}(d^3)$				$\mathrm{Box}(d^3)\cap\mathbb{Z}^5_+$	$\mathscr{R}_0(d^3)$		
FR	0	0	0	0	0	0	0	0
R	−1	−1	0	0	0	−1	−1	0
P_{fr}	0	1	0	1	0	0	1	1
P_r	0	−1	0	−1	0	0	−1	−1
Sp	0	0	0	0	0	0	0	0

Using the reactions from $\mathscr{R}_0(d^3)$ only yields the two representations

$$d^3 = \begin{pmatrix} 0 \\ -1 \\ 1 \\ -1 \\ 0 \end{pmatrix} = \begin{pmatrix} 0 \\ -1 \\ 0 \\ 0 \\ 0 \end{pmatrix} + \begin{pmatrix} 0 \\ 0 \\ 1 \\ -1 \\ 0 \end{pmatrix}$$

with $\mathscr{R}_{d^3,\lambda^1} = \{r^{17}\}$ and $\mathscr{R}_{d^3,\lambda^2} = \{r^{15}, r^{21}\}$. The third representation from Example 14.4 is not valid anymore as the involved reaction r^{13} does not belong to $\mathscr{R}_0(d^3)$.

Note that all representations from Example 14.4 not occurring anymore contradict the monotonicity for P_{fr} and P_r, as mentioned in Remark 14.2. Hence, taking monotonicity into account decreases only the number of inappropriate solution alternatives, while preserving all suitable solutions.

The ideas from [105] are generalized in [106] further as follows. Define the set

$$\mathscr{X}'_T = \left\{x^k \in \mathscr{X}\left(x^1\right) = \left\{x^0; x^1, \dots, x^k\right\} : x^1 \in \text{Ini}\right\} \cup \{\mathbf{0}\}$$

of *terminal states* in $\mathscr{X}'$ where no further reaction has been observed. Denoting by $T(x^k)$ the set of transitions enabled at state x^k, we have $T(x^k) = \emptyset$ for all $x^k \in \mathscr{X}'_T$ due to the experimental observation, and thus none of the reconstructed networks must contain a reaction from $T(x^k)$ for some $x^k \in \mathscr{X}'_T$. The impact is that we can reduce, for each $d \in \mathscr{D}$, the set of considered reaction vectors further by excluding all vectors from $\text{Box}(d)$ enabled at any terminal state, that is, all vectors from the set

$$\mathscr{R}_T(d) = \left\{r \in \text{Box}(d) : \exists x^k \in \mathscr{X}'_T \text{ with } r \in T\left(x^k\right)\right\}.$$

Remark 14.4 The vectors in $\text{Box}(d) \cap \mathbb{Z}_+^{|P|}$ correspond to interface reactions producing something without consuming any token. They can also be seen as reactions being enabled at the special terminal state $\mathbf{0}$ (where no action can occur apart from inputs to the system). Here, the set of vectors in $\text{Box}(d) \cap \mathbb{Z}_+^{|P|}$ is generalized to all vectors in $\mathscr{R}_T(d)$, and $\mathscr{R}_T(d) = \text{Box}(d) \cap \mathbb{Z}_+^{|P|}$ holds if and only if $\mathscr{X}'_T = \{\mathbf{0}\}$.

Furthermore, in [106] it is established that also none of the vectors $d - r$ with $r \in \mathscr{R}_T(d)$ can occur in any representation of d.

Theorem 14.1 [106] *For monotone data we have that, for all $d \in \mathscr{D}$, the set $\Lambda(d)$ consists of all integral solutions λ of the system*

$$d = \sum_{r^t \in \mathscr{R}(d)} \lambda_t r^t, \quad \lambda_t \in \mathbb{Z}_+ \tag{14.4}$$

using vectors from $\mathscr{R}(d) = \text{Box}(d) \setminus \{r, d - r : r \in \mathscr{R}_T(d)\}$ only.

As consequence, we deduce assertions for the existence and uniqueness of a representation for $d \in \mathscr{D}$.

Lemma 14.1 *For monotone data $\mathscr{X}'$, a vector $d \in \mathscr{D}$ with $d \neq \mathbf{0}$ has*

- *a representation if and only if $d \notin \mathscr{R}_T(d)$ holds;*
- *a unique representation if $d \notin \mathscr{R}_T(d)$ but $\mathscr{R}_T(d)$ contains a vector r differing from d in one component $q \in P$ by one, i.e., for $r = d + e^q$ or $r = d - e^q$.*

Remark 14.5 Note that the first assertion is equivalent to require $\mathscr{R}(d) \neq \emptyset$, as $d \in \mathscr{R}(d)$ holds whenever d is not enabled at any terminal state. In case of a unique representation, we have $\mathscr{R}(d) = \{d\}$ and d has only itself as representation. In particular, this happens in the special case that d has a unique negative entry $d_q = -1$, as the vector $d + e^q$ has only nonnegative entries and is enabled at $\mathbf{0} \in \mathscr{X}'_T$.

Corollary 14.1 *For monotone data $\mathscr{X}'$, there is a conformal network if and only if none of the vectors in $\mathscr{D}$ is enabled at any terminal state in $\mathscr{X}'_T$. If all vectors have a representation, then the vectors in $\mathscr{D}$ already represent a conformal network.*

Corollary 14.1 implies an efficient *feasibility test* for both the existence of a representation for $d \in \mathscr{D}$ and a conformal network: whenever a vector $d \in \mathscr{D}$ is enabled at a state in $\mathscr{X}'_T$, the problem is not solvable with the considered set P of components. This shows that some further components are involved in the studied system which have not been taken into account yet. Marwan et al. [245] proposed to use *additional components* in this situation.

We extend the component set accordingly by $P \cup P_A$ where P still contains the original places $p_1, \dots, p_n$ and P_A the additional places $p_{n+1}, \dots, p_{n+a}$. The n-dimensional vectors $x^j \in \mathscr{X}'$ and $d \in \mathscr{D}$ have to be extended to vectors $\overline{x}^j$ and $\overline{d}$ of dimension $n + a$, starting with unknown values for the additional components (as those components were not subject to experimental observation). We use an upper bound $u_p = 1$ for the capacity of each $p \in P_A$ (as we can only deal with the availability of the additional components).

We next outline, how the values of the additional components have to be determined according to [106]. More precisely, all the n-dimensional state vectors $x^j \in \mathscr{X}'$ are extended to suitable $(n + a)$-dimensional vectors

$$\overline{x}^j = \begin{pmatrix} x^j \\ z^j \end{pmatrix}$$

with $z^j \in [0, 1]^a$ (which then implies the corresponding extensions for the vectors in $\mathscr{D}$). Hence, we extend the potential state space accordingly to

$$\overline{\mathscr{X}} = \left\{\bar{x} = (x, z)^\top \in \mathbb{Z}^{n+a} : \mathbf{0} \le x \le u, \mathbf{0} \le z \le \mathbf{1}\right\}.$$

For that, we interpret the experimental data $\mathscr{X}' = \{\mathscr{X}(x^1) : x^1 \in \text{Ini}\}$ as a directed graph $D(\mathscr{X}') = (\mathscr{X}', A_D \cup A_P)$ having the measured states $x^j \in \mathscr{X}'$ as nodes and two kinds of arcs:

- $(x^j, x^{j+1}) \in A_D$ corresponding to measured differences if x^j, x^{j+1} are consecutive states in a sequence $\mathscr{X}(x^1) = (x^0; x^1, \ldots, x^j, x^{j+1}, \ldots, x^k\}$ with $j > 0$,
- $(x^0, x^1) \in A_P$ corresponding to perturbations if the initial state $x^1 \in \text{Ini}$ of some sequence is obtained by stimulating the system in some other state $x^0 \in \mathscr{X}'$.

The studied extensions $\overline{x}^j$ of the states $x^j \in \mathscr{X}'$ correspond to *labelings* of $D(\mathscr{X}')$:

- if $a = 1$, to (0,1)-labelings, where label i is assigned to node x^j if $\overline{x}^j_{n+1} = z^j = i$ is selected for $i \in \{0, 1\}$;
- if $a = 2$, to (0,1,2,3)-labelings, where the labels are assigned to the four different states $(0,0)^T$, $(0,1)^T$, $(1,0)^T$, and $(1,1)^T$,
- if $a \geq 3$, to labelings using similar encodings for all 2^a different 0/1-vectors.

However, not every labeling L of $D(\mathscr{X}')$ is appropriate, as some resulting vector $\overline{d}^j$ might not have a representation, a state $\overline{x}^k$ with $x^k \in \mathscr{X}'_T$ might have a successor state, a state $\overline{x}^j$ might have multiple successor states (recall that we assume reproducible data $\mathscr{X}'$), or some stimulation changes more than the target input component(s).

Definition 14.3 A labeling L of $D(\mathscr{X}')$ is *valid* if it satisfies the following conditions:

- for each $x^k \in \mathscr{X}'_T$, also $\overline{x}^k$ is a terminal state,
- none of the resulting differences $\overline{d}^j$ is enabled at any terminal state,
- there are no two paths from one state to different terminal states, and
- any stimulation preserves the values on the additional component(s).

Due to [106], the validity of a labeling L is ensured as follows:

- If $(x^j, x^{j+1}) \in A_D$ and $\bar{x}^j \neq \bar{x}^{j+1}$, then $\bar{x}^j \not\leq \bar{x}^{j+1}$ follows (to ensure that $\bar{d}^j = \bar{x}^{j+1} - \bar{x}^j$ has a negative entry).
- If $(x^j, x^{j+1}) \in A_D$ with $\bar{x}^j \neq \bar{x}^{j+1}$ and $x^k \in \mathscr{X}'_T$, then $\bar{x}^k + \bar{d}^j = \bar{x}^k + \bar{x}^{j+1} - \bar{x}^j \notin \overline{\mathscr{X}}$.
- If there are two directed paths $(x^i, \ldots, x^k)$ and $(x^j, \ldots, x^l)$ in A_D to different terminal states $x^k \neq x^l \in \mathscr{X}'_T$, then $\bar{x}^i \neq \bar{x}^j$.
- If $(x^i, x^j) \in A_S$, then $z^i = z^j$.

This implies some rules for the additional states z^j represented by a valid labeling.

Corollary 14.2 [106] *For a valid labeling L of $D(\mathscr{X}')$, we have*:

1. *If $(x^j, x^{j+1}) \in A_D$ and $x^j < x^{j+1}$ or $x^j = x^{j+1}$ but $z^j \neq z^{j+1}$ holds, then $z^j \neq \mathbf{0}$, $z^j \neq z^{j+1}$ and $z^{j+1} \neq \mathbf{1}$ follows.*
2. *Let $(x^j, x^{j+1}) \in A_D$ and $x^k \in \mathscr{X}'_T$ with $x^k + d^j = x^k + x^{j+1} - x^j \in \mathscr{X}$. Then $x^j \neq x^{j+1}$ or $z^j \neq z^{j+1}$ implies $z^k \neq z^j$, $z^k \neq \mathbf{1} - z^{j+1}$ and $z^j \neq z^{j+1}$.*
3. *Consider two directed paths $(x^i, \ldots, x^k)$ and $(x^j, \ldots, x^l)$ in A_D to different terminal states $x^k, x^l \in \mathscr{X}'_T$. If $x^k \neq x^l$ or $z^k \neq z^l$ then $z^i \neq z^j$ holds.*

Remark 14.6 We obtain even stronger implications for up to two additional components as Corollary 14.2 characterizes in that case the extensions as follows:

$$z^j \neq \mathbf{0}, \quad z^j \neq z^{j+1}, \quad z^{j+1} \neq \mathbf{1} \iff z^j \nleq z^{j+1}$$
$$z^j \neq z^k, \quad z^j \neq z^{j+1}, \quad z^{j+1} \neq \mathbf{1} - z^k \iff z^k + z^{j-1} + z^j \in \mathrm{Box}(\mathbf{1})$$

which can be observed by enumerating all combinations $z^k, z^j, z^{j+1} \in \mathrm{Box}(\mathbf{1})$.

For one additional component and $(x^j, x^{j+1}) \in A_D$ with $x^j < x^{j+1}$ resulting in a positive difference vector d^j, we can even directly derive $\overline{x}^j_{n+1} = z^j = 1$ and $\overline{x}^{j+1}_{n+1} = z^{j+1} = 0$ (as d^j requires as unique extension a negative entry in the new component).

We illustrate the construction of valid labelings with the help of an example.

Example 14.6 We reconsider the experimental time series data $\mathscr{X}'$ from Example 14.1. The previous example showed that the difference vector $d^2 = (0, 0, 0, 0, 1)^T$ has no representation. This can now be verified with the help of the feasibility test according to Lemma 14.1 and Corollary 14.1, as d^2 is enabled at the terminal state x^0 of the sequence $\mathscr{X}'(x^4) = \{x^2; x^4, x^0\}$.

We extend the component set by one additional element Z and choose as upper bound $u_Z = 1$. The 5-dimensional vectors $x^j \in \mathscr{X}'$ and $d^j \in \mathscr{D}$ have to be extended to appropriate vectors $\overline{x}^j$ and $\overline{d}^j$ of dimension 6.

As d^2 has a unique extension $\overline{d}^2 = (0, 0, 0, 0, 1, -1)^T$ by Remark 14.6, the extensions of the corresponding vectors from $\mathscr{X}'$ are fixed to $\overline{x}^2_Z = 1$ and $\overline{x}^3_Z = 0$. As any stimulation effects input components only, the value of the additional element $\overline{x}^4_Z$ equals the unchanged value of $\overline{x}^2_Z = 1$. Furthermore, we have $\overline{x}^0_Z = 0$ (as otherwise sporulation could occur without stimulation, a contradiction to the experimental dark control $\mathscr{X}'(x^0) = \{x^0\}$) and, thus, also $\overline{x}^1_Z = 0$. This leads to the scheme in Fig. 14.2, using vectors of the form $\overline{x}^T = (\overline{x}_{FR}, \overline{x}_R, \overline{x}_{P_{fr}}, \overline{x}_{P_r}, \overline{x}_{Sp}, \overline{x}_Z)^T$.

For the sake of minimality (i.e., in order to avoid an unnecessarily large number of network alternatives which are caused by artificial effects in the additional components), we shall further consider only those valid labelings L of $D(\mathscr{X}')$ (resp. extensions of the states in $\mathscr{X}'$) with a *minimal* number of label changes (or, equivalently, with a *maximal* number of resulting extensions $\overline{d}^j$ of vectors $d^j \in \mathscr{D}$ having $\overline{d}^j_{p_{n+i}} = 0$ for all $p_{n+i} \in P_A$). In other words, the additional elements should only be used during the reconstruction process if they are indeed required to explain an observation.

In order to find all valid labelings of $D(\mathscr{X}')$ (and hence all suitable extensions of the states in $\mathscr{X}'$) being *optimal* in this respect, Durzinsky et al. [106] proposed to set up an integer linear program encoding all the above rules and having a suitable objective function. Then all feasible solutions of this program correspond to all valid labelings of $D(\mathscr{X}')$, and the optimal solutions to the studied labelings with a minimal number of label changes.

The reconstruction process has to be applied to every optimal valid labeling.

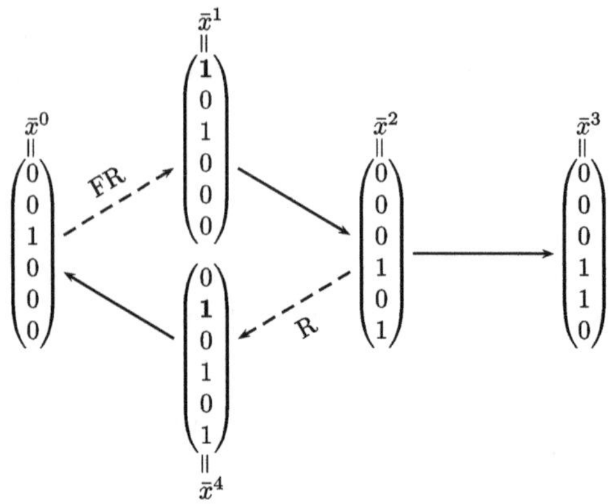

Fig. 14.2 A scheme illustrating the extended experimental setting (*dashed arrows* stand for stimulations, *solid arrows* for the observed responses of the system)

14.3 Results: A Reconstruction Algorithm

In this section, we outline the algorithm for reconstructing networks from monotone experimental time series data presented in [106]. This algorithm combines all the findings and results discussed in the previous section.

We first describe the input and the initial steps of the algorithm.

RECONSTRUCTION_FROM_MONOTONE_DATA

Input:
Set P of components and family $\mathscr{P} \subset 2^P$ of P-invariants;
Monotone time series data $\mathscr{X}' = \{\mathscr{X}(x^1), x^1 \in \text{Ini}\}$.

Initialization:
Explore $\mathscr{X}'$ and determine
Set $\mathscr{D} := \{d^j = x^{j+1} - x^j \;:\; j > 0,\; x^j, x^{j+1} \in \mathscr{X}'\}$ of difference vectors;
Set $\mathscr{X}'_T = \{x^k : \mathscr{X}(x^1) = \{x^0; x^1, \ldots, x^k\}, x^1 \in \text{Ini}\}$ of terminal states.

Feasibility Test:
IF no $d \in \mathscr{D}$ is enabled at a state in $\mathscr{X}'_T$
Call CONSTRUCT_NETWORKS($\mathscr{X}'_T, \mathscr{D}$).
ELSE
Call ADD_COMPONENT($\mathscr{X}', \mathscr{X}'_T, \mathscr{D}$).

The algorithm RECONSTRUCTION_FROM_MONOTONE_DATA takes as input a set P of components together with a family $\mathscr{P} \subset 2^P$ of subsets $P' \subseteq P$ corresponding to P-invariants, and experimental time series data $\mathscr{X}' = \{\mathscr{X}(x^1), x^1 \in \text{Ini}\}$ which are supposed to be monotone.

In an initialization step, it determines the set

$$\mathscr{D} = \left\{d^j = x^{j+1} - x^j : j > 0, x^j, x^{j+1} \in \mathscr{X}'\right\}$$

of difference vectors and, in addition, the set

$$\mathscr{X}_T' = \left\{x^k \in \mathscr{X}\left(x^1\right) = \left\{x^0; x^1, \ldots, x^k\right\} : x^1 \in \mathrm{Ini}\right\}$$

of terminal states in $\mathscr{X}'$ where no further reaction has been observed.

In a next step, we check for feasibility with the help of Lemma 14.1 or Corollary 14.1: If no vector $d \in \mathscr{D}$ is enabled at any terminal state in $\mathscr{X}_T'$, then no additional component is required and, hence, we can directly proceed with the subroutine CONSTRUCT_NETWORKS. Otherwise, at least one additional component is required and we call the subroutine ADD_COMPONENT to adjust the setting accordingly.

If the subroutine ADD_COMPONENT is called, we have to extend the component set $P = \{p_1, \ldots, p_n\}$ by additional components as follows.

```
ADD_COMPONENT (𝒳', 𝒳'_T, 𝒟)
Construct the directed graph D(𝒳').
Set a := 1
REPEAT
   Find the set ℒ of all optimal valid labelings L of D(𝒳')
   in dimension a.
   FOR ALL labelings L ∈ ℒ:
      Determine 𝒟(L) := {d̄^j : d^j ∈ 𝒟},
      Determine 𝒳'_T(L) := {x̄^k : x^k ∈ 𝒳'_T}.
      IF ∃d̄^j ∈ 𝒟(L) with d̄^j ∈ T(x̄) for some x̄ ∈ 𝒳'_T(L)
         Set ℒ := ℒ \ {L}
      ELSE
         Call CONSTRUCT_NETWORKS (𝒳'_T(L), 𝒟(L)).
   Set a := a + 1
UNTIL ℒ ≠ ∅
```

The subroutine ADD_COMPONENT takes as input the experimental data $\mathscr{X}'$, the sets of terminal states $\mathscr{X}_T'$ and of difference vectors $\mathscr{D}$. The subroutine first determines the directed graph $D(\mathscr{X}')$ corresponding to $D(\mathscr{X}') = (\mathscr{X}', A_D \cup A_P)$ having the measured states $x^j \in \mathscr{X}'$ as nodes and two kinds of arcs:

- $(x^j, x^{j+1}) \in A_D$ corresponding to measured differences
- $(x^0, x^1) \in A_P$ corresponding to perturbations.

The extensions $\overline{x}^j$ of the states $x^j \in \mathscr{X}'$ are determined according to Corollary 14.2 in terms of the set $\mathscr{L}$ of all *valid labelings* L of $D(\mathscr{X}')$ being optimal in the above sense (i.e., with the minimal number of label changes).

For each optimal valid labeling $L \in \mathscr{L}$, the corresponding extensions

$$\mathscr{D}(L) := \left\{\overline{d}^j : d^j \in \mathscr{D}\right\}$$

and

$$\mathscr{X}'_T(L) := \{\overline{x}^k : d^k \in \mathscr{X}'_T\}$$

of difference vectors and terminal states are constructed.

We also assume monotonicity for the additional components (to further avoid a large number of representations caused only by artificial effects due to the freedom of the values in the additional components). For each $L \in \mathscr{L}$, we call the subroutine CONSTRUCT_NETWORKS $(\mathscr{X}'_T(L), \mathscr{D}(L))$ separately.

The subroutine CONSTRUCT_NETWORKS $(\mathscr{X}'_T, \mathscr{D})$ is called either with the original sets $\mathscr{X}'_T$ of terminal states and $\mathscr{D}$ of difference vectors, or from the subroutine ADD_COMPONENT for each feasible case of extended settings $(\mathscr{X}'_T, \mathscr{D})$ resulting from an optimal valid labeling, where $\mathscr{X}'_T$ contains the extended terminal states and $\mathscr{D}$ the extended difference vectors. (This is possible as we assume monotonicity also for the additional components and, thus, there is no need to distinguish between the original and an extended setting in the reconstruction process.)

```
CONSTRUCT_NETWORKS(𝒳'_T, 𝒟)

Let C_o := ∅ and 𝒞 := ∅.

Representation of difference vectors:
FOR ALL d ∈ 𝒟:
   IF (d has a unique representation by Lemma 14.1)
      Let C_o := C_o ∪ {d} and 𝒟 := 𝒟 − {d}.
   ELSE
      Determine ℛ(d) = Box(d) \ {r, d − r : r ∈ ℛ_T(d)},
      Determine Λ(d) = {λ ∈ ℤ_+^{|ℛ(d)|} : d = Σ_{r^t ∈ ℛ(d)} λ_t r^t}.
      FOR ALL λ ∈ Λ(d):
         Determine ℛ_{d,λ} = {r^t ∈ ℛ(d) : λ_t > 0}.
      IF Λ(d) = {λ}:
         Let C_o := C_o ∪ ℛ_{d,λ} and 𝒟 := 𝒟 − {d}.

Combination of matrices:
FOR ALL κ ∈ ℤ_+^{|𝒟|} with 1 ≤ κ_d ≤ |Λ(d)|:
   Determine C(κ) = C_o ∪ ⋃_{d∈𝒟} ℛ_{d,λ_{κ_d}},
   Let 𝒞 := 𝒞 ∪ C(κ).

RETURN 𝒞.
```

The reconstruction is performed in the subroutine CONSTRUCT_NETWORKS as follows. For each $d \in \mathscr{D}$, we first check whether d has a unique representation with the help of Lemma 14.1.

If this is the case, we append d to the set of vectors C_o which are used in any conformal network, and remove d from $\mathscr{D}$. Otherwise, we determine the set $\mathscr{R}(d)$

of reactions which can be used to represent d. We have

$$\mathscr{R}(d) = \text{Box}(d) \setminus \{r, d - r : r \in \mathscr{R}_T(d)\}$$

as only vectors from $\text{Box}(d)$ can be used and all vectors r and $d - r$ with $r \in \mathscr{R}_T(d)$ can be excluded further by Theorem 14.1. (Note that the restrictions coming from P-invariants of the system are obeyed in the construction of $\text{Box}(d)$.)

In order to determine $\Lambda(d)$, we have to find all $\lambda \in \mathbb{Z}_+^{|\mathscr{R}(d)|}$ with $d = \sum_{r^t \in \mathscr{R}(d)} \lambda_t r^t$. For each $\lambda \in \Lambda(d)$, we select the set $\mathscr{R}_{d,\lambda} = \{r^t \in \mathscr{R}(d) : \lambda_t > 0\}$ of used reactions. If d has only one representation $\Lambda(d) = \{\lambda\}$, we append $\mathscr{R}_{d,\lambda}$ to the set of vectors C_o which are used in any conformal network, and remove d from $\mathscr{D}$.

In order to combine the incidence matrices of the resulting conformal networks, we first define a selection vector $\kappa \in \mathbb{Z}_+^{|\mathscr{D}|}$ with $1 \leq \kappa_d \leq |\Lambda(d)|$ for the difference vectors which remained in $\mathscr{D}$. For each $\kappa \in [1, |\Lambda(d)|]^{|\mathscr{D}|}$, we determine

$$C(\kappa) = C_o \cup \bigcup_{d \in \mathscr{D}} \mathscr{R}_{d,\lambda_{\kappa_d}}$$

as the incidence matrix of the network corresponding to the selected solutions $\lambda_{\kappa_d} \in \Lambda(d)$ for each $d \in \mathscr{D}$ (recall that all reaction vectors required for difference vectors with a unique representation are collected in C_o). We append $C(\kappa)$ to $\mathscr{C}$ and finally return $\mathscr{C}$ as complete list of all conformal networks w.r.t. $\mathscr{X}'_T$ and $\mathscr{D}$).

In [106], it is shown that the above algorithm indeed solves the Network Reconstruction Problem as it reconstructs all networks which are able to explain the observed phenomena.

Theorem 14.2 [106] *Given the experimental setting $(P, \mathscr{X}')$, the Network Reconstruction Algorithm generates all the networks $G = (P \cup T, F, W)$ being conformal with $\mathscr{X}'$.*

Note that all resulting conformal networks are minimal in the sense that only effects have been taken into account which are indeed crucial to explain the observed phenomenon. This avoids in fact to produce solutions which differ only due to some artificial effects, caused by uncertainty of the experimental data.

We finally illustrate the steps of the reconstruction algorithm with the help of our running example.

Example 14.7 As the experimental time series data $\mathscr{X}'$ described in Example 14.1 satisfy the monotonicity property, we can apply the above algorithm to reconstruct the complete list of (minimal) networks being conformal with $\mathscr{X}'$.

As initialization step, the algorithm explores $\mathscr{X}'$ in order to extract the sets $\mathscr{D} = \{d^1 = x^2 - x^1, d^2 = x^3 - x^2, d^3 = x^0 - x^4\}$ of difference vectors and $\mathscr{X}'_T = \{x^0, x^3\}$

of terminal states as

$$d^1 = \begin{pmatrix} -1 \\ 0 \\ -1 \\ 1 \\ 0 \end{pmatrix}, \quad d^2 = \begin{pmatrix} 0 \\ 0 \\ 0 \\ 0 \\ 1 \end{pmatrix}, \quad d^3 = \begin{pmatrix} 0 \\ -1 \\ 1 \\ -1 \\ 0 \end{pmatrix}, \quad \text{and}$$

$$x^0 = \begin{pmatrix} 0 \\ 0 \\ 1 \\ 0 \\ 0 \end{pmatrix}, \quad x^3 = \begin{pmatrix} 0 \\ 0 \\ 0 \\ 1 \\ 1 \end{pmatrix}.$$

In the feasibility test, the algorithm detects that d^2 is enabled at the terminal state x^0 and calls, therefore, the subroutine ADD_COMPONENT. Here, the component set is extended by one additional element Z and the 5-dimensional vectors $x^j \in \mathscr{X}'$ and $d^j \in \mathscr{D}$ are extended according to the above rules. For that, the directed graph $D(\mathscr{X}')$ is constructed (yielding the scheme shown in Fig. 14.1). As d^2 has a unique extension by Remark 14.6, we set $d^2_Z := -1$ and, thus, $x^2_Z := 1$ and $x^3_Z := 0$. Furthermore, we have $x^0_Z := 0$ by $x^0 \in \mathscr{X}'_T$ and $d^2 \in T(x^0)$. As a stimulation does not change values different from these of input components, this yields $x^1_Z := 0$ and $x^4_Z := 1$. Hence, applying the rules yields a unique valid labeling (corresponding to the scheme in Fig. 14.2) which is, thus, also optimal.

The algorithm calls the subroutine CONSTRUCT_NETWORKS with the following sets $\mathscr{D}$ of difference vectors and $\mathscr{X}'_T$ of terminal states:

$$d^1 = \begin{pmatrix} -1 \\ 0 \\ -1 \\ 1 \\ 0 \\ 1 \end{pmatrix}, \quad d^2 = \begin{pmatrix} 0 \\ 0 \\ 0 \\ 0 \\ 1 \\ -1 \end{pmatrix}, \quad d^3 = \begin{pmatrix} 0 \\ -1 \\ 1 \\ -1 \\ 0 \\ -1 \end{pmatrix}, \quad \text{and}$$

$$x^0 = \begin{pmatrix} 0 \\ 0 \\ 1 \\ 0 \\ 0 \\ 0 \end{pmatrix}, \quad x^3 = \begin{pmatrix} 0 \\ 0 \\ 0 \\ 1 \\ 1 \\ 0 \end{pmatrix}.$$

To find all representations for d^1, the algorithm tests first for a unique representation. As d^1 does not satisfy the conditions of Lemma 14.1, the set $\mathscr{R}(d^1)$ is determined:

	Box(d^1)						$\mathcal{R}_T(d^1)$		$\overline{\mathcal{R}}_T(d^1)$		$\mathcal{R}(d^1)$
FR	−1	−1	−1	−1	0	0	0	0	−1	−1	−1
R	0	0	0	0	0	0	0	0	0	0	0
P_{fr}	−1	−1	0	0	−1	−1	−1	−1	0	0	−1
P_r	1	1	0	0	1	1	1	1	0	0	1
Sp	0	0	0	0	0	0	0	0	0	0	0
Z	0	1	0	1	0	1	0	1	0	1	1

Here, we use the notation $\overline{\mathcal{R}}_T(d^i) = \{d - r : r \in \mathcal{R}_T(d^i) \setminus \{\mathbf{0}\}\}$. From $\mathcal{R}(d^1) = \{d^1\}$ it follows that d^1 has itself as unique representation, we append d^1 to the set of vectors C_o which are used in any conformal network, and remove d^1 from $\mathcal{D}$.

For d^2, the algorithm tests for a unique representation according to Lemma 14.1. As d^2 satisfies the conditions, we append d^2 to C_o and remove d^2 from $\mathcal{D}$.

To find all representations for d^3, the algorithm tests again for a unique representation. As d^3 does not satisfy the conditions, the set $\mathcal{R}(d^3)$ is determined as follows:

	Box(d^3)							$\mathcal{R}_T(d^3)$	$\overline{\mathcal{R}}_T(d^3)$	$\mathcal{R}(d^3)$				
FR	0	0	0	0	0	0	0	0	0	0	0	0	0	0
R	−1	−1	−1	−1	0	0	0	0	−1	−1	−1	−1	0	0
P_{fr}	0	0	1	1	0	1	1	1	0	0	1	1	0	1
P_r	0	0	−1	−1	0	−1	−1	−1	0	0	−1	−1	0	−1
Sp	0	0	0	0	0	0	0	0	0	0	0	0	0	0
Z	−1	0	−1	0	−1	−1	0	0	−1	0	−1	0	−1	−1

Then, the algorithm computes all representations of d^3 using vectors from $\mathcal{R}(d^3)$ which yields the following three alternatives:

$$d^3 = \begin{pmatrix} 0 \\ -1 \\ 1 \\ -1 \\ 0 \\ -1 \end{pmatrix} = \begin{pmatrix} 0 \\ -1 \\ 0 \\ 0 \\ 0 \\ 0 \end{pmatrix} + \begin{pmatrix} 0 \\ 0 \\ 1 \\ -1 \\ 0 \\ -1 \end{pmatrix} = \begin{pmatrix} 0 \\ -1 \\ 1 \\ -1 \\ 0 \\ 0 \end{pmatrix} + \begin{pmatrix} 0 \\ 0 \\ 0 \\ 0 \\ 0 \\ -1 \end{pmatrix}.$$

Hence, after the representation step of the subroutine CONSTRUCT_NETWORKS, we end up with $C_o = \{d^1, d^2\}$, $\mathcal{D} = \{d^3\}$, and the above three alternative representations of d^3. In the last step of the algorithm, C_o is finally combined with the three sets $\mathcal{R}_{d^3,\lambda}$, $\lambda \in \Lambda(d^3)$ to the incidence matrices of the three conformal networks shown in Fig. 14.3 having one transition for each used reaction vector.

This algorithm effectively solves the problem to reconstruct all conformal networks using up to two additional elements, if the experimental data are monotone (the running time of the algorithm exponentially grows for more than two additional components as determining the optimal valid labelings is the most expensive step).

For nonmonotone data, applying the above approach would mean that only these dynamical features of the studied system are taken into account which have become

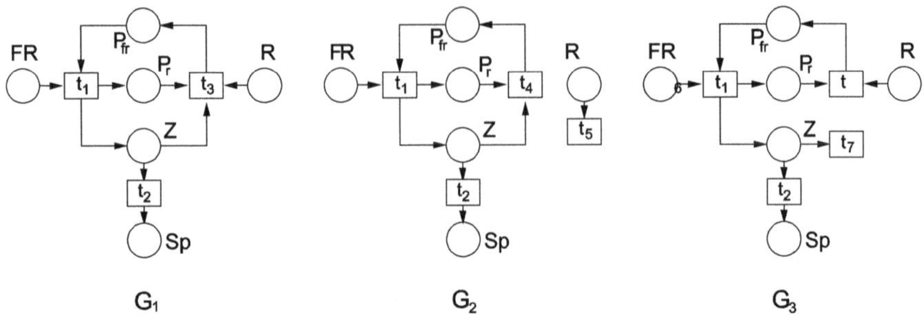

Fig. 14.3 The three reconstructed conformal networks

visible by the experimental observations. However, it is possible to adapt the original approach from [245] along the lines from the previous algorithm and further ideas from [105]. This should include to use monotonicity for all components where it can be guaranteed and to consider terminal states in order to keep the potential reaction sets $\mathscr{R}(d)$ as small as possible, and to discard homogeneous solutions from the consideration.

14.4 Related Work

Besides predicting the structure of the studied biological system in terms of networks being conformal with given experimental data, it is also interesting to consider the dynamic behavior of the studied system.

Petri nets describe dynamics in terms of reachability, by generating all possible sequences $x^0, \ldots, x^k$ of states which can be reached from an initial state x^0 by switching, in each state x^j, one transition among all transitions enabled at x^j in order to reach x^{j+1}. For states where at least two transitions are enabled, the decision between the different alternatives is typically taken nondeterministically.

If, however, the underlying biological system shows a deterministic behavior as a certain stimulation always results in the same response, additional activation rules are required in order to force the switch from a state x^j to its observed successor x^{j+1} (see [395, 396], or [408] for more details).

For this purpose, priorities between the transitions of the network can be used to determine which of the enabled transitions has to be taken. Note that these priorities typically reflect the rate of the corresponding reactions where the fastest reaction has highest priority. In [245], it is proposed to model such priorities with the help of partial orders on the set T of transitions of the network G. Here, a *partial order* O on T is a relation $\leq$ between pairs of elements of T respecting

- reflexivity (i.e., $t \leq t$ holds for all $t \in T$),
- transitivity (i.e., from $t \leq t'$ and $t' \leq t''$ follows $t \leq t''$ for all $t, t', t'' \in T$), and
- anti-symmetry (i.e., $t \leq t'$ and $t' \leq t$ implies $t = t'$).

In Torres et al. [397], it is shown that appropriate partial orders correspond to acyclic orientations of a *conflict graph* $C(T) = (T, E)$, where each node represents a transition $t_i \in T$ and edges conflicts between pairs of transitions in the sense that

$$tt' \in E \iff \exists x \text{ with } t, t' \in T(x)$$

holds. Let $B \subseteq P$ be the set of places of the network G having bounded capacities and u_B be the capacity vector, where u_p is the maximum number of tokens allowed for place $p \in B$. Further, recall that the set of pre-places of a transition t is $^\bullet(t)$, and let $(t)^\bullet_B$ denote the set of bounded post-places of t. Then the conflicts between pairs of transitions can be characterized as follows.

Lemma 14.2 [396] *In a network $G = (P \cup T, F, W)$ with capacity vector u_B, two transitions $t, t' \in T$ are in conflict if and only if*

$$W(p,t) \leq u_p^B - W(t', p) \ \forall p \in^\bullet (t) \cap (t')^\bullet_B \quad \textit{and}$$
$$W(p,t') \leq u_p^B - W(t, p) \ \forall p \in^\bullet (t') \cap (t)^\bullet_B \quad \textit{holds.}$$

In the special case $B = P$ and $u_p = 1$ for all $p \in P$, two transitions $t, t' \in T$ are in conflict if and only if $^\bullet(t) \cap (t')^\bullet_B = \emptyset$ and $^\bullet(t') \cap (t)^\bullet_B = \emptyset$. Based on these characterizations, an algorithm to create $C(T)$ is discussed in Torres et al. [397].

Using this concept, we derive a stronger version of conformality with given data:

Definition 14.4 A network $G = (P \cup T, F, W)$ with incidence matrix $C \in \mathbb{Z}^{|P|\times|T|}$ and conflict graph $C(T)$ is *strongly conformal* with experimental time series data $\mathscr{X}'$ if, for any two consecutive states $x^j, x^{j+1} \in \mathscr{X}'$, the linear equation system

$$x^{j+1} - x^j = d = C\lambda$$

has one integral solution $\lambda \in \mathbb{Z}_+^{|T|}$ and the edges of $C(T)$ can be oriented in such a way that $t \leq t'$ holds for all $t \in \mathscr{R}_{d,\lambda}$ and $t' \in T(x^j) - \mathscr{R}_{d,\lambda}$.

Thus, a strongly conformal network does not only contain appropriate transitions to reach x^{j+1} from the previously measured state x^j in $\mathscr{X}'$, but uses in each intermediate step this transition with highest priority.

If a biological system shows a deterministic behavior in the experimental data $\mathscr{X}'$, then a conformal network $G = (P \cup T, F, W)$ is *not* strongly conformal with $\mathscr{X}'$ if for some pair of transitions in T, opposite priorities are required to fit the experimental observations. The impact of considering this stronger version of conformality is, therefore, that we can rule out conformal networks which are not strongly conformal to be appropriate models of systems with a deterministic behavior.

Example 14.8 Consider the three networks $G_i = (P \cup T_i, F, W)$ from Fig. 14.3 with

$$T_1 = \{t_1, t_2, t_3\}, \qquad T_2 = \{t_1, t_2, t_4, t_5\}, \qquad T_3 = \{t_1, t_2, t_6, t_7\}$$

all being conformal with the experimental data $\mathcal{X}'$ described in Example 14.1. For all three networks, we check the experiments $x^1 \to x^2 \to x^3$ and $x^4 \to x^0$.

In all three networks, $T(x^1) = \{t_1\}$ holds and switching t_1 indeed yields x^2.

In the case of network G_1, we obtain the following. $T(x^2) = \{t_2\}$ holds and switching t_2 indeed yields x^3. Furthermore, we have $T(x^4) = \{t_2, t_3\}$ and t_3 is forced to switch in order to reach x^0, which requires that $t_3 > t_2$. As no other priority is needed, G_1 is strongly conformal with $\mathcal{X}'$.

In the case of network G_2, $T(x^2) = \{t_2, t_5\}$ holds and t_2 is forced to switch in order to reach x^3, which implies $t_2 > t_5$. On the other hand, we have $T(x^4) = \{t_2, t_4, t_5\}$ and t_4, t_5 have to switch in order to reach x^0, which requires $t_4 > t_2$ and $t_5 > t_2$. As opposite priorities for the pair t_2, t_5 are needed to simulate the experiments, G_2 is not strongly conformal with $\mathcal{X}'$.

For network G_3, $T(x^2) = \{t_2, t_7\}$ holds and t_2 is again forced to switch in order to reach x^3, which implies $t_2 > t_7$. We further have $T(x^4) = \{t_2, t_6, t_7\}$ where t_6, t_7 have to switch in order to reach x^0, which requires $t_6 > t_2$ and $t_7 > t_2$. As opposite priorities for the pair t_2, t_7 are needed to simulate the experiments, G_3 is not strongly conformal with $\mathcal{X}'$ as well.

Hence, only one of the three conformal networks is strongly conformal and explains the experimentally observed behavior in a suitable way. The remaining network G_1 is exactly the expected one, but comes here with the certificate of being the unique explanation (which is not possible using heuristical reasoning).

14.5 Summary

We addressed in this chapter different ways to solve the Network Reconstruction Problem, the challenging task to generate all models that explain the observed phenomena. Using the framework of Petri nets to describe models for the regulatory mechanisms of biological systems, this means to predict all the possible network structures being conformal with the given experimental data.

In Sect. 14.2, we discussed two approaches for solving the Network Reconstruction Problem, depending on the kind and quality of the given data $\mathcal{X}'$, which both construct the complete set of networks being conformal with the given data.

We first outlined in Sect. 14.2.1 the principle ideas of a combinatorial reconstruction approach proposed by Marwan et al. [245] for the general case, where no information is known on the intermediate states between two consecutively measured states in $\mathcal{X}'$. Here, all possible changes of the values in the intermediate states have to be taken into account which typically leads to a large number of conformal networks, even if homogeneous solutions are not considered for the sake of minimality. In addition, it turned out that this approach is not appropriate as soon as the intermediate sequences linking two consecutively measured states of the experimental data are restricted (Example 14.4 and Remark 14.2).

In Sect. 14.2.2, we discussed the approach from Durzinsky et al. [105] for the case of monotone data, where the network elements have been measured so accurately that an oscillation of their values between two measured states can be excluded a priori. This restricts the complexity of the problem as for the representations of a difference vector d, exclusively reactions from the restricted set $\mathscr{R}_0(d)$ have to be used which can even be restricted further to $\mathscr{R}(d)$ by taking terminal states into account. In addition, we have a *feasibility test* for the existence of a conformal network: the problem is not solvable with the considered set P of components if and only if $\mathscr{D}$ contains a vector being enabled at a terminal state. In this situation, additional nonobserved elements are required to provide us with a meaningful model. Thus, the additional elements help to construct conformal networks. The drawback is that the total number of conformal networks increases due to artificial effects caused by the freedom in the additional components.

In order to avoid such effects, the reconstruction algorithm from Durzinsky et al. [106] presented in Sect. 14.3 employs some further assumptions and conditions in order to keep the number of those conformal networks as small as possible which only show some artificial effects or can be ruled out with the help of a deeper analysis of the experimental data. For that,

- reactions are excluded from the reconstruction process that are enabled in a terminal state of $\mathscr{X}'$,
- if additional elements are required, only valid labelings with a minimal number of label changes are considered, and
- monotonicity is also assumed for the additional elements.

The resulting conformal networks are minimal in the sense that only effects have been taken into account which are indeed crucial to explain the observed phenomenon (Example 14.7).

In Sect. 14.4, we further discussed a stronger version of conformality which also takes some dynamic aspects into account, as suggested in Marwan et al. [245] and refined by Torres et al. [395] and Torres and Wagler [396]. That way, we can rule out further networks as appropriate models for the studied system which are conformal but not strongly conformal with the given data (Example 14.8).

Thus, we conclude that the proposed approach provides a powerful tool for effectively predicting network structures for biological systems. The art is, however, to find suitable assumptions and conditions for the reconstruction process to exclude as many artificial aspects as possible, but still generating all meaningful network alternatives being (strongly) conformal with the given data.

14.6 Exercises and Solutions

Exercise 14.1 Explain why the difference vectors in Example 14.3 were represented using a subset of appropriate reaction vectors from the set $\mathscr{R}$ constructed in Example 14.2.

Exercise 14.2 Consider the different representations for the difference vectors d^1, d^2, d^3 presented in Example 14.4. Determine how many different networks out of the 144 possible selections indeed occur. How many different networks remain if the representations involving the homogeneous solution are discarded?

(Hint: start with the second question and then determine which variants of these solutions can be obtained by reconsidering homogeneous solutions.)

Exercise 14.3 Which rules have been applied to obtain the valid labeling in Example 14.6?

Exercise 14.4 Consider the difference vector d^2 and its representation in Example 14.7. First, construct the sets $\mathrm{Box}(d^2)$, $\mathscr{R}_T(d^2)$, and $\overline{\mathscr{R}_T}(d^2)$ explicitly to verify that $\mathscr{R}(d^2) = \emptyset$ holds. Then, check for its extension $\overline{d}^2$ that $\mathscr{R}(\overline{d}^2) = \{\overline{d}^2\}$ holds by explicitly constructing the sets $\mathrm{Box}(\overline{d}^2)$, $\mathscr{R}_T(\overline{d}^2)$, and $\overline{\mathscr{R}_T}(\overline{d}^2)$.

Exercise 14.5 Consider the extended difference vectors $d^1, d^2, d^3 \in \mathbb{Z}^6$ from Example 14.7 and construct the set $\mathscr{R}(d^j)$ for each of them without taking the P-invariant $P' = \{P_{fr}, P_r\}$ into account. Discuss the effect on the number of resulting reaction vectors and representations for the difference vectors as well as the number of resulting conformal networks.

Exercise 14.6 Check the new network alternatives from Exercise 14.5 obtained by dropping the P-invariant $P' = \{P_{fr}, P_r\}$ for strong conformality.

Glossary

Here, we sum up what has been compiled in this book. Notions and notations presented here will be applied all over the rest of this book.

A. Static Structure

Petri net $N = (P, T, F, W, m_0)$:

P finite set of *places*	graphically: circles
T finite set of *transitions*	graphically: squares
$F \subseteq (P \times T) \cup (T \times P)$ *flow relation*	graphically: arrows
$W : F \to \mathbb{N}$ *arc weight*	graphically: inscriptions
$m_0 : P \to \mathbb{N}$ *initial marking*	graphically: dots ("tokens")

Derived Notions for N:

$E =_{\text{def}} P \cup T$ set of *elements*

For $e \in E$: ${}^{\bullet}e =_{\text{def}} \{d \mid (d, e) \in F\}$ *pre-set* of e
$e^{\bullet} =_{\text{def}} \{d \mid (e, d) \in F\}$ *post-set* of e

$\overline{W} : E \times E \to \mathbb{N}$ "*extended arc weight*" with

$$\overline{W}(d, e) = \begin{cases} W(d, e), & \text{if } (d, e) \in F \\ 0, & \text{otherwise} \end{cases}$$

B. Dynamics

$m : P \to \mathbb{N}$ *marking*.
m *enables* t iff for all $p \in {}^{\bullet}t$ $W(p, t) \leq m(p)$.
$m \xrightarrow{t} m'$ is a *step* iff

(a) m enables t, and
(b) for all $p \in P$ $m'(p) - \overline{W}(p, t) + \overline{W}(t, p)$.

$\sigma : m_0 \xrightarrow{u_1} m_1 \xrightarrow{u_2} \cdots \xrightarrow{u_n} m_n$ is a *computation of* N iff $m_{i-1} \xrightarrow{u_i} m_i$ are steps ($i = 1, 2, \ldots, n$).

I. Koch et al. (eds.), *Modeling in Systems Biology*, Computational Biology 16,
DOI 10.1007/978-1-84996-474-6, © Springer-Verlag London Limited 2011

A marking m is *reachable in* N iff some computation ends at m.
σ *reproduces* m_0 if $m_n = m_0$.
The *counting vector* $c : T \to \mathbb{N}$ of σ assigns each t the number of occurrences of t (i.e., of indices i with $u_i = t$) in σ.

C. Linear-Algebraic Representation

Assume *any* order on P and on T. Typically: along indices in $P = \{p_1, \ldots, p_k\}$, $T = \{t_1, \ldots, t_l\}$.
For $m : P \to T$ and $t \in T$ let

$$\underline{m} : \begin{pmatrix} m(p_1) \\ \vdots \\ m(p_k) \end{pmatrix} \quad \text{and} \quad \underline{t} : \begin{pmatrix} \overline{W}(t, p_1) - \overline{W}(p_1, t) \\ \vdots \\ \overline{W}(t, p_k) - \overline{W}(p_k, t) \end{pmatrix}$$

If t is *enabled at* m, then $\underline{m} \xrightarrow{t} \underline{m} + \underline{t}$ is a step.
$\underline{N} : (\underline{t_1} \ldots \underline{t_k})$ is the *matrix* of N.

D. Analysis Techniques

(a) $i : P \to \mathbb{Z}$ is a *place invariant* of N iff i solves $x \cdot \underline{N} = 0$.
Theorem 3.1 For each reachable m, $i \cdot m = i \cdot m_0$.
(b) $j : T \to \mathbb{N}$ is a *transition invariant* of N iff j solves $\underline{N} \cdot x = 0$.
Theorem 3.2 If j is the counting vector of a step sequence, it reproduces the initial marking.
(c) A set Q of places is a *trap* iff $Q^{\bullet} \subseteq {}^{\bullet}Q$.
Observation A trap never runs empty.
(d) The *marking graph* of N has the reachable states and steps as vertices and edges.
Observation If finite, this graph decides termination, divergence, liveness, weak liveness, boundedness and reversibility.

References

1. Ackermann, J., Wlotzka, B., McCaskill, J.S.: In vitro DNA-based predator-prey system with oscillatory kinetics. Bull. Math. Biol. **60**, 329–354 (1998)
2. Agapakis, C., Silver, P.: Synthetic biology: exploring and exploiting genetic modularity through the design of novel biological networks. Mol. Biosyst. **7**(5), 704–713 (2009)
3. Akashi, M., Takumi, T.: The orphan nuclear receptor Rora regulates circadian transcription of the mammalian core-clock Bmal1. Nat. Struct. Mol. Biol. **12**, 441–448 (2005)
4. Akutsu, T., Miyano, S., Kuhara, S.: Identification of genetic networks from small number of gene expression patterns under the Boolean network model. Proc. Pac. Symp. Biocomp. **4**, 17–28 (1999)
5. Albert, R., Othmer, H.G.: The topology of the regulatory interactions predicts the expression pattern of the segment polarity genes in drosophila melanogaster. J. Theor. Biol. **223**(1), 1–18 (2003)
6. Alberts, B., Johnson, A., Lewis, J., Raff, M., Roberts, K., Walter, P.: Molecular Biology of the Cell, 4th edn. Taylor & Francis, London (2002)
7. Aldridge, B.B., Saez-Rodriguez, J., Muhlich, J.L., Sorger, P.K., Lauffenburger, D.A.: Fuzzy logic analysis of kinase pathway crosstalk in TNF/EGF/insulin-induced signaling. Comput. Biol. **5**(4), e1000340 (2009)
8. Alexopoulos, C.: Statistical analysis of simulation output: state of the art. In: WSC'07: Proceedings of the 39th conference on Winter simulation, pp. 150–161. IEEE Press, New York (2007)
9. Alfarano, C., Andrade, C.E., Anthony, K., et al.: The Biomolecular Interaction Network Database and related tools 2005 update. Nucleic Acids Res. **33**, D418–D424 (2005)
10. Alla, H., David, R.: Continuous and hybrid Petri nets. JCSC **8**, 159–188 (1998)
11. Alon, U.: An Introduction to Systems Biology. CRC Press, Boca Raton (2006)
12. Alon, U.: Network motifs: theory and experimental approaches. Nat. Rev. Genet. **8**(6), 450–461 (2007)
13. Alvarez-Buylla, E.R., Benítez, M., Dávila, B., Chaos, Á., Espinosa-Soto, C., Padilla-Longoria, P.: Gene regulatory network models for plant development. Curr. Opin. Plant Biol. **10**, 83–91 (2007)
14. Alves, R., Antunes, F., Salvador, A.: Tools for kinetic modelling of biochemical networks. Nat. Biotechnol. **24**, 667–672 (2006)
15. Arkin, A.J., Ross, J., McAdams, H.H.: Gene Ontology: tool for the unification of biology. Genetics **149**(4), 1633–1648 (1998)
16. Ashburner, M., Ball, C.A., Blake, J.A., et al.: Gene Ontology: tool for the unification of biology. Nat. Genet. **25**, 25–29 (2000)
17. Ashyraliyev, M., Fomekong-Nanfack, Y., Kaandorp, J.A., Blom, J.G.: Systems biology: parameter estimation for biochemical models. FEBS J. **276**, 886–902 (2009)
18. Atkins, P.W., De Paula, J.: Physical Chemistry, 7th edn. John Wiley, New York (2002)

I. Koch et al. (eds.), *Modeling in Systems Biology*, Computational Biology 16,
DOI 10.1007/978-1-84996-474-6, © Springer-Verlag London Limited 2011

19. Backhaus, K., Erichson, B., Plinke, W., Weiber, R.: Multivariate analysis methods. An application-oriented introduction, 10th edn. Springer, Berlin (2000) (in German)
20. Bagheri, N., Stelling, J., Doyle, F.J. III: Quantitative performance metrics for robustness in circadian rhythms. Bioinformatics **23**(3), 358–364 (2007)
21. Bahi-Jaber, N., Pontier, D.: Modeling transmission of directly transmitted infectious diseases using colored stochastic Petri nets. Math. Biosci. **185**(1), 1–13 (2003)
22. Banks, R., Steggles, L.J.: A high-level Petri net framework for genetic regulatory networks. J. Integr. Bioinform. **4**(3), 1–12 (2007)
23. Banks, R., Khomenko, V., Steggles, L.J.: A case for using signal transition graphs for analysing and refining genetic networks. Electron. Notes Theor. Comput. Sci. **227**, 3–19 (2008)
24. Banks, R.: Qualitatively modelling genetic regulatory networks: Petri net techniques and tools. Ph.D. thesis, School of Computing Science, Newcastle University, upon Tyne (2009)
25. Barabasi, A.-L., Oltvai, Z.N.: Network biology: understanding the cell's functional organization. Nat. Genet. **5**(2), 101–113 (2004)
26. Bard, J., Rhee, S.Y., Ashburner, M.: An ontology for cell types. Genome Biol. **6**, R21 (2005)
27. Bassett, D.E.Jr., Eisen, M.B., Boguski, M.S.: Gene expression informatic—it's all in your mine. Nat. Genet. **21**, 51–55 (1999)
28. Basu, S., Gerchman, Y., Collins, C.H., Arnold, F.H., Weiss, R.: A synthetic multicellular system for programmed pattern formation. Nature **434**(7037), 1130–1134 (2005)
29. Bause, F., Kemper, P., Kritzinger, P.: Abstract Petri net notation. Forschungsbericht **563**, Fachbereichs Informatik, Universität Dortmund (Germany) (1994). Also in: Petri Net Newsletter **49**, 9–27 (1995)
30. Becker-Weimann, S., Wolf, J., Herzel, H., Kramer, A.: Modeling feedback loops of the mammalian circadian oscillator. Biophys. J. **87**, 3023–3034 (2004)
31. Becker-Weimann, S., Wolf, J., Kramer, A., Herzel, H.: A model of the mammalian circadian oscillator including the REV-ERBα module. Genome Inform. **15**(1), 3–12 (2004)
32. Behre, J., Schuster, S.: Modelling signal transduction in enzyme cascades with the concept of elementary flux modes. J. Comput. Biol. **16**(6), 829–844 (2009)
33. Behre, J., Wilhelm, T., von Kamp, A., Ruppin, E., Schuster, S.: Structural robustness of metabolic networks with respect to multiple knockouts. J. Theor. Biol. **252**, 433–441 (2008)
34. Bhalla, U.S., Ram, P.T., Iyengar, R.: MAP kinase phosphatase as a locus of flexibility in a mitogen-activated protein kinase signaling network. Science **297**(5583), 1018–1023 (2002)
35. Bhutkar, A.: Synthetic biology: navigating the challenges ahead. J. BioLaw Bus. **8**, 19–29 (2005)
36. Biebricher, Ch.K., Eigen, M., Gardiner, W.C. Jr.: Kinetics of RNA replication. Biochemistry **22**, 2544 (1983)
37. Biedl, T.C., Madden, B., Tollis, I.G.: The three-phase method: a unified approach to orthogonal graph drawing. In: Proceedings international symposium on graph drawing. Lecture Notes in Computer Science, vol. 1353, pp. 391–402. Springer, Berlin (1997)
38. Bisswanger, H.: Enzyme Kinetics: Principles and Methods. Wiley-VHC, New York (2008)
39. Blazewicz, J., Formanowicz, D., Formanowicz, P., Sackmann, A., Sajkowski, M.: Modeling the process of human body iron homeostasis using a variant of timed Petri nets. Discrete Appl. Math. **157**(10), 2221–2231 (2009)
40. Bock, R.: Cell and molecular biology of plastids. In: Topics in Current Genetics, vol. 19. Springer, Berlin (2007)
41. Bortfeldt, R., Schuster, S., Koch, I.: Exhaustive analysis of the modular structure of the spliceosomal assembly network—a Petri net approach. In Silico Biol. **10**, 0007 (2010)
42. Bosl, W.: Systems biology by the rules: hybrid intelligent systems for pathway modeling and discovery. BMC Syst. Biol. **1**(1), 13 (2007)
43. Bower, J.M., Bolouri, H.: Computational Modelling of Genetic and Biochemical Networks. MIT Press, Cambridge (2001)
44. Bowtell, D.D.: Options available—from start to finish—for obtaining expression data by microarray. Nat. Genet. **21**, 25–32 (1999)

45. Box, G.E.P., Draper, N.R.: Empirical Model-Building and Response Surfaces. Wiley Series in Probability and Statistics, p. 424. Wiley, New York (1987)
46. Boyle, J.: Gene-Expression Omnibus integration and clustering tools in SeqExpress. Bioinformatics **21**, 2550–1551 (2005)
47. Brazma, A., Hingamp, P., Quackenbush, J., et al.: Minimum information about a microarray experiment (MIAME)-toward standards for microarray data. Nat. Genet. **29**, 365–371 (2001)
48. Breeding, K.J.: Digital Design Fundamentals. Prentice Hall, New York (1992)
49. Breitling, R., Gilbert, D., Heiner, M., Orton, R.: A structured approach for the engineering of biochemical network models, illustrated for signalling pathways. Brief. Bioinform. **9**(5), 404–421 (2008)
50. Brondello, J.-M., Pouysségur, J., McKenzie, F.R.: Reduced MAP kinase phosphatase-1 degradation after p42/p44MAPK-dependent phosphorylation. Science **286**(5449), 2514–2517 (1999)
51. Broy, M.: The structure and function of plastids. In: Wise, R.R., Hoober, K.J. (eds.) Advances in Photosynthesis and Respiration. Springer, Dordrecht (2006)
52. Bryant, R.E.: Graph-based algorithms for boolean function manipulation. IEEE Trans. Comput. **35**(8), 677–691 (1986)
53. Burgard, A.P., Nikolaev, E.V., Schilling, C.H., Maranas, C.D.: Flux coupling analysis of genome-scale metabolic network reconstructions. Genome Res. **14**(2), 301–312 (2004)
54. Cantor, C.R., Schimmel, P.R.: Biophysical Chemistry. Part III, The Behavior of Biological Macromolecules. Freeman, New York (1980)
55. Cao, Y., Li, H., Petzold, L.: Efficient formulation of the stochastic simulation algorithm for chemically reacting systems. J. Chem. Phys. **121**(9), 4059–4067 (2004)
56. Cao, Y., Gillespie, D.T., Petzold, L.R.: Avoiding negative populations in explicit Poisson tau-leaping. J. Chem. Phys. **123**(5), 054104–054108 (2005)
57. Caspi, R., Foerster, H., Fulcher, C.A., Kaipa, P., Krummenacker, M., Latendresse, M., Paley, S., Rhee, S.Y., Shearer, A.G., Tissier, C., Walk, T.C., Zhang, P., Karp, P.D.: The MetaCyc Database of metabolic pathways and enzymes and the BioCyc collection of Pathway/Genome Databases. Nucleic Acids Res. **36**, D623–D631 (2008)
58. Chiola, G., Franceschinis, G., Gaeta, R., Ribaudo, M.: GreatSPN 1.7: graphical editor and analyzer for timed and stochastic Petri nets. Perform. Eval. **24**, 47–68 (1995)
59. Cell Illustrator Online 4.0: http://cionline.hgc.jp/
60. Cell Illustrator Online: http://www.cellillustrator.org
61. Cell System Markup Language: http://www.csml.org/
62. Chang, S., Johnston, R.J.Jr., Frokjaer-Jensen, C., Lockery, S., Hobert, O.: MicroRNAs act sequentially and asymmetrically to control chemosensory laterality in the nematode. Nature **430**, 785–789 (2004)
63. Chaouiya, C.: Petri net modelling of biological networks. Brief. Bioinform. **8**(4), 210–219 (2007)
64. Chaouiya, C., Remy, E., Mossé, B., Thieffry, D.: Qualitative analysis of regulatory graphs: a computational tool based on a discrete formal framework. LNCIS **294**, 119–126 (2003)
65. Chaouiya, C., Remy, E., Ruet, P., Thieffry, D.: Qualitative modelling of genetic networks: from logical regulatory graphs to standard Petri nets. In: LNCS, vol. 3099, pp. 137–156. Springer, Berlin (2004)
66. Chaouiya, C., Remy, E., Thieffry, D.: Qualitative Petri net modelling of genetic networks. In: Trans. Comp. Syst. Biol. VI. LNCS, vol. 4220, pp. 95–112. Springer, Berlin (2006)
67. Chaouiya, C., Remy, E., Thieffry, D.: Petri net modelling of biological regulatory networks. J. Discret. Algorithms **6**(2), 165–177 (2008)
68. Chaouiya, C., Naldi, A., Remy, E., Thieffry, D.: Petri net representation of multi-valued logical regulatory networks. Nat. Comput. (2010). doi:10.1007/s1104-010-9178-0
69. Chaves, M., Albert, R., Sontag, E.D.: Robustness and fragility of boolean models for genetic regulatory networks. J. Theor. Biol. **235**(3), 431–449 (2005)
70. Chen, M., Hofestädt, R.: A Petri net application of metabolic processes. Syst. Anal. Mod. Simul. **16**, 113–122 (1994)

71. Chen, M., Hofestädt, R.: Quantitative Petri net model of gene regulated metabolic networks in the cell. In Silico Biol. **3**, 347–365 (2003)
72. Chickarmane, V., Paladugu, S.R., Bergmann, F., Sauro, H.M.: Bifurcation discovery tool. Bioinformatics **21**(18), 3688–3690 (2005)
73. Chiola, G., Dutheillet, C., Franceschinis, G., Haddad, S.: Stochastic well-formed colored nets and symmetric modeling applications. IEEE Trans. Comput. **42**(11), 1343–1360 (1993)
74. Chiola, G., Franceschinis, G., Gaeta, R., Ribaudo, M.: Greatspn 1.7: Graphical editor and analyzer for timed and stochastic Petri nets. Perform. Eval. **24**, 47–68 (1995)
75. Chu, T.A.: Synthesis of self-timed VLSI circuits from graph-theoretic specifications. MIT/LCS/TR-393, Lab. for Comp. Sci., MIT, Cambridge (1987)
76. Ciardo, G., Muppala, J.K., Trivedi, K.S.: Stochastic Petri net package. In: Intern. Workshop on Petri Nets and Performance Models (PNPM'89), pp. 142–151 (1989)
77. Clark, G., Courtney, T., Daly, D., Deavours, D., Derisavi, S., Doyle, J.M., Sanders, W.H., Webster, P.: The Möbius modeling tool. In: Intern. Workshop on Petri Nets and Performance Models (PNPM'01), pp. 241–250 (2001)
78. Clarke, B.L.: Complete set of steady states for the general stoichiometric dynamical system. J. Chem. Phys. **75**, 4970–4979 (1981)
79. Clarke, E.M., Grumberg, O., Peled, D.: Model Checking. MIT Press, Cambridge (1999)
80. Colom, J.M., Silva, M.: Convex geometry and semiflows in P/T nets. A comparative study of algorithms for computation of minimal P-semiflows. In: Rozenberg, G. (ed.) Advances in Petri Nets, pp. 79–112. Springer, Berlin (1990)
81. Comet, J.-P., Klaudel, H., Liauzu, S.: Modeling multi-valued genetic regulatory networks using high-level Petri nets. In: LNCS, vol. 3536, pp. 208–227. Springer, Berlin (2005)
82. Comparot-Moss, S., Denyer, K.: The evolution of the starch biosynthetic pathway in cereals and other grasses. J. Exp. Bot. **60**, 2481–2492 (2009)
83. Cortadella, J., Kishinevsky, M., Kondratyev, A., Lavagno, L., Yakovlev, A.: PETRIFY: a tool for manipulating concurrent specifications and synthesis of asynchronous controllers. IEICE Trans. Inform. Syst **E80-D**(3), 315–325 (1997)
84. Cortadella, J., Kishinevsky, M., Kondratyev, A., Lavagno, L., Yakovlev, A.: Logic Synthesis of Asynchronous Controllers and Interfaces. Springer Series in Advanced Microelectronics, vol. 8. Springer, Berlin (2002)
85. Cowan, K.J., Storey, K.B.: Mitogen-activated protein kinases: new signaling pathways functioning in cellular responses to environmental stress. J. Exp. Biol. **206**, 1107–1115 (2003)
86. Dahlquist, K.D., Salomonis, N., Vranizan, K., et al.: GenMAPP, a new tool for viewing and analyzing microarray data on biological pathways. Nat. Genet. **31**, 19–22 (2002)
87. von Dassow, G., Odell, G.M.: Design and constraints of the drosophila segment polarity module: robust spatial patterning emerges from intertwined cell state switches. J. Exp. Zool. **294**(3), 179–215 (2002)
88. David, R., Alla, H.: Discrete, Continuous, and Hybrid Petri Nets. Springer, Berlin (2005)
89. de Figueiredo, L.F., Schuster, S., Kaleta, C., Fell, D.A.: Can sugars be produced from fatty acids? A test case for pathway analysis tools. Bioinformatics **25**, 152–158 (2009)
90. Delaunay, F., Laudet, V.: Circadian clock and microarrays: mammalian genome gets rhythm. TREN. Genet. **18**(12), 565–597 (2002)
91. Deuflhard, P., Bornemann, F.: Numerische Mathematik 2: Gewöhnliche Differentialgleichungen, 3rd edn. de Gruyter, Berlin (2008)
92. Diniz, S.C., Voss, I., Steinbüchel, A.: Optimization of cyanophycin production in recombinant strains of *Pseudomonas putida* and *Ralstonia eutropha* employing elementary mode analysis and statistical experimental design. Biotechnol. Bioeng. **93**, 698–717 (2006)
93. Dittrich, P., di Fenizio, P.S.: Chemical organisation theory. Bull. Math. Biol. **69**, 1199–1231 (2007)
94. Doi, A., Nagasaki, M., Fujita, S., Matsuno, H., Miyano, S.: Abstract Genomic Object Net: II. Modelling biopathways by hybrid functional Petri net with extension. Appl. Bioinform. **2**(3), 185–188 (2003)
95. Doi, A., Fujita, S., Matsuno, H., Nagasaki, M., Miyano, S.: Constructing biological pathway models with hybrid functional Petri nets. In Silico Biol. **4**(3), 271–291 (2004)

96. Doi, A., Nagasaki, M., Matsuno, H., Miyano, S.: Simulation based validation of the p53 transcriptional activity with hybrid functional Petri net. In Silico Biol. **6**(1–2), 1–13 (2006)
97. Doi, A., Nagasaki, M., Ueno, K., Matsuno, H., Miyano, S.: A combined pathway to simulate CDK-dependent phosphorylation and ARF-dependent stabilization for p53 transcriptional activity. Genome Inform. **17**(1), 112–123 (2006)
98. Doob, J.L.: Stochastic Processes. Wiley, New York (1953)
99. Drath, R.: Hybrid object nets: an object oriented concept for modeling complex hybrid systems. In: Proc. 3rd International Conference on Automation of Mixed Processes: Hybrid Dynamical Systems (ADPM), pp. 437–442. Shaker Verlag, Aachen (1998)
100. Drubin, D.A., Way, J.C., Silver, P.A.: Designing biological systems. Genes Dev. **21**(3), 242–254 (2007)
101. DSSZ-MC: Tools for the symbolic analysis of bounded Petri nets. http://www-dssz.informatik.tu-cottbus.de/index.html?/software/mc.html
102. Dumont, J.E., Dremier, S., Pirson, I., Maenhaut, C., et al.: Cross signaling, cell specificity, and physiology. Am. J. Physiol. Cell Physiol. **283**, C2–C28 (2002)
103. Dunlap, J.C.: Molecular bases for circadian clocks. Cell **96**, 271–290 (1999)
104. Durzinsky, M., Marwan, W., Wagler, A., Weismantel, R.: Automatic reconstruction of molecular and genetic networks from experimental time series data. Biosystems **93**, 181–190 (2008)
105. Durzinsky, M., Wagler, A., Weismantel, R.: A combinatorial approach to reconstruct Petri nets from experimental data. In: Heiner, M., Uhrmacher, A.M. (eds.) Proc. of CMSB 2008. LNBI, vol. 5307, pp. 328–346. Springer, Berlin (2008)
106. Durzinsky, M., Wagler, A., Weismantel, R.: An algorithmic framework for network reconstruction. J. Theor. Comput. Sci. (2009, to appear)
107. Dwyer, T., Marriott, K., Schreiber, F., et al.: Exploration of networks using overview + detail with constraint-based cooperative layout. IEEE Trans. Vis. Comput. Graph. **14**(6), 1293–1300 (2008)
108. International Union of Biochemistry and Moleculare Biology, Nomenclature Committee: Enzyme Nomenclature. Academic Press, San Diego (1992)
109. Elowitz, M.B., Leibler, S.: A synthetic oscillatory network of transcriptional regulators. Nature **403**(6767), 335–338 (2000)
110. Enright, A.J., Ouzounis, C.A.: BioLayout-an automatic graph layout algorithm for similarity visualization. Bioinformatics **17**(9), 853–854 (2001)
111. Érdi, P., Tóth, J.: Mathematical Models of Chemical Reactions. Manchester University Press, Manchester (1989)
112. Erhard, F., Friedel, C.C., Zimmer, R.: FERN—a Java framework for stochastic simulation and evaluation of reaction networks. BMC Bioinform. **9**, 356 (2008)
113. Fages, F., Soliman, S.: From reaction models to influence graphs and back: a theorem. In: Formal Methods in Systems Biology. LNCS, vol. 5054, pp. 90–102. Springer, Berlin (2008)
114. Fall, C., Marland, E., Wagner, J., Tyson, J.: Computational Cell Biology. Springer, Berlin (2002)
115. Fauré, A., Naldi, A., Chaouiya, C., Thieffry, D.: Dynamical analysis of a generic Boolean model for the control of the mammalian cell cycle. Bioinformatics **22**(14), 124–131 (2006)
116. Fauré, A., Naldi, A., Chaouiya, C., Ciliberto, A., Thieffry, D.: Modular logical modelling of the budding yeast cell cycle. Mol. Biosyst. **5**, 1787–1796 (2009)
117. Forger, D.B., Peskin, C.S.: A detailed predictive model of the mammalian circadian clock. Proc. Natl. Acad. Sci. USA **100**(25), 14806–14811 (2003)
118. Fowler, J.H., Dawes, C.T., Christakis, N.A.: Model of genetic variation in human social networks. Proc. Natl. Acad. Sci. USA **106**(6), 1720–1724 (2009)
119. Fruchterman, T., Reingold, E.: Graph Drawing by Force-directed Placement. Software—Practice and Experience **21**(11), 1129–1164 (1991)
120. Funahashi, A., Tanimura, N., Morohashi, M., Kitano, H.: CellDesigner: a process diagram editor for gene-regulatory and biochemical networks. Biosilico **1**, 159–162 (2003)
121. Functional genomics: Lipidomics In: The Scientific Athlete (2009). http://technorganiac.wordpress.com/category/definitions/

122. Gardiner, C.W.: Handbook of Stochastic Methods. Springer, Berlin (1997)
123. Garg, A., Di Cara, A., Xenarios, I., Mendoza, L., De Micheli, G.: Synchronous versus asynchronous modelling of gene regulatory networks. Bioinformatics **24**(17), 1917–1925 (2008)
124. Genrich, H.J., Küffner, R., Voss, K.: Executable Petri net models for the analysis of metabolic pathways. Int. J. STTT **3**, 394–404 (2001)
125. Gershenson, C.: Classification of random boolean networks. In: Standish, R.K., et al. (eds.) Proc. of the 8th Int. Conf. on Artificial Life, pp. 1–8. MIT Press, Cambridge (2002)
126. Gevorgyan, A., Poolman, M.G., Fell, D.A.: Detection of stoichiometric inconsistencies in biomolecular models. Bioinformatics **24**, 2245–2251 (2008)
127. Gibson, M.A., Bruck, J.: Efficient exact stochastic simulation of chemical systems with many species and many channels. J. Phys. Chem. **104**(9), 1876–1889 (2000)
128. Gilbert, S.F.: Developmental Biology, 8th edn. Sinauer, Sunderland (2006)
129. Gilbert, D., Heiner, M.: From Petri nets to differential equations—an integrative approach for biochemical network analysis. In: Petri Nets and Other Models of Concurrency. LNCS, vol. 4024, pp. 181–200. Springer, Berlin (2006)
130. Gillespie, D.T.: A general method for numerically simulating the stochastic time evolution of coupled chemical reactions. J. Comput. Phys. **22**(4), 403–434 (1976)
131. Gillespie, D.T.: Exact stochastic simulation of coupled chemical reactions. J. Phys. Chem. **81**(25), 2340–2361 (1977)
132. GINsim: Gene Interaction Network simulation, http://gin.univ-mrs.fr/GINsim
133. Glossary In: Biotechnology Industry Organization (2009), http://www.bio.org/speeches/pubs/er/glossary_s.asp
134. Glossary In: The New Genetics. National Institute of General Medical Sciences (2006). Publication No. 07-662. http://publications.nigms.nih.gov/thenewgenetics/glossary.html. Cited 20 March 2010
135. Glossop, N.R.J., Lyons, L.C., Hardin, P.E.: Interlocked feedback loops within the *Drosophila* circadian oscillator. Science **286**, 766–768 (1999)
136. Goldberg, D.E., Holland, J.H.: Genetic algorithms and machine learning. Mach. Learn. **3**(2), 95–99 (1988)
137. Goldbeter, A.: Computational approaches to cellular rhythms. Nature **420**(6912), 238–245 (2002)
138. Gonzalez, M., Bagatolli, L.A., Echabe, I., Arrondo, J.L.R., Argaran, C.A., Cantor, C.R., Fidelio, G.D.: Interaction of Biotin with Streptavidin. J. Biol. Chem. **272**, 11288–11294 (1997)
139. Goodrich, J.A., Kugel, J.F.: Binding and Kinetics for Molecular Biologists. Cold Spring Harbor Laboratory Press, Cold Spring Harbor (2007)
140. Goss, P.J.E., Peccoud, J.: Quantitative modeling of stochastic systems in molecular biology by using stochastic Petri nets. Proc. Natl. Acad. Sci. USA. **95**(12), 6750–6755 (1998)
141. Goss, P.J.E., Peccoud, J.: Analysis of the stabilizing effect of Rom on the genetic network controlling ColE1 plasmid replication. Proc. Pac. Symp. Biocomput. **4**, 65–76 (1999)
142. Grafahrend-Belau, E.: Classification of t-Invariants in biochemical Petri nets on the basis of various cluster analysis methods. Master's thesis, Technical University of Applied Sciences (TFH). Berlin (2006) (in German)
143. Grafahrend-Belau, E., Schreiber, F., Heiner, M., Sackmann, A., Junker, B.H., Grunwald, S., Speer, A., Winder, K., Koch, I.: Modularization of biochemical networks based on classification of Petri net t-Invariants. BMC Bioinform. **9**, 90 (2008)
144. Grahlmann, B.: The PEP tool. In: Proc. of 9th Int. Conf. on Computer Aided Verification. LNCS, vol. 1254, pp. 440–443. Springer, Berlin (1997)
145. Gross, D., Miller, D.R.: The randomization technique as a modeling tool and solution procedure for transient Markov processes. Oper. Res. **32**(2), 343–361 (1984)
146. Grossman, P.: Discrete Mathematics for Computing, 2nd edn. Palgrave MacMillan, Basingstoke (2002)
147. Grunwald, S., Speer, A., Ackermann, J., Koch, I.: Petri net modelling of gene regulation of the Duchenne muscular dystrophy. Biosystems **92**(2), 189–205 (2008)
148. Haken, H.: Synergetics. Springer, Berlin (2004)

149. Hardy, S., Robillard, P.N.: Modeling and simulation of molecular biology systems using Petri nets: modeling goals of various approaches. J. Bioinform. Comput. Biol. **2**(4), 619–637 (2004)
150. Hassel, M.P., Comins, H.N., May, R.M.: Spatial structure and chaos in insect population dynamics. Nature **353**, 255–258 (1991)
151. Hastings, M.H.: Circadian clockwork: two loops are better than one. Nat. Rev. Neurosci. **1**(2), 143–146 (2000)
152. Heiner, M., Koch, I.: Petri net based model validation in systems biology. In: Applications and Theory of Petri Nets. LCNS, vol. 3099, pp. 216–237. Springer, Berlin (2004)
153. Heiner, M., Koch, I., Will, J.: Model validation of biological pathways using Petri nets—demonstrated for apoptosis. Biosystems **75**(1), 15–28 (2004)
154. Heiner, M., Gilbert, D., Donaldson, R.: Petri nets for systems and synthetic biology. In: Formal Methods for Computational Systems Biology. LNCS, vol. 5016, pp. 215–264. Springer, Berlin (2008)
155. Heiner, M., Richter, R., Schwarick, M.: Snoopy: a tool to design and animate/simulate graph-based formalisms. In: Proc. 1st Intern. Conf. Simulation Tools and Techniques for Communications, Networks and Systems & Workshops, pp. 1–10 (2008)
156. Heiner, M., Lehrack, S., Gilbert, D., Marwan, W.: Extended stochastic Petri nets for modelbased design of wetlab experiments. In: Computational Models for Cell Processes. Transactions on Computational Systems Biology XI. LNCS, vol. 5750, pp. 138–163. Springer, Berlin (2009)
157. Heinrich, R., Schuster, S.: The Regulation of Cellular Systems. Chapman and Hall, London (1996)
158. Heinrich, R., Neel, B.G., Rapoport, T.A.: Mathematical models of protein kinase signal transduction. Mol. Cell **9**, 957–970 (2002)
159. Hengge-Aronis, R.: The general stress response in *Escherichia coli*. In: Storz, G., Hengge-Aronis, R. (eds.) Bacterial Stress Responses, pp. 161–178. ASM, Materials Park (2000)
160. Hill, A.V.: The possible effects of the aggregation of the molecules of huemoglobin on its dissociation curves. In: Proceedings of the Physiological Society, pp. 4–7 (1910)
161. Hitchcock, S.E.: Extinction probabilities in predator-prey models. J. Appl. Probab. **23**(1), 1–13 (1986)
162. Hofestädt, R.: A petri net application to model metabolic processes. Syst. Anal. Mod. Simul. **16**(2), 113–122 (1994)
163. Hofestädt, R., Thelen, S.: Quantitative modeling of biochemical networks. In Silico Biol. **1**, 39–53 (1998)
164. Huang, B., Wu, H., Bhaya, D., Grossman, A., Granier, S., Kobilka, B.K., Zare, R.N.: Count ing low-copy-number proteins in a single cell. Science **315**, 81–84 (2007)
165. Ikeo, K., Ishi-i, J., Tamura, T., et al.: CIBEX: center for information biology gene expression database. C. R. Biol. **326**, 1079–1082 (2003)
166. INA: Integrated Net Analyzer, http://www2.informatik.hu-berlin.de/~starke/ina.html
167. Indic, P., Gurdziel, K., Kronauer, R.E., Klerman, E.B.: Development of a two-dimension manifold to represent high dimension mathematical models of the intracellular Mammalian circadian clock. J. Biol. Rhythms. **21**(3), 222–232 (2006)
168. Ingolia, N.T.: Topology and robustness in the drosophila segment polarity network. PLoS Biol. **2**(6), e123 (2004)
169. Ingram, P., Stumpf, M., Stark, J.: Network motifs: structure does not determine function. BMC Genom. **7**(1), 108 (2006)
170. Jeong, H., Tombor, B., Albert, R., Oltvai, Z.N., Barabási, A.L.: The large-scale organization of metabolic networks. Nature **407**, 651–654 (2000)
171. Jeong, E., Nagasaki, M., Saito, A., Miyano, S.: Cell system ontology: representation for modeling, visualizing, and simulating biological pathways. In Silico Biol. **7**(6), 623–638 (2007)
172. Johnston, R.J.Jr., Hobert, O.: A microRNA controlling left/right neuronal asymmetry in *Caenorhabditis elegans*. Nature **426**, 845–849 (2003)

173. Johnston, R.J.Jr., Hobert, O.: A novel *C. elegans* zinc finger transcription factor, *lsy-2*, required for the cell type-specific expression of the *lsy-6* microRNA. Development **132**(24), 5451–5460 (2005)
174. Johnston, R.J.Jr., Chang, S., Etchberger, J.F., Ortiz, C.O., Hobert, O.: MicroRNAs acting in a double-negative feedback loop to control a neuronal cell fate decision. Proc. Natl. Acad. Sci. USA **102**(35), 12449–12454 (2005)
175. de Jong, H.: Modeling and simulation of genetic regulatory systems: a literature review. J. Computat. Biol. **9**, 67–103 (2002)
176. Joshi-Tope, G., Gillespie, M., Vastrik, I., et al.: Reactome: a knowledgebase of biological pathways. Nucleic Acids Res. **33**, D428–D432 (2005)
177. Junker, B.H., Klukas, C., Schreiber, F.: VANTED: A system for advanced data analysis and visualization in the context of biological networks. BMC Bioinform. **7**, 109 (2006)
178. Jurica, M.S., Moore, M.J.: Pre-mRNA splicing: awash in a sea of proteins. Mol. Cell **12**, 5–14 (2003)
179. Kam, T., Villa, T., Brayton, R.K., Sangiovanni-Vincentelli, A.L.: Multi-valued decision diagrams: theory and applications. Int. J. Multiple-Valued Logic **4**, 9–62 (1998)
180. Kanehisa, M., Goto, S., Kawashima, S., et al.: The KEGG resource for deciphering the genome. Nucleic Acids Res. **32**, D277–D280 (2004)
181. Karlebach, G., Shamir, R.: Modelling and analysis of gene regulatory networks. Nat. Rev. Mol. Cell Biol. **9**, 770–780 (2008)
182. Karlebach, G., Shamir, R.: Modelling and analysis of gene regulatory networks. Nat. Rev. Mol. Cell Biol. **9**(10), 770–780 (2008)
183. Karp, P.D., Riley, M., Paley, S.M., Pellegrini-Toole, A., Krummenacker, M.: EcoCyc: Encyclopedia of *Escherichia coli* genes and metabolism. Nucleic Acids Res. **26**, 50–53 (1998)
184. Karp, P.D., Riley, M., Saier, M., Paulsen, I.T., Paley, S.M., Pellegrini-Toole, A.: The EcoCyc and MetaCyc databases. Nucleic Acids Res. **28**, 56–59 (2000)
185. Kauffman, S.: Metabolic stability and epigenesis in randomly constructed genetics nets. J. Theor. Biol. **22**, 437–467 (1969)
186. Keener, J., Sneyd, J.: Mathematical Physiology. Springer, Berlin (1998)
187. Kholodenko, B.N.: Negative feedback and ultrasensitivity can bring about oscillations in the mitogen-activated protein kinase cascades. Eur. J. Biochem. **267**(6), 1583–1588 (2000)
188. Kholodenko, B.N., Kiyatkin, A., Bruggeman, F.J., Sontag, E., Westerhoff, H.V., Hoek, J.B.: Untangling the wires: a strategy to trace functional interactions in signaling and gene networks. Proc. Natl. Acad. Sci. USA **99**(20), 12841–12846 (2002)
189. Khomenko, V.: Model checking based on prefixes of Petri net unfoldings. Ph.D. thesis, School of Computing Science, Newcastle University (2003)
190. Kielbassa, J., Bortfeldt, R., Schuster, S., Koch, I.: Modeling of the U1 snRNP assembly pathway in alternative splicing in human cells using Petri nets. Comput. Biol. Chem. **33**(1), 46–61 (2009)
191. Kinniment, D.J.: Synchronization and Arbitration in Digital Systems. Wiley, New York (2007)
192. Kirkpatrick, S., Gelatt, C.D., Vecchi, M.P.: Optimization by simulated annealing. Science **220**(4598), 671–680 (1983)
193. Kitano, H.: Systems biology: a brief overview. Science **295**, 1662–1664 (2002)
194. Klamt, S.: Generalized concept of minimal cut sets in biochemical networks. Biosystems **83**, 233–247 (2006)
195. Klamt, S., Saez-Rodriguez, J., Lindquist, J.A., Simeoni, L., Gilles, E.D.: A methodology for the structural and functional analysis of signaling and regulatory networks. BMC Bioinform. **7**, 56 (2006)
196. Klamt, S., Gilles, E.D.: Minimal cut sets in biochemical reaction networks. Bioinformatics **20**(2), 226–234 (2004)
197. Klamt, S., Gagneur, J., von Kamp, A.: Algorithmic approaches for computing elementary modes in large biochemical reaction networks. IEE Proc. Syst. Biol. **152**, 249–255 (2005)
198. Klamt, S., Saez-Rodriguez, J., Gilles, E.D.: Structural and functional analysis of cellular networks with CellNetAnalyzer. BMC Syst. Biol. **1**, 2 (2007)

199. Klamt, S., Haus, U.U., Theis, F.: Hypergraphs and cellular networks. PLoS Comput. Biol. **5**, e1000385 (2009)
200. Klipp, E., Liebermeister, W., Helbig, A., Kowald, A., Schaber, J.: System biology standards—the community speaks. Nat. Biotechnol. **25**, 390–391 (2007)
201. Kindler, E., Pales, C.: 3D-Visualization of Petri net models: concept and realization. In: Proceedings of the 25th International Conference on Applications and Theory of Petri Nets, Bologna. Lecture Notes in Computer Science, vol. 3099, pp. 464–473. Springer, Berlin (2004)
202. Klaudel, H., Pommereau, F.: M-nets: a survey. Acta Inform. **45**, 537–564 (2008)
203. Klipp, E., Herwig, R., Kowald, A., Wierling, C., Lehrach, H.: Systems Biology in Practice. Wiley VCH, Weinheim (2006)
204. Klukas, C., Spies, K., Lange, M., et al.: Comparison of pathway visualisation tools. J. Integr. Bioinform. (2010, to appear)
205. Klipp, E., Liebermeister, W., Wierling, C., Kowald, A.: Systems Biology: A Textbook. Wiley VCH, Weinheim (2009)
206. Ko, C.H., Takahashi, S.: Molecular components of the mammalian circadian clock. Hum. Mol. Genet. **15**(2), R271–277 (2006)
207. Koch, I.: Petri nets and GRN Models. In: Das, S., Caragea, D., Hsu, W.H., Welch, S.M. (eds.): Handbook of Research on Computational Methodologies in Gene Regulatory Networks, pp. 604–637. IGI Global, Hershey–New York (2010). Chapter 25
208. Koch, I., Heiner, M.: Petri nets. In: Junker, B.H., Schreiber, F. (eds.) Biological Network Analysis, Wiley Book Series on Bioinformatics, pp. 139–180. Wiley, New York (2008). Chapter 7
209. Koch, I., Kaden, F., Selbig, J.: Analysis of protein sheet topologies by graph-theoretical methods. Protein Struct. Funct. Genet. **12**, 314–324 (1992)
210. Koch, I., Schüler, M., Heiner, M.: STEPP—search tool for exploration of Petri net paths: a new tool for Petri net-based path analysis in biochemical networks. In Silico Biol. **5**(2), 129–137 (2005)
211. Koch, I., Junker, B.H., Heiner, M.: Application of Petri net theory for modelling and validation of the sucrose breakdown pathway in the potato tuber. Bioinformatics **21**(7), 1219–1226 (2005)
212. Koshland, D.: Application of a theory of enzyme specificity to protein synthesis. Proc. Natl. Acad. Sci. USA **44**(2), 98–104 (1958)
213. Krishna, K., Guo, S.: A partial Granger causality approach to explode causal networks derived from multi-parameter data. In: Heiner, M., Uhrmacher, A.M. (eds.) Proc. of CMSB 2008. LNBI, vol. 5307, pp. 9–27. Springer, Berlin (2008)
214. Kruger, N.J., von Schaewen, A.: The oxidative pentose phosphate pathway: structure and organisation. Curr. Opin. Plant Biol. **6**, 236–246 (2003)
215. Krull, M., Pistor, S., Voss, N., Kel, A., Reuter, I., Kronenberg, D., Michael, H., Schwarzer, K., Potapov, A., Choi, C., Kel-Margoulis, O., Wingender, E.: TRANSPATH: an information resource for storing and visualizing signaling pathways and their pathological aberrations. Nucleic Acids Res. **34**, D546–D551 (2006)
216. Lamparter, T., Marwan, W.: Spectroscopic detection of a phytochrome-like photoreceptor in the Myxomycete *Physarum polycephalum* and the kinetic mechanism for the photocontrol of sporulation by P_{fr}. Photochem. Photobiol. **73**, 697–702 (2001)
217. Larhlimi, A., Bockmayr, A.: A new approach to flux coupling analysis of metabolic networks. In: Computational Life Sciences II. LNCS, vol. 4216, pp. 205–215. Springer, Berlin (2006)
218. Laubenbacher, R., Stigler, B.: A computational algebra approach to reverse engineering of gene regulatory networks. J. Theor. Biol. **229**, 523–537 (2005)
219. Lautenbach, K.: Exact liveness conditions of a Petri net class. GMD Report 82, German National Research Center for Information Technology, Sankt Augustin, Germany (1973) (in German)
220. Lautenbach, K.: Linear algebraic techniques for place/transition nets. In: Petri Nets: Central Models and Their Properties. LNCS, vol. 254, pp. 142–167. Springer, Berlin (1987)

221. Leadbeater, B.S.C., Green, J.C.: The Flagellates. Unity, Diversity and Evolution, pp. 110–123. Routledge, London (2000)
222. Lee, C.C.: Fuzzy logic in control systems: fuzzy logic controller. I. IEEE Trans. Syst. Man Cybern. **20**(2), 404–418 (1990)
223. Lee, C., Etchegaray, J.P., Cagampang, F.R., Loudon, A.S., Reppert, S.M.: Posttranslation mechanism regulate the mammalian circadian clock. Cell **107**, 855–867 (2001)
224. Lehninger, A.L., Nelson, D.L., Cox, M.M.: Principles of Biochemistry. Worth Publisher, New York (2004)
225. Leloup, J.C., Gonze, D., Goldbeter, A.: Limit cycle models for circadian rhythms based on transcriptional regulation in *Drosophila* and *Neurospora*. J. Biol. Rhythms **14**(6), 433–448 (1999)
226. Leloup, J.C., Goldbeter, A.: Modeling the molecular regulatory mechanism of circadian rhythms in *Drosophila*. BioEssays **22**(1), 84–93 (2000)
227. Leloup, J.C., Goldbeter, A.: Toward a detailed computational model for the mammalian circadian clock. Proc. Natl. Acad. Sci. USA **100**(12), 7051–7056 (2003)
228. Leloup, J.C., Goldbeter, A.: Modeling the circadian clock: from molecular mechanism to physiological disorders. BioEssays **30**, 590–600 (2008)
229. Le Nov'ere, N., Hucka, M., Mi, H., et al.: The systems biology graphical notation. Nat. Biotechnol. **27**, 735–741 (2009)
230. Li, C., Suzuki, S., Ge, Q.-W., Nakata, M., Matsuno, H., Miyano, S.: Structural modeling and analysis of signaling pathways based on Petri nets. J. Bioinf. Comput. Biol. **4**(5), 1119–1140 (2006)
231. Li, C., Nagasaki, M., Ueno, K., Miyano, S.: Simulation-based model checking approach to cell fate specification during *Caenorhabditis elegans* vulval development by hybrid functional Petri net with extension. BMC Syst. Biol. **3**, 42 (2009)
232. Lide, R.D.: CRC Handbook of Chemistry and Physics, 90th edn. Taylor & Francis, London (2009)
233. Lipshtat, A., Purushothaman, S.P., Iyengar, R., Ma'ayan, A.: Functions of bifans in context of multiple regulatory motifs in signaling networks. Biophys. J. **94**(7), 2566–2579 (2008)
234. Loew, L.M., Schaff, J.C.: The virtual cell: a software environment for computational cell biology. Trends Biotechnol. **19**(10), 401–406 (2001)
235. Logan, D.C.: The mitochondrial compartment. J. Exp. Bot. **57**(6), 1225–1243 (2006)
236. Lotka, A.J.: Undamped oscillations derived from the laws of mass action. J. Am. Chem. Soc. **42**, 1595–1599 (1920)
237. Ma'ayan, A.: Network integration and graph analysis in mammalian molecular systems biology. IET Syst. Biol. **2**(5), 206–221 (2008)
238. Ma'ayan, A., Iyengar, R.: From components to regulatory motifs in signalling networks. Brief. Funct. Genomics Proteomics **5**(1), 57–61 (2006)
239. Ma'ayan, A., Jenkins, S.L., Neves, S., Hasseldine, A., Grace, E., Dubin-Thaler, B., Eungdamrong, N.J., Weng, G., Ram, P.T., Rice, J.J., Kershenbaum, A., Stolovitzky, G.A., Blitzer, R.D., Iyengar, R.: Formation of regulatory patterns during signal propagation in a mammalian cellular network. Science **309**(5737), 1078–1083 (2005)
240. Mangan, S., Alon, U.: Structure and function of the feed-forward loop network motif. Proc. Natl. Acad. Sci. USA **100**(21), 11980–11985 (2003)
241. Marbach, D., Mattiussi, C., Floreano, D.: Combining multiple results of a reverse-engineering algorithm: application to the DREAM five-gene network challenge. Ann. NY Acad. Sci. **1158**, 102–113 (2009)
242. Marbach, D., Schaffter, T., Mattiussi, C., Floreano, D.: Generating realistic in silico gene networks for performance assessment of reverse engineering methods. J. Comput. Biol. **16**(2), 229–239 (2009)
243. Marsan, A.M., Conte, G., Balbo, G.: A class of generalized stochastic Petri nets for the performance evaluation of multiprocessor systems. ACM Trans. Comput. Syst. **2**(2), 93–122 (1984)

244. Marwan, W., Sujatha, A., Starostzik, C.: Reconstructing the regulatory network controlling commitment and sporulation in physarum polycephalum based on hierarchical petri net modelling and simulation. J. Theor. Biol. **236**(4), 349–365 (2005)
245. Marwan, W., Wagler, A., Weismantel, R.: A mathematical approach to solve the network reconstruction problem. Math. Methods Oper. Res. **67**, 117–132 (2008)
246. Mason, O., Verwoerd, M.: Graph theory and networks in Biology. IET Syst. Biol. **1**(2), 89–119 (2007)
247. Materna, S.C., Davidson, E.H.: Logic of gene regulatory networks. Curr. Opin. Biotechnol. **18**, 351–354 (2007)
248. Matsuno, H., Doi, A., Nagasaki, M., Miyano, S.: Hybrid Petri net representation of gene regulatory network. Proc. Pac. Symp. Biocomput. **5**, 341–352 (2000)
249. Matsuno, H., Fujita, S., Doi, A., Nagasaki, M., Miyano, S.: Towards biopathway modeling and simulation. In: Proc. ICATPN 2003. LNCS, vol. 2679, pp. 3–22. Springer, Berlin (2003)
250. Matsuno, H., Murakami, R., Yamane, R., Yamasaki, N., Fujita, S., Yoshimori, H., Miyano, S.: Boundary formation by notch signaling in *Drosophila multicellular* systems: experimental observations and gene network modeling by Genomic Object Net. Pac. Symp. Biocomput. **8**, 152–163 (2003)
251. Matsuno, H., Tanaka, Y., Aoshima, H., Doi, A., Matsui, M., Miyano, S.: Biopathways representation and simulation on hybrid functional Petri net. In Silico Biol. **3**(3), 389–404 (2003)
252. May, P., Kreuchwig, A., Steinke, T., Koch, I.: PTGL: A data base for secondary structure-based protein topologies. Nucleic Acids Res. **38**, D326–D330 (2010)
253. Matsuno, H., Doi, A., Nagasaki, M., Miyano, S.: Hybrid Petri Net Representation of Gene Regulatory Network. In: Proceedings Pacific Symposium Biocomputing 2000, vol. **5**, pp. 338–349 (2000)
254. Matsuno, H., Tanaka, Y., Aoshima, H., Doi, A., Matsui, M., Miyano, S.: Biopathway representation and simulation on hybrid functional Petri net. In Silico Biology **3**(3), 389–404 (2003)
255. Matsuno, H., Inouye, S.-I.T., Okitsu, Y., Fujii, Y., Miyano, S.: A new regulatory interaction suggested by simulations for circadian genetic control mechanism in mammals. J. Bioinf. Comput. Biol. **4**(1), 139–153 (2006)
256. McColluma, J.M., Peterson, G.D., Cox, C.D., Simpson, M.L., Samatova, N.F.: The sorting direct method for stochastic simulation of biochemical systems with varying reaction execution behavior. Comput. Biol. Chem. **30**(1), 29–39 (2006)
257. Mendel, J.M.: Fuzzy logic systems for engineering: a tutorial. Proc. IEEE **83**(3), 345–377 (1995)
258. Mendoza, L.: A network model for the control of the differentiation process in Th cells. Biosystems **84**(2), 101–114 (2006)
259. Mertens, E.: Pyrophosphate-dependent phosphofructokinase, an anaerobic glycolytic enzyme? FEBS Lett. **285**, 1–5 (1991)
260. Meyer, J.F., Movaghar, A., Sanders, W.H.: Stochastic Activity Networks: Structure, Behavior, and Application. In: International Workshop on Timed Petri Nets, pp. 106–115. IEEE Comput. Soc., Los Alamitos (1985)
261. Michaelis, M., Menten, M.L.: Die Kinetik der Invertin-Wirkung. Biochem. Z. **49**, 333–369 (1913)
262. MGED society: Microarray Gene Expression Data Society. http://www.mged.org/
263. Michels, K., Klawonn, F., Kruse, R., Nürnberger, A.: Fuzzy-Regelung. Grundlagen, Entwurf, Analyse. Springer, Berlin (2003)
264. Milo, R., Shen-Orr, S., Itzkovitz, S., Kashtan, N., Chklovskii, D., Alon, U.: Network motifs: simple building blocks of complex networks. Science **298**(5594), 824–827 (2002)
265. Mirschel, S., Steinmetz, K., Rempel, M., Ginkel, M., Gilles, E.D.: ProMoT: ModularModeling for Systems Biology. Bioinformatics **25**(5), 687–689 (2009)
266. Mito, N., Ikegami, Y., Matsuno, H., Miyano, S., Inouye, S.: Simulation analysis for the effect of light-dark cycle on the entrainment in circadian rhythm. Genome Inform. **21**, 212–223 (2008)

267. Molloy, M.K.: Performance analysis using stochastic Petri nets. IEEE Trans. Comput. **31**(9), 913–917 (1982)
268. Muller, D.E., Bartky, W.S.: A theory of asynchronous circuits. In: Proc. Int. Symp. of the Theory of Switching, Part I, pp. 204–243. Harvard University Press, Cambridge (1959)
269. Muppala, J.K., Ciardo, G., Trivedi, K.S.: Stochastic reward nets for reliability prediction. In: Communications in Reliability, Maintainability and Serviceability, pp. 9–20 (1994)
270. Mura, I., Csiksz-Nagy, A.: Stochastic Petri Net extension of a yeast cell cycle model. J. Theor. Biol. **254**(4), 850–860 (2008)
271. Murata, T.: Petri nets: Properties, analysis and applications. Proc. IEEE **77**(4), 541–580 (1989)
272. Murray, J.D.: Mathematical Biology. Springer, Berlin (2008)
273. MVSIS. UC Berkeley. http://www-cad.eecs.berkeley.edu/mvsis
274. Nagasaki, M., Doi, A., Matsuno, H., Miyano, S.: Genomic object net: I, a platform for modeling and simulating biopathways. Appl. Bioinform. **2**(3), 181–184 (2003)
275. Nagasaki, M., Doi, A., Matsuno, H., Miyano, S.: A versatile Petri net based architecture for modeling and simulation of complex biological processes. Genome Inform. **15**(1), 180–197 (2004)
276. Nagasaki, M., Doi, A., Matsuno, H., Miyano, S.: Petri net based description and modeling of biological pathways. In: Algebraic Biology 2005: Proc. of the 1st International Conference on Algebraic Biology—Computer Algebra in Biology, pp. 19–31. Universal Acad. Press, Tokyo (2005)
277. Nagasaki, M., Doi, A., Matsuno, H., Miyano, S.: Computational modeling of biological processes with Petri net based architecture. In: Chen, Y.P. (ed.) Bioinformatics Technologies, pp. 179–243. Springer, Berlin (2005)
278. Nagasaki, M., Yamaguchi, R., Yoshida, R., Imoto, S., Doi, A., Tamada, Y., Matsuno, H., Miyano, S., Higuchi, T.: Genomic data assimilation for estimating hybrid functional Petri net from time-course gene expression data. Genome Inform. **17**(1), 46–61 (2006)
279. Nagasaki, M., Saito, A., Li, C., Jeong, E., Miyano, S.: Systematic reconstruction of TRANSPATH data into Cell System Markup Language. BMC Syst. Biol. **2**, 53 (2008)
280. Nagasaki, M., Saito, A., Doi, A., Matsuno, H., Miyano, S.: Foundations of Systems Biology—Using Cell Illustrator and Pathway Databases. Springer, Berlin (2009)
281. Nakamura, K., Yoshida, R., Nagasaki, M., Miyano, S., Higuchi, T.: Parameter estimation of in silico biological pathways with particle filtering towards a petascale computing. Pac. Symp. Biocomput. **14**, 227–238 (2009)
282. Naldi, A., Thieffry, D., Chaouiya, C.: Decision diagrams for the representation of logical models of regulatory networks. In: LNBI, vol. 4695, pp. 233–247. Springer, Berlin (2007)
283. Naldi, A., Berenguier, D., Fauré, A., Lopez, F., Thieffry, D., Chaouiya, C.: Logical modelling of regulatory networks with GINsim 2.3. Biosystems **97**(2), 134–139 (2009)
284. Naldi, A., Remy, E., Thieffry, D., Chaouiya, C.: A reduction of logical regulatory graphs preserving essential dynamical properties. In: LNBI, vol. 5688, pp. 266–280. Springer, Berlin (2009)
285. Nuño, J.C., Sánchez-Valdenebro, I., Pérez-Iratxeta, C., Meléndez-Hevia, E., Montero, F.: Network organization of cell metabolism: monosaccharide interconversion. Biochem. J. **324**, 103–111 (1997)
286. Ng, A., Bursteinas, B., Gao, Q., Mollison, E., Zvelebil, M.: pSTIING: a 'systems' approach towards integrating signalling pathways, interaction and transcriptional regulatory networks in inflammation and cancer. Nucleic Acids Res. **34**, D527–D534 (2006)
287. Ng, A., Bursteinas, B., Gao, Q., Mollison, E., Zvelebil, M.: Resources for integrative systems biology: fromdata through databases to networks and dynamic systemmodels. Brief. Bioinform. **7**(4), 318–330 (2006)
288. Oliveira, J.S., Jones-Oliveira, J.B., Dixon, D.A., Bailey, C.G., Gull, D.W.: Hyperdigraph-theoretical analysis of the EGFR signaling network: initial steps leading to GTP: Ras complex formation. J. Comput. Biol. **11**(5), 812–842 (2004)
289. Oppenheim, A.B., Kobiler, O., Stavans, J., Court, D.L., Adhya, S.L.: Switches in bacteriophage λ development. Annu. Rev. Genet. **39**, 4470–4475 (2005)

290. Palsson, B.Ø.: Systems Biology: Properties of Reconstructed Networks. Cambridge University Press, Cambridge (2006)
291. Papin, J.A., Stelling, J., Price, N.D., Klamt, S., Schuster, S., Palsson, B.Ø.: Comparison of network-based pathway analysis methods. Trends Biotechnol. **22**(8), 400–405 (2004)
292. Parkinson, H., Sarkans, U., Shojatalab, M., et al.: ArrayExpress—a public repository for microarray gene expression data at the EBI. Nucleic Acids Res. **33**, D553–D555 (2005)
293. Pearson, J.E.: Complex patterns in a simple system. Science **261**, 189–192 (1993)
294. Peccoud, P.: Stochastic Petri nets for genetic networks. Med. Sci. **14**(8–9), 991–993 (1998)
295. Peleg, M., Rubin, D., Altman, R.B.: Using Petri net tools to study properties and dynamics of biological systems. J. Am. Med. Inform. Assoc. **12**(2), 181–199 (2005)
296. PEP homepage: http://theoretica.informatik.uni-oldenburg.de/~pep
297. Pérès, S.: Analysis of the structure of metabolic networks: Application to mitochondrial energy metabolism. Ph.D. thesis, Université de Bordeaux 2 (2005) (in French)
298. Pérès, S., Beurton-Aimar, M., Mazat, J.P.: Pathway classification of TCA cycle. IEE Proc. Syst. Biology **153**(5), 369–371 (2006)
299. Petri, C.A.: Communication with automata. Institut für Instrumentelle Mathematik, Bonn: Schriften des IIM **3** (1962) (in German)
300. Petri, C.A., Reisig, W.: Scholarpedia **3**(4), 6477 (2008). http://www.scholarpedia.org/article/Petri_net
301. Petri nets World: http://www.informatik.uni-hamburg.de/TGI/PetriNets
302. Pfeiffer, T., Sánchez-Valenebro, I., Nuño, J.C., Montero, F., Schuster, S.: METATOOL: for studying metabolic networks. Bioinformatics **15**(3), 251–257 (1999)
303. Pinney, J.W., Westhead, R.D., McConkey, G.A.: Petri net representations in systems biology. Biochem. Soc. Trans. **31**, 1513–1515 (2003)
304. Planes, F.J., Beasley, J.E.: A critical examination of stoichiometric and pathfinding approaches to metabolic pathways. Brief. Bioinform. **9**, 422–436 (2008)
305. Plaxton, W.C.: The organization and regulation of plant glycolysis. Annu. Rev. Plant Physiol. Plant Mol. Biol. **47**, 185–214 (1996)
306. PNML.org: The reference site for the Petri Net Markup Language. http://www.pnml.org/
307. Pommereau, F.: Quickly prototyping Petri nets tools with SNAKES. ACM Digital Library, pp. 1–10, ACM (2008)
308. Popova-Zeugmann, L., Heiner, M., Koch, I.: Time Petri nets for modelling and analysis of biochemical networks. Fundam. Inform. **67**, 149–162 (2005)
309. Preitner, N., Domiola, F., Molina, L.L., Zakany, J., Doboule, D., Albrecht, U., Schibler, U.: The orphan nuclear receptor REV-ERBα controls circadian transcription within the positive limb of the mammalian circadian oscillator. Cell **110**, 251–260 (2002)
310. Prill, R.J., Iglesias, P.A., Levchenko, A.: Dynamic properties of network motifs contribute to biological network organization. PLoS Biol. **3**(11), e343 (2005)
311. Ptashne, M.: A Genetic Switch: Phage Lambda Revisited, 3rd edn. Cold Spring Harbor Laboratory Press, Cold Spring Harbor (2004)
312. Puchałka, J., Kierzek, A.M.: Bridging the gap between stochastic and deterministic regimes in the kinetic simulations of the biochemical reaction networks. Biophys. J. **86**(3), 1357–1372 (2004)
313. Python Software Foundation: Python programming language. http://www.python.org
314. Reddy, V.N., Mavrovouniotis, M.L., Liebman, M.N.: Petri net representation in metabolic pathways. In: Proc. Int. Conf. Intell. Syst. Mol. Biol., vol. 1, pp. 328–336. MIT Press, Cambridge (1993)
315. Reddy, V.N., Liebmann, M.N., Mavrovouniotis, M.L.: Qualitative analysis of biochemical reaction systems. Comput. Biol. Med. **26**(1), 9–24 (1996)
316. Reder, C.: Metabolic control theory: a structural approach. J. Theor. Biol. **135**, 175–201 (1988)
317. Reed, J.L., Palsson, B.O.: Genome-scale in silico models of *E-coli* have multiple equivalent phenotypic states: assessment of correlated reaction subsets that comprise network states. Genome Res. **14**, 1797–1805 (2004)

318. Reisig, W.: Petri nets: an introduction. In: Brauer, W., et al. (eds.) EATCS Monographs on Theoretical Computer Science. Springer, Berlin (1985)
319. Reisig, W., Rozenberg, G. (eds.): Lectures on Petri Nets I: Basic Models. LNCS, vol. 1491. Springer, Berlin (1998)
320. Remy, E., Ruet, P., Mendoza, L., Thieffry, D., Chaouiya, C.: From logical regulatory graphs to standard Petri nets: dynamical roles of functionality of feedback circuits. Concurr. Models Molec. Biol. **4230**, 52–72 (2004)
321. Rino, J., Carvalho, T., Braga, J., Desterro, J.M.P., Lührmann, R., Carmo-Fonseca, M.: A stochastic view of spliceosome assembly and recycling in the nucleus. PLoS Comput. Biol. **3**, 2019–2031 (2007)
322. Rohwer, J.M., Botha, F.C.: Analysis of sucrose accumulation in the sugar cane culm on the basis of in vitro kinetic data. Biochem. J. **358**, 437–445 (2001)
323. Ropers, D., de Jong, H., Page, M., Schneider, D., Geiselmann, J.: Qualitative simulation of the nutritional stress response in *Escherichia coli*. Biosystems **84**(2), 124–152 (2006)
324. Ribeiro, J.M., Fontes, R., Sillero, A.: Enzyme inhibition as visualized with the reservoir model: relationships between I50 and inhibition constant(s) of an enzyme inhibitor. Comput. Biol. Med. **24**(2), 129–144 (1994)
325. Rosenblum, L., Yakovlev, A.: Signal graphs: from self-timed to timed ones. In: Proc. of the Int. Workshop on Timed Petri Nets, pp. 199–206. IEEE Comput. Soc., Los Alamitos (1985)
326. Rusak, B., Zucker, I.: Neural regulation of circadian rhythms. Physiol. Rev. **59**, 449–526 (1979)
327. Sackmann, A., Heiner, M., Koch, I.: Application of Petri net based analysis techniques to signal transduction pathways. BMC Bioinform. **7**(1), 482 (2006)
328. Sackmann, A., Formanowicz, D., Formanowicz, P., Koch, I., Blazewicz, J.: An analysis of the Petri net based model of the human body iron homeostasis process. Comput. Biol. Chem. **31**, 1–10 (2007)
329. Sackmann, A., Formanowicz, D., Formanowicz, P., Blazewicz, J.: New insights into the human body iron metabolism analyzed by a Petri net based approach. Biosystems **96**(1), 104–113 (2009)
330. Saez-Rodriguez, J., Simeoni, L., Lindquist, J.A., Hemenway, R., Bommhardt, U., Arndt, B., Haus, U.U., Weismantel, R., Gilles, E.D., Klamt, S., Schraven, B.: A logical model provides insights into T cell receptor signaling. PLoS Comput. Biol. **3**(8), e163 (2007)
331. Saito, A., Nagasaki, M., Doi, A., Ueno, K., Miyano, S.: Cell fate simulation model of gustatory neurons with microRNAs double-negative feedback loops by hybrid functional Petri net with extension. Genome Inform. **17**(1), 100–111 (2006)
332. Salwinski, L., Miller, C.S., Smith, A.J., et al.: The Database of Interacting Proteins: 2004 update. Nucleic Acids Res. **32**, D449–D451 (2004)
333. Sánchez, L., Thieffry, D.: A logical analysis of the Drosophila gap-gene system. J. Theor. Biol. **211**(2), 115–141 (2001)
334. Sánchez, L., Thieffry, D.: Segmenting the fly embryo: a logical analysis of the pair-rule cross-regulatory module. J. Theor. Biol. **224**(4), 517–537 (2003)
335. Sánchez, L., Chaouiya, C., Thieffry, D.: Segmenting the fly embryo: a logical analysis of the segment polarity cross-regulatory module. Int. J. Dev. Biol. **52**(8), 1059–1075 (2008)
336. Sano, T., Cantor, C.R.: Cooperative biotin binding by streptavidin. J. Biol. Chem. **265**, 3369–3373 (1990)
337. Sato, T.K., Panda, S., Miraglia, L.J., Reyes, T.M., Rudic, R.D., McNamara, K.A., FitzGerald, G.A., Kay, S.A., Hogenesch, J.B.: A functional genomics strategy reveals Rora as a component of the mammalian circadian clock. Neuron **43**, 527–537 (2004)
338. Sato, Y., Hashiguchi, Y., Nishida, M.: Evolution of multiple phosphodiesterase isoforms in stickleback involved in cAMP signal transduction pathway. BMC Syst. Biol. **3**, 23 (2009)
339. Systems Biology Ontology (2010), http://www.ebi.ac.uk/sbo/
340. Schaefer, G., Nakashima, T., Ishibuchi, H.: Gene expression analysis by fuzzy and hybrid fuzzy classification. In: Fuzzy Systems in Bioinformatics and Computational Biology, pp. 127–140 (2009)

341. Schibler, U., Sassone-Corsi, P.: A web of circadian pacemakers. Cell **111**, 919–922 (2002)
342. Schilling, C.H., Letscher, D., Palsson, B.Ø.: Theory for the systemic definition of metabolic pathways and their use in interpreting metabolic function from a pathway-oriented perspective. J. Theor. Biol. **203**(3), 229–248 (2000)
343. Schlitt, T., Brazma, A.: Current approaches to gene regulatory network modelling. BMC Bioinform. **8**(Suppl. 6), S9 (2007)
344. Schreiber, F., Dwyer, T., Marriott, K., Wybrow, M.: A generic algorithm for layout of biological networks. BMC Bioinform. **10**, 375 (2009)
345. Schuster, S., Höfer, T.: Determining all extreme semi-positive conservation relations in chemical reaction systems. A test criterion for conservativity. J. Chem. Soc. Faraday Trans. **87**, 2561–2566 (1991)
346. Schuster, S., Hilgetag, C.: On elementary flux modes in biochemical reaction systems at steady state. J. Biol. Syst. **2**, 165–182 (1994)
347. Schuster, S., Hilgetag, C.: What information about the conserved-moiety structure of chemical reaction systems can be derived from their stoichiometry? J. Phys. Chem. **99**, 8017–8023 (1995)
348. Schuster, S., Hilgetag, C., Schuster, R.: Determining elementary modes of functioning in biochemical reaction networks at steady state. In: Proc. Second Gauss Symposium 1993, pp. 101–114 (1996)
349. Schuster, S., Dandekar, T., Fell, D.A.: Detection of elementary flux modes in biochemical networks: A promising tool for pathway analysis and metabolic engineering. Trends. Biotechnol. **17**, 53–60 (1999)
350. Schuster, S., Fell, D., Dandekar, T.: A general definition of metabolic pathways useful for systematic organization and analysis of complex metabolic networks. Nat. Biotechnol. **18**(3), 326–332 (2000)
351. Schuster, S., Hilgetag, C., Woods, J.H., Fell, D.A.: Reaction routes in biochemical reaction systems: algebraic properties, validated calculation procedure and example from nucleotide metabolism. J. Math. Biol. **45**, 153–181 (2002)
352. Schuster, S., Klamt, S., Weckwerth, W., Moldenhauer, F., Pfeiffer, T.: Use of network analysis of metabolic systems in bioengineering. Bioproc. Biosyst. Eng. **24**, 363–372 (2002)
353. Schuster, S., Marhl, M., Höfer, T.: Modelling of simple and complex calcium oscillations. Eur. J. Biochem. **269**, 1333–1355 (2002)
354. Schuster, S., Pfeiffer, T., Moldenhauer, F., Koch, I., Dandekar, T.: Exploring the pathway structure of metabolism: decomposition into subnetworks and application to *Mycoplasma pneumoniae*. Bioinformatics **18**, 351–361 (2002)
355. Schuster, S., von Kamp, A., Pachkov, M.: Understanding the roadmap of metabolism by pathway analysis. In: Weckwerth, W. (ed.) Metabolomics, Methods and Protocols, pp. 199–226. Humana Press, Totowa (2007)
356. Schwender, J., Goffman, F., Ohlrogge, J.B., Shachar-Hill, Y.: Rubisco without the Calvin cycle improves the carbon efficiency of developing green seeds. Nature **432**, 779–782 (2004)
357. Shaffer, C.A., Randhawa, R., Tyson, J.J.: The role of composition and aggregation in modeling macromolecular regulatory networks. In: Proc. of the Winter Simulation Conf., pp. 1628–1636 (2006)
358. Shannon, P., Markiel, A., Ozier, O., Baliga, N.S., Wang, J.T., Ramage, D., Amin, N., Schwikowski, B., Ideker, T.: Cytoscape: a software environment for integrated models of biomolecular interaction networks. Genome Res. **13**(11), 2498–2504 (2003)
359. Sharov, A.A.: Self-reproducing systems: structure, niche evolution and evolution. Biosystems **25**, 237–249 (1991)
360. Shearman, L.P., Sriram, S., Weaver, D.R., et al.: Interacting molecular loops in the mammalian circadian clock. Science **288**, 1013–1019 (2000)
361. Shen-Orr, S.S., Milo, R., Mangan, S., Alon, U.: Network motifs in the transcriptional regulation network of *Escherichia coli*. Nat. Genet. **31**(1), 64–68 (2002)
362. Simão, E., Remy, E., Thieffry, D., Chaouiya, C.: Qualitative modelling of regulated metabolic pathways: application to the tryptophan biosynthesis in *E. Coli*. Bioinformatics **21**(2), ii190–ii196 (2005)

363. Sivakumaran, S., Hariharaputran, S., Mishra, J., Bhalla, U.S.: The database of quantitative cellular signaling: management and analysis of chemical kinetic models of signaling networks. Bioinformatics **19**(3), 408–415 (2003)
364. Smith, B., Ashburner, M., Rosse, C., Bard, J., Bug, W., Ceusters, W., Goldberg, L.J., Eilbeck, K., Ireland, A., Mungall, C.J., The OBI Consortium, Leontis, N., Rocca-Serra, P., Ruttenberg, C.J., Sansone, S.-A., Scheuermann, R.H., Shah, N., Whetzel, P.L., Lewis, S.: The OBO Foundry: coordinated evolution of ontologies to support biomedical data integration. Nat. Biotechnol. **25**, 1251–1255 (2007)
365. Smolen, P., Baxter, D.A., Byrne, J.H.: Modeling circadian oscillations with interlocking positive and negative feedback loops. J. Neurosci. **21**(17), 6644–6656 (2001)
366. Snel, B., Lehmann, G., Bork, P., Huynen, M.A.: STRING: a web-server to retrieve and display the repeatedly occurring neighbourhood of a gene. Nucleic Acids Res. **28**(18), 3442–3444 (2000)
367. Spellman, P.T., Miller, M., Stewart, J., et al.: Design and implementation of microarray gene expression markup language (MAGE-ML). Genome Biol. **3**(9), 46.1–46.9 (2002)
368. Sriram, K., Gopinathan, M.S.: A two variable delay model for the circadian rhythm of Neurospora crassa. J. Theor. Biol. **231**(1), 23–38 (2004)
369. Srivastava, R., Peterson, M.S., Bentley, W.E.: Stochastic kinetic analysis of the *Escherichia coli* stress circuit using $\sigma-32$ targeted antisense. Biotechnol. Bioeng. **75**(1), 120–129 (2001)
370. Starke, P.H.: Analysis of Petri Net Models. Teubner-Verlag, Stuttgart (1990) (in German)
371. Starke, P.: INA—Integrated Net Analyzer. Manual. Humboldt University Berlin, Dept. Computer Science (1998)
372. Starostzik, C., Marwan, W.: Functional mapping of the branched signal transduction pathway that controls sporulation in *Physarum polycephalum*. Photochem. Photobiol. **62**, 930–933 (1995)
373. Steggles, L.J., Banks, R., Wipat, A.: Modelling and analysing genetic networks: from Boolean networks to Petri nets. In: Priama, C. (ed.) Computational Methods in Systems Biology. Intern. Conf. CMSB 2006. LNBI, vol. 4210, pp. 127–141. Springer, Berlin (2006)
374. Steggles, L.J., Banks, R., Shaw, O., Wipat, A.: Qualitatively modelling and analysing genetic regulatory networks: a Petri net approach. Bioinformatics **23**(3), 336–343 (2007)
375. Stelling, J., Klamt, S., Bettenbrock, K., Schuster, S., Gilles, E.D.: Metabolic network structure determines key aspects of functionality and regulation. Nature **420**, 190–193 (2002)
376. Steuer, R., Junker, B.H.: Computational models of metabolism: stability and regulation in metabolic networks. Adv. Chem. Phys. **142**, 105–251 (2009)
377. Strehl, K., Thiele, L.: Interval diagrams for efficient symbolic verification of process networks. IEEE Trans.-Comput., Aided Des. Integr. Circuits Syst. **19**(8), 939–956 (2000)
378. Berg, J.M., Tymoczko, J.L., Stryer, L.: Biochemistry, 6th edn. Freeman, New York (2006)
379. Suderman, M., Hallett, M.: Tools for visually exploring biological networks. Bioinformatics **23**(10), 2651–2659 (2007)
380. Sugiyama, K., Tagawa, S., Toda, M.: Methods for visual understanding of hierarchical system structures. IEEE Trans. Syst. Man Cybern. **SMC-11**(2), 109–125 (1981)
381. Sun, H., Charles, C.H., Lau, L.F., Tonks, N.K.: MKP-1 (3CH134), an immediate early gene product, is a dual specificity phosphatase that dephosphorylates MAP kinase in vivo. Cell **75**(3), 487–493 (1993)
382. Systems Biology in Wikipedia (2009), http://technorganiac.wordpress.com/2007/10/14/functional-genomics-lipidomics/
383. Szallai, Z., Stelling, J., Periwal, V.: System Modeling in Cellular Biology. MIT Press, Cambridge (2006)
384. Takigawa-Imamura, H., Mochizuki, A.: Transcriptional autoregulation by phosphorylated and non-phosphorylated KaiC in cyanobacterial circadian rhythms. J. Theor. Biol. **241**(2), 178–192 (2006)
385. Takigawa-Imamura, H., Mochizuki, A.: Predicting regulation of the phosphorylation cycle of KaiC clock protein using mathematical analysis. J. Biol. Rhythms **21**(5), 405–416 (2006)

386. Tasaki, S., Nagasaki, M., Oyama, M., Hata, H., Ueno, K., Yoshida, R., Higuchi, T., Sugano, S., Miyano, S.: Modeling and estimation of dynamic EGFR pathway by data assimilation approach using time series proteomic data. Genome Inform. **17**(2), 226–238 (2006)
387. Terzer, M., Stelling, J.: Large-scale computation of elementary flux modes with bit pattern trees. Bioinformatics **24**, 2229–2235 (2008)
388. Thirumalai, D., Woodson, S.A.: Kinetics of folding of proteins and RNA. Acc. Chem. Res. **29**, 433–439 (1996)
389. Thomas, R.: Boolean formalisation of genetic control circuits. J. Theor. Biol. **42**, 565–583 (1973)
390. Thomas, R., D'Ari, R.: Biological Feedback. CRC Press, Boca Raton (1990)
391. Thomas, R.: Regulatory networks seen as asynchronous automata: a logical description. J. Theor. Biol. **153**, 1–23 (1991)
392. Thomas, R., Thieffry, D., Kaufman, M.: Dynamical behaviour of biological regulatory networks, I. Biological role of feedback loops and practical use of the concept of the loop-characteristic state. Bull. Math. Biol. **57**, 247–276 (1995)
393. TINA: TIme petri Net Analyzer, http://www.laas.fr/tina/
394. Tolba, C., Lefebvre, D., Thomas, P., El Moudni, A.: Continuous and timed Petri nets for the macroscopic and microscopic traffic flow modelling. Simul. Modell. Pract. Theory **13**, 407–436 (2005)
395. Torres, L.M., Wagler, A., Weismantel, R.: Modeling the dynamic behavior of deterministic biological systems. In: Proc. of ALIO/EURO Workshop on Appl. Combin. Optim., Buenos Aires (2008)
396. Torres, L.M., Wagler, A.: Encoding the dynamics of deterministic systems. Math. Methods Oper. Res. **36**, 175–182 (2010)
397. Trinh, C.T., Wlaschin, A., Srienc, F.: Elementary mode analysis: a useful metabolic pathway analysis tool for characterizing cellular metabolism. Appl. Microbiol. Biotechnol. **81**(5), 813–826 (2009)
398. Troncale, S., Tahi, F., Campard, D., Vannier, J.-P., Guespin, J.: Modeling and simulation with hybrid functional Petri nets of the role of interleukin-6 in human early haematopoiesis. Pac. Symp. Biocomput. **11**, 427–438 (2006)
399. Tsavachidou, D., Liebman, M.N.: Modeling and simulation of pathways in menopause. J. Am. Med. Inform. Assoc. **9**(5), 461–471 (2002)
400. Ueda, H.R., Hagiwara, M., Kitano, H.: Robust oscillations within the interlocked feedback model of *Drosophila* circadian rhythm. J. Theor. Biol. **210**, 401–406 (2001)
401. Ueki, K., Matsuda, S., Tobe, K., Gotoh, Y., Tamemoto, H., Yachi, M., Akanuma, Y., Yazaki, Y., Nishida, E., Kadowaki, T.: Feedback regulation of mitogen-activated protein kinase kinase kinase activity of c-Raf-1 by insulin and phorbol ester stimulation. J. Biol. Chem. **269**(22), 15756–15761 (1994)
402. Urbanczik, R., Wagner, C.: An improved algorithm for stoichiometric network analysis: theory and applications. Bioinformatics **21**, 1203–1210 (2005)
403. van Dien, S.J., Lidstrom, M.E.: Stoichiometric model for evaluating the metabolic capabilities of the facultative methylotroph *Methylobacterium extorquens* AM1, with application to reconstruction of C3 and C4 metabolism. Biotechnol. Bioeng. **78**, 296–312 (2002)
404. Vayttaden, S.J., Bhalla, U.S.: Developing complex signaling models using GENESIS/Kinetikit. Sci. STKE **2004**(219), pl4 (2004)
405. Voet, D.J., Voet, J.G.: Biochemistry, 3rd edn. Wiley, New York (2004)
406. von Kamp, A., Schuster, S.: Metatool 5.0: fast and flexible elementary modes analysis. Bioinformatics **22**, 1930–1931 (2006)
407. Voss, K., Heiner, M., Koch, I.: Steady state analysis of metabolic pathways using Petri nets. In Silico Biol. **3**, 367–387 (2003)
408. Wagler, A., Weismantel, R.: The combinatorics of modeling and analyzing biological systems. Nat. Comput. (2009, to appear)
409. Wang, J.: Timed Petri nets. Theory and Applications. Kluwer Academic, Dordrecht (1998)

410. Windhager, L., Zimmer, R.: Intuitive modeling of dynamic systems with Petri nets and fuzzy logic. In: Proceedings German Conference on Bioinformatics. Lecture Notes in Informatics, pp. 106–115. Gesellschaft für Informatik, Bonn (2008)
411. Wlotzka, B., McCaskill, J.S.: A molecular predator and its prey: coupled isothermal amplification of nucleic acids. Biol. Chem. **4**, 25 (1997)
412. Weinman, E.O., Strisower, E.H., Chaikoff, I.L.: Conversion of fatty acids to carbohydrate: application of isotopes to this problem and role of the Krebs cycle as a synthetic pathway. Physiol. Rev. **37**, 252–272 (1957)
413. Weng, G., Bhalla, U.S., Iyengar, R.: Complexity in biological signaling systems. Science **284**(5411), 92–96 (1999)
414. Westerhoff, H.V., van Dam, K.: Thermodynamics and Control of Biological Free-Energy Transduction. Elsevier, Amsterdam (1987)
415. Willeboordse, F.H., Kaneko, K.: Pattern dynamics of a coupled map lattice for open flow. Physica D **86**, 428–455 (1995)
416. Windhager, L., Zimmer, R.: Intuitive modeling of dynamic systems with Petri nets and fuzzy logic. In: Proc. German Conf. Bioinform., pp. 106–115 (2008)
417. Wolpert, L., Beddington, R., Brockes, J., Jessell, T., Lawrence, P., Meyerowitz, E.: Principles of Development, 3rd edn. Oxford University Press, London (2006)
418. WormBase: http://www.wormbase.org/db/gene/variation?name=pas40332;class=Variation
419. Wu, J.Q., Pollard, T.D.: Counting cytokinesis proteins globally and locally in fission yeast. Science D **310**, 310–314 (2005)
420. Yamada, S., Shiono, S., Joo, A., Yoshimura, A.: Control mechanism of JAK/STAT signal transduction pathway. FEBS Let. **534**(1–3), 190–196 (2003)
421. Yook, S.H., Jeong, H., Barabasi, A.L.: Modeling the Internet's large-scale topology. Proc. Natl. Acad. Sci. USA **99**, 13382–13386 (2002)
422. Zadeh, L.A.: Fuzzy sets. Inf. Control **8**, 338–353 (1965)
423. Zaitsev, D.A.: Decomposition of Petri nets. Cybern. Syst. Anal. **40**(5), 739–746 (2004)
424. Zaitsev, D.A.: Compositional analysis of Petri nets. Cybern. Syst. Anal. **42**(1), 126–136 (2006)
425. Zanzoni, A., Montecchi-Palazzi, L., Quondam, M., et al.: MINT: a Molecular INTeraction database. FEBS Lett. **513**, 135–140 (2002)
426. Zeigarnik, A.V., Temkin, O.N.: A graph-theoretical model of complex reaction mechanisms: bipartite graphs and the stoichiometry of complex reactions. Kinet. Catal. **35**, 647–655 (1994)
427. Zevedei-Oancea, I., Schuster, S.: Topological analysis of metabolic networks based on Petri net theory. In Silico Biol. **3**(3), 323–345 (2003)
428. Zevedei-Oancea, I., Schuster, S.: A theoretical framework for detecting signal transfer routes in signalling networks. Comput. Chem. Eng. **29**, 597–617 (2005)

Index

A
Abstract Petri net notation, 264
Active site, 24
Activity, 8, 109
Activity function, 109
Additional component, 323–325, 328
Adequateness, 8
Algebraic equation system, 213
Allosteric effects, 165
Alternative splicing, 222
Amplification, 244
Amyloplasts, 22
Anabolism, 23
Anti-coincidence, 61
APNN, 264
Apoptosis, 104, 221
Arbiter, 86, 90
Arc, 44, 45, 309
Arc weights, 44, 309
ArrayExpress, 11
ASEL cell, 115
ASER cell, 115
Asynchronous Boolean network, 86
Asynchronous circuit, 86
Attractor, 162
Autotrophic cell, 29
Avogadro number, 156

B
Berkely Madonna, 174
Bhalla–Iyengar model, 225, 227, 229 231–233, 235, 238, 239, 241–243, 248, 249
Bifan, 226
Bifurcation, 158
Bimolecular reaction, 157, 158, 211
Binary decision diagram, 258, 259
Binary Decision Tree, BDD, 258
Binding, 103, 104
BINDplus, 12
Biochemical Petri net, 51
BioLayout, 15
Biological pathway, 103
Biological system, 42
BioPAX, 175
BIOSPICE, 174
BioTapestry, 175
Biotin, 169–172
BioUML, 175
Bipartite graph, 223
Bistable behavior, 238
Boolean function, 258, 259
Boolean network, 74, 76, 264
  example of, 76
  neighbourhood state, 76
  synchronous or asynchronous behavior, 76
  synchronous Petri net model construction, 78
Boolean regulatory graph, 264
Boolean variable, 259
Bounded, 53
Bounded capacities, 333

C
C-element, 88, 91
C. elegans, 114, 116
CADLIVE, 175
Calcium3D, 174
CAMP, 226
Capacity vector, 333
Carbohydrate metabolism, 23
Catabolism, 23
Catalytic activity, 166
Catalytic rate, 166
Catalyzed reactions, 154

I. Koch et al. (eds.), *Modeling in Systems Biology*, Computational Biology 16,
DOI 10.1007/978-1-84996-474-6,

Cell biology, 19
Cell Illustrator, 101, 102, 114–116, 119, 174, 288
Cell system markup language, 115, 119
Cell system ontology, 119
CellDesigner, 15, 174
Cellerator, 174
CellML, 175
CellNetAnalyzer, 220
Cellular automata, 165
Cellular organization, 20
CellWare, 174
Central metabolic pathways, 29
Chapman–Kolmogorov equation, 136
Chemesis, 174
Chemical master equation, 124
Chemical reaction, 154
Chemical substance, 38
CIBEX, 11
Circadian rhythms, 282
CKS, 174
Cleavage, 104
Clock genes, 285
Clustering of T-invariants, 216
Collection of interconnected Logical Regulatory Modules, CLRM, 265–267, 269, 271, 277
Combinatorial reconstruction approach, 313
Combined glucose and pentose phosphate pathway, 37
Competitive inhibition, 25, 28, 167, 168
Concurrent, 37, 53, 54
Confidence interval, 143
Conflict, 112
Conflict graph, 333
Conflict resolution, 112
Conformal networks, 319, 323, 334
Conformality, 335
Conservation property, 230
Conservation relation, 217
Conservative component, 230
Conservativity, 218
Continuous Petri net, 165
Continuous place, 106, 113, 286
Continuous timed Petri net, 106
Continuous token, 106
Continuous transition, 106, 113, 117, 285
Continuous-time Markov chain, 124
Cooperative binding, 169, 172
COPASI, 174
Coupled differentiation equations, 154
Covered by P-invariants, CPI, 50, 64
Covered by T-invariants, CTI, 51, 65
CPN-AMI, 219
CSML, 119
CSO, 119
CTPN, 106, 107
Cytoscape, 15, 175

D

DAG, 226
Decision diagram, 258, 259
Defuzzifier, 191
Degradation, 103, 104
Delay, 104
Delay component, 296
Depth-first search, 224
Determinism, 7
Deterministic behavior, 333
Deterministic models, 10
Difference vector, 321
Differential equation, 165
Diffusion, 164
DIP, 12
Disabled transition, 111
Discrete, continuous, and generic places and transitions, 117
Discrete, continuous and generic, 113
Discrete functional levels, 255
Discrete model, 163, 164
Discrete multi-valued functions, 258
Discrete Petri net, 59
Discrete Petri net model, 227, 228
Discrete place, 106
Discrete transition, 106, 112
Discreteness, 7
Dissociation, 103
Dizzy, 174
DNA, 32, 103, 104
DOQCS database, 238
Double negative feedback loop, 114
DREAM challenge, 200
Drosophila, 253, 254, 266, 270, 276
Dually phosphorylated, 230
Dynafit, 174
Dynamic behavior, 9, 45, 153, 227, 247
Dynamic graph, 225, 236, 239, 249
Dynamic pathway modeling, 101
Dynetica, 174

E

EcoCyc, 13
E. coli, 81, 222
  nutritional stress response, 81
Edinburgh pathway editor, 175
Educts, 154
EFMTool, 220

Elementary entities, 156
Elementary modes, 213–216, 220
Elementary Petri net, 101
Elementary reaction, 125, 155
Enabled transitions, 45
Encapsulated regulatory network, 266
Endothermic, 154
Endoplasmatic reticulum, 20
Energy metabolism, 23
Enzyme, 23, 24, 39
Enzyme activity, 27
Enzyme concentration, 166
Enzyme kinetics, 27
Enzyme subsets, 220
Enzyme–substrate complex, 165, 166
Enzyme-catalyzed reactions, 165
Equilibrium constant, 159
Equilibrium equation, 155
Equilibrium reaction, 157, 158
Ergodicity, 133
Eukaryotic cell, 19
Exothermic, 154
Experimental time series data, 310, 316
Exponential random variable, 122
External metabolites, 213
Extracellular matrix, 20

F
Fatty acid, 30, 211
Feasibility test, 323, 335
Feedforward loops, 226
Feedforward motif, 244
FEM, 176
Firing, 110, 112, 268
  condition, 103
  of continuous transition, 113
  of discrete transition, 112
  of generic transition, 113
  order enforcement, 93, 94, 97
  speed, 285
First-order reactions, 155
Flux, 218
Flux coupling analysis, 221
FOE, *see* Firing order enforcement
Functional module, 247
Functional unit, 233
Futile cycle, 214
Fuzzification, 185
Fuzzy implication, 188
Fuzzy logic system, 182, 192–194
Fuzzy sets, 181, 185

G
Gates, 226
Gene expression, 11, 32
Gene regulation, 34, 230
Gene regulatory network, 34, 102
Generic place, 107
Generic transition, 107, 113, 117, 118
GENESIS, 238
Genesis/Kinetikit, 174
Genetic interaction, 285
GenMAPP, 15, 175
GEO, 11
GEPASI, 174
GINsim, 264, 276, 277
Glycolysis, 29
Glyoxylate shunt, 212
Golgi apparatus, 20
Graph theory, 210, 223
Graphical representation, 40
GRNs, *see* Gene regulatory network

H
Heterotrophic organism, 29
HFPNe model, 117, 119
High-level Petri net , HLPN, 253, 264, 265, 267–270, 277
Hill
  coefficient, 173
  constant, 172
  curve, 170–172
  equation, 169, 173
  function, 170
  kinetics, 153, 154, 169
Holistic paradigm, 4
Homogeneous solution, 317–319, 332, 334
Homogeneous system, 317
Hopf bifurcation, 162, 164
Hybrid dynamic net, 107
Hybrid functional Petri net, HFPN, 107, 108, 284
Hybrid functional Petri net with extension, HFPNe, 101, 102, 106–108, 115, 117, 119, 284
Hybrid Petri net, HPN, 102, 107
Hybrid system, 107
Hypergraph, 223

I
INA, 217, 264, 276
Incidence matrix, 63, 219, 311, 317, 329
Indirect interaction, 295
Influence graph, 232, 233, 235, 236, 249
Inhibition, 25, 103
  competitive, 25, 28, 167, 168
  irreversible, 167

Inhibition (*cont.*)
  MAPK, 242
  noncompetitive, 25, 28
  reversible, 167
  uncompetitive, 26, 167, 168
Inhibitor arcs, 255
Inhibitory, 108
Inhibitory interaction, 295
Initial concentration, 159
Initial condition, 114
Initial marking, 45, 47, 49–51, 104, 110, 309
Initial state, 261, 274, 310
Input arc, 103, 108
Input place, 103
Input speed, 106
Integer linear programming, 219, 325
Integration function, 271, 275
Interaction graph, 230–232
Interaction network, 273
Interface, 42
Intermediate state, 40
Interval Decision Diagrams, IDD, 256, 259, 262, 263
Invariant, 49, 37, 52
Invariant analysis, 231
Irreversible inhibition, 167
Isoenzymes or isozyme, 24
Isoforms, 24

J
Jacobian matrix, 162, 163
Jarnac/JDesigner, 174
JSim, 174

K
KDBI, 176
KEGG, 13
Kinetic dynamic, 249
Kinetic model, 175
Kinetic rate constant coefficient, 158
Kinetics, 118
KINSIM/FITSIM, 174
Kohn interaction maps, 175
Krebs cycle, 29

L
Labeling, optimal, 325
Labeling, valid, 324
λ phage, *see Phage λ*
LARKIN, 176
Lifetime, 247
Limit cycle, 163, 164
Linear algebra, 47, 48
Linear algebraic equation system, 219
Linear equation, 162
Linear equation system, 312
Logic minimization, 77, 83
Logical approach, 255
Logical formalism, 254, 255
Logical functions, 257, 258, 259
Logical modeling, 254
Logical operator, 60
  conjunction, 60
  disjunction, 60
  implication, 60
  negation, 60
Logical Regulatory Graph, LRG, 254–256, 259–265, 267, 274
Logical Regulatory Module, LRM, 265, 266, 268, 269, 272, 274, 277
Lotka–Volterra model, 159, 160, 162

M
MACRON, 176
MAPK, 227, 231, 236, 241–247
MAPK inhibition, 242
MAPK response, 248
MAPK/PKC model, 247
MAPK/PKC pathway, 225–241
Marking, 44, 45, 110, 268, 309
Marking graph, 52
Marking probability distribution, 132
Marking-dependence, 127, 128
Markov process, 156
Markup language, 264
Mass action, 114, 154, 155
  kinetics, 165
  reversible, 158
Mass conservation, 159, 166, 168
Mass transfer, 218
Master equation, 163
MATHML, 175
MathSBML, 174
MatLab, 219, 238
Mauritius maps, 70
Maximal common transition set, *see* MCT-set
MCT-set, 67, 69, 70, 221, 222
MDD, 258
Metabolic network, 30, 209, 218
Metabolic network analysis, 214
Metabolic pathway, 102
Metabolic reaction, 104
Metabolism, 23
Metabolite, 24
MetaCyc, 113
METATOOL, 214, 218, 220
Michaelis–Menten, 228, 229

Michaelis–Menten (*cont.*)
constant, 27, 166
equation, 27, 167
kinetics, 106, 114, 153–155, 165
MicroRNA, 103, 114, 117
Microscopic dissociation constant, 170, 171
Minimal T-invariants, 213, 215, 216, 233
MINT, 12
MIRIAM, 175
Mitochondria, 20, 21
Mixotroph organisms, 29
Model, 5
Model checking, 81, 84
Model composition, 9
Model development, 8
Model validation, 84
Module, 264
Mole, 156
Molecular signal, 249
Monostable behavior, 238
Monotone, 319
Monotone data, 312, 319, 322, 323, 326
Monotone experimental time series data, 326
Monotonicity, 320, 322, 329, 332, 335
mRNA, 32, 103, 114–116, 231
Multi-valued decision diagram, 258
Multi-valued function, 259
Multi-valued logical formalism, 253
Multi-valued regulatory Petri net, MRPN, 261
Multigraph, 256
Mutant analysis, 85

N
Negative feedback loop, 227, 240, 242–244
Nested feedforward motif, 245
Net flux, 213
NetBuilder, 175
Network, 309
asynchronous Boolean, 86
Boolean, *see* Bolean network
conformal, 312, 319, 323, 334
diagram, 13
encapsulated regulatory, 266
endothermic, 154
gene regulatory, 34, 74, 81, 102
interaction, 273
metabolic, 30, 209, 218
reconstruction, 334
reconstruction problem, 307, 312–314, 329, 334
regulatory, 209, 218, 253–255, 264, 276
signaling, 225
strongly conformal, 333
Neutral binding, 169–171
Neutral stable steady-state, 164
Non-minimal T-invariant, 219
Noncompetitive inhibition, 25, 28
Nondeterministic models, 10
Nondeterministically, 332
Nonmonotone data, 331
Nucleus, 20, 103
Null-space, 217, 220

O
ODE model, 233, 236, 238, 240
Operations on fuzzy sets, 186, 187
Ordinary differential equation, ODE, 153, 156, 161, 213, 227, 228, 249
Oscillation, 158, 164
Oscillatory behavior, 159, 260
Output arc, 103, 108
Output place, 103
Output speed, 106

P
P-invariant, 63, 212, 217, 230, 231, 233, 237
P/T net, 261–263, 267
PARKIN, 176
Partial order, 332
Pasadena Twain, 174
PathVision, 175
Pathway, 223
Pathway databases, 12
Pathway modeling, 103
Penicillin, 167
PEP, 74, 80, 81, 83, 84, 100
Periodicity, 7
Perturbation, 162
Petri net, PN, 16, 37, 38, 43–56, 59, 103, 153, 165, 170, 174, 212, 223–226, 234, 237, 253, 276, 308, 334
continuous, 165
continuous timed, 106
discrete, 59
discrete model, 227, 228
elementary, 101
formalism, 153–154
high-level, 253, 264–270, 277
hybrid, 102, 107
hybrid functional, HFPN, 107, 284
hybrid functional with extension, HFPNe, 101–103, 106, 107, 284
library, 269
markup language, 64
model, 38, 42–47, 172, 229–231, 236
modeling application, 197
multi-valued regulatory, 261

Petri net (*cont.*)
stochastic, 122
timed, 104
theory, 240
with fuzzy logic, 194, 195
"Petri Nets World" archive, 219, 223
Phage λ, 94
lysis–lysogeny switch, 94
lysogenic cycle, 95
lytic cycle, 95
Phosphorylation, 103
Phosphorylation of MKP1, 242
Photosynthesis, 22
Physarum polycephalum, 310
Pitchfork bifurcation, 163
Place, 40, 43, 45, 103, 108, 113, 309
Place invariant, 49, 50, 55
PLAS, 174
Plastid, 21, 22
PLmaddon, 174
PNFL, 194, 195
PNMA, 197
PNML, 264
Poisson distribution, 139
Positive feedback loop, 226, 227, 240, 245, 246
Posttranslational modification, 226
Potato tuber, 210, 214, 221
Predator prey model, 159
Prediction of network structure, 307, 308
Primary metabolism, 23
Probability distribution, 163
Prokaryotes, 20
ProMoT, 277
ProMoT/DIVA/Diana, 174
Propensity function, 123, 125
Protein–protein interaction databases, 11
PSI-MI, 175
pSTIING, 12
PyBioS, 174

Q
QSSA, 166, 168
Qualitative modeling, 253, 254
Quantitative analysis, 153
Quantitative methods, 153
Quantitative modeling, 154
Quasi steady-state approximation, 166

R
Race policy, 129
Randomization algorithm, 137
Rate constant, 125
Rate constants coefficients, 157
Reachability, 260, 332
Reachable states, 274
Reachable marking, 46, 47, 52
Reactants, 154
Reaction flow ratio, 237
Reaction rate, 154
Reaction space, 314
Reaction vector, 315, 321, 322
REACTOME, 13
Read arc, 65, 231
Receptor, 36
Reconstruction algorithm, 326, 329
Red blood cells, 38
Reduced ordered multi-valued decision diagram, 258
Reductionist paradigm, 4
Regulation of metabolism, 31
Regulatory component, 256
Regulatory interaction, 255, 257
Regulatory modules, 249
Regulatory motif analysis, 227
Regulatory motif module, 248
Regulatory motif, 236, 242, 247, 248
Regulatory network, 209, 218, 253–255, 264, 276
Repetitive component, 232, 233
Repressilator, 278
Reserved marking, 110
Reversibility, 7
Reversible inhibition, 167
Reversible mass action, 158
Ribosome, 32, 33
RNA, 117, 231
RNA splicing, 222
Robustness, 8, 210
ROMDD, 258

S
Saccharomyces cerevisiae, 222
SB Toolbox, 174
SBRT, 174
SBW, 174
Second-order reactions, 157
Secondary metabolism, 23
Segment-polarity module, 273–275
Self-loop, 257
Self-regulation, 257
Semi-continuous modeling, 197, 198
Semi-discrete modeling, 195, 196
Semi-positive conservation relation, 217
SI circuit, *see* Speed-independent circuit
Signal propagation, 232
Signal transduction, 34, 230

Signal transition graph, STG, 87, 260, 262, 264
  boundedness, 89
  complete state coding, 90
  consistency, 87, 89
  CSC, *see* complete state coding
  OP, *see* output-persistency
  output-persistency, 87, 89, 96
  signal, 87
    input signal, 88
    internal signal, 88
    local signal, 88
    output signal, 88
Signaling component, 237
Signaling network, 225
Signaling pathway, 102, 104, 226
Simulation trace, 140
Singly phosphorylated, 230
Siphon, 37
SNAKES toolkit, 269
SNP, 117
Spatial instability, 164
Speed-independent circuit, 86
Spliceosome, 222
Sporulation, 310
Stability, 8, 288
Stable states, 260
State transition graph, 260, 261
State vector, 309
State, 40
Static net structure, 44
Stationary state, 10
Statistic average, 142
Steady-state, 48, 124, 133, 158, 161–163, 211, 220, 223
  concentration, 170, 173
  condition, 170, 171
  flux, 212, 213, 221
STG, *see* signal transition graph, 260, 262, 264
Stochastic and/or deterministic information, 226
Stochastic model, 164
Stochastic Petri net, 122
Stochastic simulation algorithm, 140
  direct method, 141
  first-reaction method, 141
  next reaction method, 142
Stoichiometric coefficients, 38, 154, 309
Stoichiometry, 215
Stoichiometry matrix, 48, 220
Streptavidin, 169–173
STRING, 12
Strongly conformal, 333, 335
Strongly connected components, 260
Subgraph, 233, 234
Substrate, 154
Superconservativity, 218
Switch-off time, 111
Switches, 226
Syphon, 52
System error, 112
System state, 9
Systems biology, 4
Systems Biology Graphical Notation, SBGN, 15
Systems Biology Markup Language, SBML, 175, 228, 229, 238

T
T-cluster, 68, 69, 70, 223
T-invariant, 51, 52, 64, 212, 213, 220–222, 233
  feasible, 65
  support, 65
  trivial, 65
Tanimoto similarity, 216
Terminal state, 322–325, 327, 335
Test arcs, 255
Theoretical model, 153
Thermodynamic parameters, 158
Time series data, 319
Timed bistable switch, 246
Timed Petri net, TPN, 104, 107
Token, 40, 41, 44, 103, 110, 268
Topological analysis, 209, 210
Topology, 240, 249
Transcription, 32, 103
Transcription factor, 34
Transition, 39, 41, 43–48, 51, 52, 55, 103, 108, 262, 309, 312
Transition graph, 255
Transition invariant, 51
Translation, 32, 103, 104
Translocation, 103, 104
TRANSPATH, 119
Trap, 37, 52
Trigger time, 111
True T-invariants, 213
Truth table, 259
Tryptophan, 222
Turing instability, 164
Turing mechanism, 164

U
Uncompetitive inhibition, 26, 167, 168
Unimolecular reaction, 155, 157
Universe of discourse, 185
Unphosphorylated, 230
Update function, 109

Updating function, 260
UPGMA, 222
Urbanczik–Wagner algorithm, 220

V
Valid labeling, 324, 327
VANTED, 15
Velocity of enzymatic reactions, 27
Virtual cell, 174, 238
VisANT, 175
Visualization, 13
Visualization tools, 175
Volterra, 159

W
Weakly live, 53
WebCell, 174
Weight, 103, 106

X
XML based standards, 175
XPP-Aut, 174

Z
Zero-order reactions, 155
Zinc-finger transcription factor, 114

MIX
Papier aus verantwortungsvollen Quellen
Paper from responsible sources
FSC® C105338

If you have any concerns about our products,
you can contact us on
ProductSafety@springernature.com

In case Publisher is established outside the EU,
the EU authorized representative is:
Springer Nature Customer Service Center GmbH
Europaplatz 3, 69115 Heidelberg, Germany

Printed by Libri Plureos GmbH
in Hamburg, Germany